Designing and Developing Innovative Mobile Applications

Debabrata Samanta
Rochester Institute of Technology, Kosovo

A volume in the Advances in Wireless Technologies and Telecommunication (AWTT) Book Series

IGI Global
PUBLISHER of TIMELY KNOWLEDGE

Published in the United States of America by
IGI Global
Information Science Reference (an imprint of IGI Global)
701 E. Chocolate Avenue
Hershey PA, USA 17033
Tel: 717-533-8845
Fax: 717-533-8661
E-mail: cust@igi-global.com
Web site: http://www.igi-global.com

Library of Congress Cataloging-in-Publication Data

Names: Samanta, Debrata, 1987- editor.
Title: Designing and developing innovative mobile applications / edited by Debrata Samanta.
Description: Hershey, PA : Information Science Reference, [2023] | Includes bibliographical references and index. | Summary: "This in turn prompted several developers to consider creating mobile applications. Many IT enthusiasts are drawn to developing mobile apps because of the market's rapid growth, which is another essential element. Each developer is expected to conduct an extensive study and innovate in this era of growing interest in mobile app development. This book is unique in that it gives students, academicians, researchers, and developers knowledge on topics ranging from the history of the Android OS to more complex issues that will help them comprehend basic ideas and keep up with the most recent developments in the field, which will encourage them to create ever-more-innovative mobile apps. This book will undoubtedly assist someone who wants to start the journey of developing mobile applications by providing them with fundamental concepts, practical exercises, and advanced database handling"-- Provided by publisher.
Identifiers: LCCN 2023003374 (print) | LCCN 2023003375 (ebook) | ISBN 9781668485828 (h/c) | ISBN 9781668485835 (s/c) | ISBN 9781668485842 (eISBN)
Subjects: LCSH: Mobile apps. | Cell phones--Programming.
Classification: LCC QA76.59 .D446 2023 (print) | LCC QA76.59 (ebook) | DDC 005.35--dc23/eng/20230201
LC record available at https://lccn.loc.gov/2023003374
LC ebook record available at https://lccn.loc.gov/2023003375

This book is published in the IGI Global book series Advances in Wireless Technologies and Telecommunication (AWTT) (ISSN: 2327-3305; eISSN: 2327-3313)

British Cataloguing in Publication Data
A Cataloguing in Publication record for this book is available from the British Library.

Advances in Wireless Technologies and Telecommunication (AWTT) Book Series

Xiaoge Xu
University of Nottingham Ningbo China, China

ISSN:2327-3305
EISSN:2327-3313

MISSION

The wireless computing industry is constantly evolving, redesigning the ways in which individuals share information. Wireless technology and telecommunication remain one of the most important technologies in business organizations. The utilization of these technologies has enhanced business efficiency by enabling dynamic resources in all aspects of society.

The **Advances in Wireless Technologies and Telecommunication Book Series** aims to provide researchers and academic communities with quality research on the concepts and developments in the wireless technology fields. Developers, engineers, students, research strategists, and IT managers will find this series useful to gain insight into next generation wireless technologies and telecommunication.

COVERAGE

- Wireless Sensor Networks
- Network Management
- Global Telecommunications
- Cellular Networks
- Telecommunications
- Broadcasting
- Radio Communication
- Mobile Communications
- Digital Communication
- Wireless Technologies

IGI Global is currently accepting manuscripts for publication within this series. To submit a proposal for a volume in this series, please contact our Acquisition Editors at Acquisitions@igi-global.com or visit: http://www.igi-global.com/publish/.

The Advances in Wireless Technologies and Telecommunication (AWTT) Book Series (ISSN 2327-3305) is published by IGI Global, 701 E. Chocolate Avenue, Hershey, PA 17033-1240, USA, www.igi-global.com. This series is composed of titles available for purchase individually; each title is edited to be contextually exclusive from any other title within the series. For pricing and ordering information please visit http://www.igi-global.com/book-series/advances-wireless-technologies-telecommunication/73684. Postmaster: Send all address changes to above address.

Titles in this Series

For a list of additional titles in this series, please visit: http://www.igi-global.com/book-series/

Multidisciplinary Applications of Computer-Mediated Communication
Hung Phu Bui (University of Economics Ho Chi Minh City, Vietnam) and Raghvendra Kumar (GIET University, India)
Information Science Reference • © 2023 • 315pp • H/C (ISBN: 9781668470343) • US $250.00

Role of 6G Wireless Networks in AI and Blockchain-Based Applications
Malaya Dutta Borah (National Institute of Technology, Silchar, India) Steven A. Wright (Georgia State University, USA) Pushpa Singh (GL Bajaj Institute of Technology and Management, Greater Noida, India) and Ganesh Chandra Deka (Ministry of Skill Development and Entrepreneurship, Government of India, India)
Information Science Reference • © 2023 • 285pp • H/C (ISBN: 9781668453766) • US $250.00

Innovative Smart Materials Used in Wireless Communication Technology
Ram Krishan (Mata Sundri University Girls College, Mansa, India) Manpreet Kaur (Yadavindra Department of Engineering, Guru Kashi College, Punjabi University, Talwandi Sabo, India) and Shilpa Mehta (Auckland University of Technology, New Zealand)
Information Science Reference • © 2023 • 369pp • H/C (ISBN: 9781668470008) • US $250.00

Economic and Social Implications of Information and Communication Technologies
Yilmaz Bayar (Bandirma Onyedi Eylul University, Turkey) and Lina Karabetyan (Independent Researcher, Turkey)
Information Science Reference • © 2023 • 318pp • H/C (ISBN: 9781668466209) • US $235.00

Modelling and Simulation of Fast-Moving Ad-Hoc Networks (FANETs and VANETs)
T.S. Pradeep Kumar (Vellore Institute of Technology, India) and M. Alamelu (Kumaraguru College of Technology, India)
Information Science Reference • © 2023 • 251pp • H/C (ISBN: 9781668436103) • US $240.00

IGI Global
PUBLISHER of TIMELY KNOWLEDGE

701 East Chocolate Avenue, Hershey, PA 17033, USA
Tel: 717-533-8845 x100 • Fax: 717-533-8661
E-Mail: cust@igi-global.com • www.igi-global.com

To my parents, Mr. Dulal Chandra Samanta, Mrs. Ambujini Samanta; my elder sister, Mrs. Tanusree Samanta; brother-in-law, Mr. Soumendra Jana; and daughter, Ms. Aditri Samanta.

Editorial Advisory Board

Table of Contents

Detailed Table of Contents

When it comes to software or mobile application development, it is only possible to make it successful with the help of a good architecture and the principles that govern it. It has been found that applications that use the principles of architecture are good at scalability, maintainability, availability, interoperability, and so on. The development of mobile applications should be based on SOLID principles which leads to high-quality code without any additional effort on the part of the developer. The architecture process focuses on the design of both functional and non-functional requirements for a system. By designing the architecture of a system, one can gain a deeper understanding of the bigger picture of the system as a whole. Well-designed architecture and principles play a critical role in enabling developers to build scalable and high-quality applications.

The suggested wellness mobile application uses many processes, both automatic and manual, to record various parameters on a daily basis. The wellness app also calculates and monitors the appropriate indicators monitoring nutrition, physical activity, and sleep quality to motivate the user to alter their lifestyle. An inference reasoning module analyses these signs to identify the types of recommendations the subject needs to adopt the desired behaviours. These recommendations might include, for example, maintaining the Mediterranean diet correctly, which is acknowledged as the suitable diet for preventing cardiovascular disease, and/or improving the activity plan. Guidance, exercise, sleep, and mindfulness are among linked aspects that Flutter, a wellness app for Android and iOS, takes care of.

Mobile architecture is an essential but often overlooked aspect of mobile app development. It determines the structure and organization of an app, and it plays a crucial role in managing the complexity of code base and enabling collaboration among developers. In today's rapidly evolving mobile environment, the importance of mobile architecture cannot be overstated. But what exactly is mobile architecture, and how can it be used to create successful mobile apps? One of the key components of mobile architecture is the use of version control systems, which are powerful tools that help developers track and manage changes to the source code of an app. In this chapter, the authors will explore the role of version control systems in mobile architecture, and discuss how these systems can help developers create high-quality and efficient mobile applications that are able to handle the challenges of the mobile environment. So, any developers looking to create the next big mobile app, read on to learn more about the importance of mobile architecture and the role of version control systems.

Mobile phones are now one of the most important elements of our lives. We know that they are helpful and enable us to perform different services for our requirements online. But there are limits to everything. It is excessive that we cannot even eat or go to sleep without our mobile phones. According to some statistics, a person uses their mobile phone on average 3 hours and 15 minutes a day. The part of the population that is affected most by this problem is teenagers. Most of them suffer from "nomophobia," which is defined as "the fear of not having a mobile phone." Teenagers are the most impacted by mobile addiction because they cannot manage their screen time and they still have not developed the ability to self-control. According to some research, it is ascertained that 27% of the population who own a mobile phone are people around 11-13 years old that don't turn off their phone, not even to sleep.

There are several studies on the software development life cycle (SDLC) and various approaches that can be utilized in planning and designing a software project. The contemporary norm for the software development process across industries is to build products using iterative or incremental design. Iterative and incremental development (IID) is useful for different types of software. The authors have provided a literature review of the current definitions of iterative and incremental design, and also discuss the most common pitfalls for each approach. The authors also analysed which approach is better for different stages of the development of mobile applications. At the end of this research, it was concluded that the iterative approach is the best approach for front-end development, while the incremental approach is a better fit for back-end development. Finally, the authors provided an analysis of utilizing IID for applications in the banking sector.

People nowadays use their phones constantly throughout the day, for a variety of reasons. It seems reasonable that they may be used for educational purposes as well. The process of education has become much easier with mobile applications. Through these mobile apps, students can access all kinds of books and information. They can save a lot of time without the need to go to the library to acquire it. Thanks to these mobile apps, classes and lectures are not confined to physical classrooms only. Students no longer have to wait until class to ask questions; instead, they may get in touch with their professors whenever the need appears. When schools were closed and lessons were done online during Covid-19, mobile applications were also incredibly helpful.

Mobility is one of the most important problems for totally and partially disabled people in daily life. Disabled people need some assistive stick or device to avoid obstacles before an accident. This research work proposes to assist disabled people in their daily routines without human guidance. There are four main sections in this paper. Firstly, in the obstacle detection system, five ultrasonic sensors sense the obstacles in the front side, left side, and right side of the user's way on the ground and another ultrasonic sensor senses the overhanging obstacles near the user's head level. Secondly, in an obstacle avoidance system, the servo steering motor is equipped with the front wheel of the stick's base. The wheel with servo motor will turn easily to the left direction or right direction depending on the detecting or sensing of the ultrasonic sensors. Thirdly, in the voice-guided system, there are six recorded human voice messages in the Myanmar language for disabled users in the author's country, Myanmar.

The aim is to design a 3D augmented reality (AR) weather visualization to investigate future tragedy and daily forecasting. Imagine a world where all physical objects could be made, destroyed, customized, and used in endless ways. Augmented reality (AR) and virtual reality (VR) can create such a place. The background for the study is thus: to analyze, learn implementing and building of an AR Weather application using the OpenWeatherMap Weather API. OpenWeatherMap provides access to local and global weather data via application programming interface (API). OpenWeatherMap serves millions of customers around the world by providing a wide selection of weather-related goods with varying degrees of detail and precision. The research question is if smartphone- and tablet-based 3D AR weather visualizations can be effective tools to enhance current weather forecasting techniques. The methods used in this chapter were making use of AR weather App using unity, AR Foundation, and OpenWeatherMap to stimulate the visualization.

This chapter gives a broad outline of machine learning and artificial intelligence, and introduces the reader to many novel and latest developments in the field of machine learning. The first half of this compilation provides a comprehensive view of the classical concepts of machine learning. Subsequently, examples of machine learning frameworks are discussed. Deep learning, concepts, models, types, and algorithms in machine learning are elaborated in the subsequent section, followed by a detailed introduction to neural networks, concepts of weights, propagation, and initialization. The final section of this chapter introduces the reader to the fascinating and latest world of cutting-edge applications of machine learning like convolutional neural networks (CNNs), bidirectional long short-term memory (BLSTM), artistic, image-generating AI engines like DALL-E and stable diffusion, music and drama writing AI engines, human-like chatbot ChatGPT, art generation with AI, generative neural network concepts, regenerative neural network, and natural language processing (NLP).

With the increased mobile usage throughout the world, there is an enormous demand for mobile applications that provide not only good functionality but also good experience to the user, which is why it is important to have a good user interface for a mobile application. In this chapter, various components of an app are designed and illustrated in Figma software to make the work easier for a developer to create them. It also demonstrates the usage of Fig-ma. Meanwhile it shows the design thinking behind adding the correct colour schemes, right font, using proper white space, proper placement of buttons, and other design principles.

Digital image processing is currently used in various fields of research. One of them is in the field of medicine. In fact, experienced radiologists have difficulty distinguishing the cancerous portions of the blood vessels in the lung or detecting fine nodules that suggest lung cancer on X-ray images. Previous studies have shown that doctors and radiologists fail to detect cancerous patches in 30% of positive cases. Implementation of CAD system to classify and detect parts of cancer has been developed, but the results obtained from this implementation are that there are still many errors in the classification results. Therefore, this study will develop android app image technique to perform the classification process of lung cancer. With this research, it is hoped that the developed algorithm can help doctors and radiologists to detect cancer in a short time with more accuracy. Finally, after 20 iterations, a percentage of 90.65% was attained for the test results' performance in classifying 10 X-ray pictures.

Class models are an essential part of mobile game architecture, providing a framework for organizing and structuring the code. These models help developers to understand the relationships and dependencies between different objects in the game, allowing for more efficient and effective code development. The use of class models is likely to continue to evolve and improve in the future, as mobile games become more complex and sophisticated. As AI and ML technologies become more advanced, developers will need to create new classes and models to represent these technologies in mobile games. Additionally, the use of cloud computing and other distributed technologies is likely to become more prevalent, requiring the creation of new classes and models to represent these technologies in the game. Overall, class models will play a critical role in enabling developers to create more engaging and immersive mobile gameplay experiences.

Class models are an essential part of mobile game architecture, providing a framework for organizing and structuring the code. These models help developers to understand the relationships and dependencies between different objects in the game, allowing for more efficient and effective code development. The use of class models is likely to continue to evolve and improve in the future, as mobile games become more complex and sophisticated. As AI and ML technologies become more advanced, developers will need to create new classes and models to represent these technologies in mobile games. Additionally, the use of cloud computing and other distributed technologies is likely to become more prevalent, requiring the creation of new classes and models to represent these technologies in the game. Overall, class models will play a critical role in enabling developers to create more engaging and immersive mobile gameplay experiences.

Websites have evolved over the past few years utilizing a variety of programming languages and frameworks. There were other developers working on them concurrently. Developers deploy their projects utilizing a number of design patterns to handle these sophisticated web apps, which makes the code simpler and easier to work with. They are well known for using the model view controller approach. Model, view, and controller, or MVC, are the three components that make up an application in accordance with the MVC architectural pattern. Each of these parts is made to address particular facets of application development. This chapter focuses on MVC for mobile device website viewing. The various sections of this chapter address the introduction to MVC, MVC components, the architecture pattern, several techniques used to implement the pattern in the project, and an example of MVC architecture. Towards the end of the chapter, the advantages and disadvantages of MVC are examined.

This chapter gives a broad outline of machine learning and artificial intelligence and introduces the reader to many novel and most recent developments in the field of machine learning. The first half of this compilation provides an all-round view of the classical concepts of machine learning, namely: ensemble learning, concept of big data, handling of big data, and predictive data analytics using big data. Examples of machine learning (ML) frameworks are discussed, which are computer vision (CV), swarm algorithm, network science/graph theory and applications in machine learning, Bioinformatics using machine learning, and internet of things (IoT). A side note—R language is added as is the second most common language used worldwide for machine learning and this chapter spotlights mostly on Python language for ML. Deep learning, concepts, models, types, and algorithms in machine learning are elaborated in the subsequent section, followed by a detailed introduction to Neural networks, concepts of weight initialization, propagation, and vanishing gradient problem.

This chapter gives a broad outline of machine learning on Android mobile phones using the Scikit-learn module. The first section introduces the reader to Python language; next, Python on Android is introduced with a brief historical note on implementations of Python on Android mobile phones. Pydroid3 is introduced in the subsequent section. This is followed by instructions on setting up an Android phone for machine learning. This is followed by a description of supportive modules for machine learning that are available for Pydroid3, and some example codes, namely: os, pathlib, Pandas, NumPy, SciPy, Matplotlib, Seaborn, PySimpleGUI, NetworkX, Biopython, WordCloud, Kivy, and Jupyter Notebook. The last section of this compilation describes the Scikit-learn library, basic concepts of the Scikit-learn module, and algorithms available with this module, namely: Linear Regression, Logistic Regression, Principal Component Analysis (PCA), XGBoost, K-nearest neighbors, and support vector machine.

Computational models that simulate human emotions have witnessed a substantial development in recent years for widening the spectrum of applications. Emotional computation is becoming crucial in human-to-computer interactions with exponential growth of artificial intelligence. Normally referred to as emotion recognition, it is widely believed that the prospective detection of a person's emotional state of mind should be computed from their facial expressions. Face-movement combinations may express many different emotion types, for instance, hate, anger, panic, joy, grief, surprise, shock, to name a few. The goal and emphasis of this manuscript is the deployment of different algorithms and computation models for emotions. Considerable advancements in this domain of emotion recognition can be made through AI model development that discusses the challenges of the system and Facial Action Coding as an integral part of the models.

Design patterns are reusable solutions to common design problems that provide a consistent user experience across different apps. This article explores the best practices in mobile app design patterns and tools with a focus on the user-centered approach to design. Design patterns such as navigation bars, tab bars, list views, and card views are discussed, along with design tools such as Sketch, Figma, Adobe XD, and InVision. The problem is to ensure that mobile app design is centered around the needs and preferences of the user, rather than the designer or the technology, and that the right design patterns and tools are used to create interfaces that are familiar and easy to use. The chapter emphasizes the importance of conducting user research to understand the needs and preferences of the target audience and using design patterns and tools to create interfaces that are familiar and easy to use. Mobile apps have become an integral part of our lives, and designing a successful mobile app is a challenging task that requires a thorough understanding of user needs and preferences.

Preface

The Smart and Intelligent Technological Era is coming to the earth. People now depend more on mobile apps. Everything is monitored or controlled by a mobile device, from nighttime bedtime to morning awakening time. The most recent study area in this expanding business is mobile applications. This book's primary focus was on the fundamentals of the Android OS and its device features. Activities and intents, the two primary building blocks of Android programming, are also covered. The several layouts are essential for demonstrating how to save or store data in any application. The deployment of any Android application is finally covered in this book's conclusion. As mobile communication has become so ingrained in our daily lives, many people find it difficult to function without one. When the phone first came out, the only commonly used features were calling and sending text messages (texts). With the development of the smart mobile phone, which has shown to be a multipurpose tool that not only works best for communication but also aids in learning, earning, and having fun. This in turn prompted several developers to consider creating mobile applications.

Chapter 1: Only with the aid of a sound architecture and the guiding principles that underpin it is it feasible to create effective software or mobile applications. Applications that adhere to architectural principles have been found to perform well in terms of scalability, maintainability, availability, interoperability, and other factors. The creation of mobile applications should be built on SOLID principles which leads to high-quality code without any additional work on the part of the developer. The design of a system's functional and non-functional requirements is the main emphasis of the architectural process. You can develop a greater grasp of the wider picture of the system as a whole by defining the system's architecture. Developers may create scalable and high-quality applications thanks to well-designed architecture and principles.

Chapter 2: The suggested wellness mobile application employs numerous procedures, both automatic and manual, to continuously record different metrics. In order to encourage the user to change their lifestyle, the Wellness App also calculates and monitors the appropriate indicators for tracking nutrition, physical activity, and sleep quality. These indicators are analyzed by an inference reasoning module to determine the kinds of suggestions the subject will require to adopt the desired behaviors. These recommendations may include things like improving the activity schedule or adhering to the Mediterranean diet, which is known to be the best diet for reducing cardiovascular disease. Flutter, a wellness software for Android and iOS, takes care of guidance, exercise, sleep, and mindfulness, among other related issues.

Chapter 3: Mobile architecture is a crucial but frequently disregarded component in developing mobile applications. It establishes an app's structure and organization, and it is essential for controlling the complexity of the code base and fostering developer cooperation. It is impossible to exaggerate the significance of mobile architecture in the fast changing mobile environment of today. So what precisely is mobile architecture, and how can it be used to the development of effective mobile apps? The usage of version control systems, which are potent instruments that assist developers in tracking and managing changes to an app's source code, is one of the essential elements of mobile architecture. The function of version control systems in mobile architecture will be examined in this paper, and it will be discussed how these systems can aid developers in producing effective and high-quality mobile apps that can withstand the demands of the mobile environment. Read on to find out more about the significance of mobile architecture and the use of version control systems if you're a developer hoping to build the next big mobile app.

Chapter 4: Mobile phone is now one of the most important elements of our lives. We know that they are helpful, and enable us to perform different services for our requirements online. But, there are limits to everything. It is excessive that we cannot even eat or go to sleep without our mobile phones. According to some statistics, a person uses their mobile phone on average 3 hours and 15 minutes a day. The part of the population that is affected most by this problem is teenagers. Most of them suffer from "nomophobia", which is defined as "the fear of not having a mobile phone". Teenagers are the most impacted by mobile addiction because they cannot manage their screen time and they stillhave not developed the ability toself-control. According to some research, it is ascertained that 27% of the population who own a mobile phone are people around 11-13 years old, that don't turn off their phone, not even to sleep. There are real-life problems that affect phone addiction, such as family problems, impaired school relationships, psychological problems, and so on. The negative impacts of mobile addiction are uncountable. It can cause real physical and mental health problems. We certainly know that we need mobile phones for our everyday tasks and we cannot live without them. Therefore, the best solution to addiction is time management. We have to use our mobile phones in our everyday lives but we can at least manage and reduce the time spent on them.

There are several studies on the Software Development Life Cycle (SDLC) and various approaches that can be utilized in planning and designing a software project.

The contemporary norm for the software development process across industries is to build products using iterative or incremental design. Iterative and Incremental Development (IID) are useful for different types of software. We have provided a literature review of the current definitions of iterative and incremental design, and we also discussed the most common pitfalls for each approach. We also analysed which approach is better for different stages of the development of mobile applications. At the end of our research, we concluded that the iterative approach is the best approach for front-end development, while the incremental approach is a better fit for back-end development. Finally, we provided an analysis of utilizing IID for applications in the banking sector. This analysis was useful in asserting the importance of field knowledge for software development. We demonstrated that different industries provide a unique set of development challenges. This allows us to assert that the best way to pick between iterative and incremental development is by scrutinizing the unique problems we need to solve for the given field and designing a custom approach based on the client's needs in this Chapter 5.

People nowadays use their phones constantly throughout the day, for a variety of reasons. It seems reasonable that they may be used for educational purposes as well. The process of education has become much easier with mobile applications. Through these mobile apps, students can access all these books and information. They can save a lot of time without the need to go to the library to acquire it. Thanks to these mobile apps, classes and lectures are not confined to physical classrooms only. Students no longer have to wait until class to ask questions; instead, they may get in touch with their professors whenever the need appears. When schools were closed and lessons were done online during Covid-19, mobile applications were also incredibly helpful. Back then, "Zoom" and "Google Classroom" were the two most popular distance learning platforms. Today a variety of apps are used to help students concentrate on their studies, assist with problem-solving and be accessible at any time and location. Some apps can assist students with time management issues in creating and following a timetable. Learning has become more entertaining and engaging as a result of educational mobile applications, which have also increased student performance in Chapter 6.

Chapter 7: Mobility is one of the main important problems for the totally and partially disable people in a daily life. The disabled people need some assistive stick or device to avoid the obstacles before accident. This research work proposes to assist the disabled people in daily routine without hu-man guidance. There are four main sections in this paper. Firstly, in obstacle detection system, five ultrasonic sensors sense the obstacles in front side, left side and right side of the user's way on the ground and the another one ultrasonic sensor senses the over- hanging obstacles near the user's head level. Secondly, in obstacle avoidance system, the servo steering motor is equipped with the front wheel of the stick's base. The wheel with servo motor will turn easily to the left direction or right direction depending on the detecting or sensing of the ultrasonic sensors. Thirdly, in voice-guided system, there are six recorded human voice messages with My-anmar language for the disabled users in our country, Myanmar. The appropriate voice message among six voice ones will play automatically, if there is an obstacle in front of, left, right of the stick or upper side for the user's head. Those six messages depend on the detecting of the sensors and the turning of the front wheel with servo steering motor. And then, the vibrating motor will turn on based on the sensors. Fourthly, in lighting system, if there is no light in the room or it is at night, the LED of the stick will be ON. If not, LED will be off automatically. Although the light of LED is not useful or effective to the disabled people, it is very useful and good for the sighted people to avoid the disabled user's stick before the sighted people to avoid the disable user's stick before the sighted or ordinary people do not hit it suddenly. All these four sections are suitable for the disabled users to avoid the obstacle or disturbance and to know where the obstacle is before accident. Hence, this voice-guided system for dis-able people using mobile robot will be very valuable, useful, and effective to walk and more easily in their surroundings.

Chapter 8 examines impending disaster and daily forecasts, 3D augmented reality (AR) weather visualization was created. Imagine a society in which any physical object may be created, destroyed, altered, and put to countless uses. Such a location can be created using augmented reality (AR) and virtual reality (VR). Background: To assess, learn how to implement, and construct an AR weather application utilizing the OpenWeatherMap Weather API. Via an API, OpenWeatherMap provides access to regional and international weather data (API). Millions of users worldwide are served by OpenWeatherMap, which offers a variety of weather-related products with varied levels of precision and depth. Can 3D AR weather visualizations on smartphones and tablets be useful tools to improve current weather forecasting methods? Methods: To enhance visualization in this chapter, we'll use an AR weather app created using Unity, the AR Foundation, and OpenWeatherMap.

Chapter 9: The reader is introduced to numerous unique and recent advancements in the field of machine learning in this chapter, which also provides a wide overview of machine learning and artificial intelligence. This compilation's first half offers a comprehensive overview of the traditional machine learning ideas, including ensemble learning, the idea of large data, how to handle big data, and predictive data analytics using big data. Computer vision (C.V.), the Swarm Algorithm, network science/graph theory and applications in machine learning, bioinformatics employing machine learning, and the Internet of Things are some examples of machine learning (M.L.) frameworks that are discussed (IoT). The second most popular language used globally for machine learning, R, is added as a side note, and this chapter focuses mostly on Python for ML. The next section elaborates on deep learning, machine learning principles, models, types, and algorithms before giving a thorough introduction to neural networks, concepts including weight initialization, propagation, and the vanishing gradient problem. This chapter's introduction to neural networks covers their initialization, weights, forward and backward propagation, and initialization. The final section of this chapter introduces the reader to the fascinating and cutting-edge world of artistic, image-generating AI engines such as DALL-E Image Synthesis Program, Stable Diffusion, and Midjourney's artistic engine via Discord bot, highly responsive, human-like chatbot ChatGPT, Art generation with AI, and Generative Neural Network conception (LLM).

Chapter 10: It is crucial to have a strong user interface for a mobile application since there is a huge demand for mobile applications that offer users both good functionality and a positive user experience as a result of the rising global use of mobile devices. To make it simpler for a developer to construct them, numerous app components are built and illustrated in this article using the Figma software. Also, it shows how to use Figma. By doing so, it demonstrates the design ideas that went into using the appropriate color schemes, typeface, amount of white space, button placement, and other design elements.

Chapter 11: Several research areas currently make use of digital image processing. One of them works in the medical industry. In fact, even skilled radiologists have trouble identifying the blood artery segments in the lung that are malignant or finding tiny nodules that point to lung cancer on X-ray images. In 30% of positive cases, doctors and radiologists are unable to find cancerous areas, according to prior studies. A CAD system has been developed and put into use to identify and detect different types of cancer, however the results show that there are still a lot of errors in the classification outcomes. It is believed that this research and the algorithm it produces would enable medical professionals to more accurately and quickly identify cancer. A final percentage of 90.65% was reached for the test results' performance in classifying 10 X-ray images following 20 iterations (3 cancer image data, 3 pneumonia image data, 3 covid-19 data and 1 normal image).

Chapter 12: The architecture of mobile games must include class models since they give the code a framework for organization and structuring. These models enable more effective and efficient code development by assisting developers in understanding the connections and interdependencies between various game components. As mobile games get more sophisticated and complicated, the use of class models is anticipated to continue to develop and advance. To depict AI and ML in mobile games, developers will need to design new classes and models as these technologies improve. Moreover, the use of remote technologies such as cloud computing is projected to increase, necessitating the development of new classes and models to represent these technologies in the game. In general, class models will be essential for enabling developers to produce more deep and compelling mobile gameplay experiences.

Chapter 13: In order to organize and structure the code, class models are a crucial component of the architecture of mobile games. These models aid programmers in comprehending the connections and interdependencies among various game components, enabling more effective and efficient code development. As mobile games become more intricate and sophisticated, the use of class models will probably continue to develop and get better. The representation of AI and ML in mobile games will require the development of new classes and models as these technologies improve. Also, it is expected that cloud computing and other distributed technologies will be used more frequently, necessitating the development of new classes and models to represent them in the game. Class models will, in general, be essential in enabling developers to produce more captivating and immersive mobile gameplay experiences.

Chapter 14: Throughout the past few years, websites have developed using a range of programming languages and frameworks. Other programmers were concurrently working on them. To handle these complex web apps, developers launch their projects using a variety of design patterns, which makes the code simpler and easier to deal with. Its use of the Model View Controller (MVC) methodology is well recognized. According to the MVC architectural pattern, an application is made up of three parts: Model, View, and Controller, or MVC. Each of these components is designed to cover a certain area of application development. This chapter focuses on MVC for viewing websites on mobile devices. This chapter is divided into several sections that include an introduction to MVC, MVC components, the architecture pattern, numerous methods for putting the pattern into practice in the project, and an illustration of MVC architecture. The merits and cons of MVC are examined at the chapter's conclusion.

The reader is introduced to numerous unique and recent advancements in the field of machine learning in this chapter, which also provides a wide overview of machine learning and artificial intelligence. This compilation's first half offers a comprehensive overview of the traditional machine learning ideas, including ensemble learning, the idea of large data, how to handle big data, and predictive data analytics using big data. Computer vision (C.V.), the Swarm Algorithm, network science/graph theory and applications in machine learning, bioinformatics employing machine learning, and the Internet of Things are addressed as examples of machine learning (M.L.) frameworks (IoT). The second most popular language used globally for machine learning, R, is added as a side note, and this chapter focuses mostly on Python for ML. The next section elaborates on deep learning, machine learning ideas, models, types, and algorithms before giving a thorough introduction to neural networks, weight initialization, propagation, and the vanishing gradient problem.

This chapter's introduction to neural networks covers their initialization, weights, forward and backward propagation, and initialization in this Chapter 15.

This chapter provides a general overview of machine learning using the Scikit-learn module on Android mobile devices. The reader is introduced to the Python language in the first section. After that, a quick historical note on Python implementations on Android mobile phones is followed by an introduction to Python on Android. The following section introduces Pydroid3. Instructions for configuring an Android phone for machine learning are provided after this. os, pathlib, Pandas, NumPy, SciPy, Matplotlib, Seaborn, PySimpleGUI, NetworkX, Biopython, WordCloud, Kivy, and Jupyter Notebook are examples of supporting modules for machine learning that are available for Pydroid3. The Scikit-learn library, fundamental ideas of the Scikit-learn module, and techniques available with this module, such as Logistic and Linear Regression algorithms, data dimensionality scaling (PCA), XGBoost, K means, and Support Vector Machine, are all described in the compilation's final section in Chapter 16.

Chapter 17: Several businesses have changed their IT infrastructure and upkeep processes as a result of cloud computing, and some may even completely stop using their current technology. Traditional techniques for configuring switches and routers may be error-prone and unable to fully exploit the potential of modern network topologies. Future networks must be intellectualized, activated, and customized using as many intelligent networking architectures as possible. The rigid connection between the control plane and the data plane in conventional network topologies can be removed thanks to software-defined networking (SDN) technology, which also makes it feasible to control, secure, and optimize network resources. In this article, we look at the issues, challenges, and potential fixes related to software-defined networks (SDN), a revolutionary idea in computer networking. The network becomes programmable, fast, and adaptive through SDN because of the separation of data and the control it has over traffic. Our research's objective is to comprehend various SDN architecture ideas in order to improve knowledge. The principles of conventional networking will be covered in this chapter, along with the significant distinctions between SDN and conventional networking. Inquiring about the principles needed to develop SDN and the range of applications for which it is useful. These are a few examples of how SDN is utilized in the business world and the primary reasons why companies use SDN.

Chapter 18: We utilize hand gestures in our daily lives for a variety of tasks, such as summoning someone or giving directions, without even realizing it or consciously recognizing them as a part of our lives. But, effectively participating in society is a much more significant aspect of daily life for those who have hearing impairments. They are an excellent means of human communication, and their primary means of communication. Processes for recognizing hand gestures based on eyesight have numerous known benefits over more traditional systems. This chapter looks at several tools that can interpret hand gestures made by users of American Sign Language to communicate (ASL). The following are the results of the survey. This idea is very easily implementable in schools to incorporate ASL into their curriculum.

Chapter 19: The range of applications for computational models that mimic human emotions has expanded significantly in recent years. With the exponential rise of artificial intelligence, emotional computation is becoming increasingly important in interactions between humans and computers. It is widely accepted that the future detection of a person's emotional state of mind should be computed from their facial expressions, a practice known as emotion recognition. Face-movement combinations can convey a wide range of emotions, including, to mention a few: hate, wrath, fright, joy, grief, surprise, and astonishment. The use of various algorithms and computing models for emotions is the manuscript's focus and main objective. Through the construction of AI models that describe the system's difficulties and incorporate Facial Action Coding as a core component of the models, significant breakthroughs in this area of emotion recognition can be realized.

Chapter 20: Reusable answers to typical design issues, design patterns offer a consistent user experience across many apps. With a focus on the user-centered approach to design, this article examines the best practices in mobile app design patterns and tools. Design tools including Sketch, Figma, Adobe XD, and InVision are explored along with design patterns like navigation bars, tab bars, list views, and card views. The challenge is to make sure that the needs and preferences of the user, rather than those of the designer or the technology, are at the center of mobile app design, and that the appropriate design patterns and tools are used to produce interfaces that are recognizable and simple to use. The article stresses the significance of employing design patterns and tools to create interfaces that are recognizable and simple to use, as well as doing user research to comprehend the needs and preferences of the target audience. Our lives now revolve around mobile apps, and creating one that is effective is a difficult effort that involves a deep comprehension of user needs and preferences.

Debabrata Samanta
Rochester Institute of Technology, Kosovo

Acknowledgment

I express our great pleasure, sincere thanks, and gratitude to the people who significantly helped, contributed and supported to the completion of this book entitled with "*Designing and Developing Innovative Mobile Applications*". My sincere thanks to Fr. Benny Thomas, Professor, Department of Computer Science and Engineering, CHRIST (Deemed to be University), Bengaluru, Karnataka India, and Siddhartha Bhattacharyya,, Principal, Rajnagar Mahavidyalaya, Rajnagar, Birbhum, India for their continuous support, advice and cordial guidance from the beginning to the completion of this book.

I would also like to express our honest appreciation to my colleagues at Department of Computational Information Technology, Rochester Institute of Technology, Kosovo, Europe for their guidance and support.

I also thank all the authors who have contributed some chapters to this book. This book would not have been possible without their contribution.

I am also very thankful to the reviewers for reviewing the book chapters. This book would not have been possible without their continuous support and commitment towards completing the chapters' review on time.

To complete this book, all the publishing team members at IGI Global extended their king cooperation, timely response, expert comments, guidance, and I am very thankful to them.

Finally, I sincerely express our special and heartfelt respect, gratitude, and gratefulness to my family members and parents for their endless support and blessings.

Chapter 1
Architecture Principles for Enterprise Software and Mobile Application Development

Tapan Kumar Behera
https://orcid.org/0000-0003-2524-9171
Forrester Research, USA

ABSTRACT

When it comes to software or mobile application development, it is only possible to make it successful with the help of a good architecture and the principles that govern it. It has been found that applications that use the principles of architecture are good at scalability, maintainability, availability, interoperability, and so on. The development of mobile applications should be based on SOLID principles which leads to high-quality code without any additional effort on the part of the developer. The architecture process focuses on the design of both functional and non-functional requirements for a system. By designing the architecture of a system, one can gain a deeper understanding of the bigger picture of the system as a whole. Well-designed architecture and principles play a critical role in enabling developers to build scalable and high-quality applications.

INTRODUCTION

For the Enterprise application or mobile application development irrespective of its Android or IoS based application the Architecture principles are independent, and can be used anywhere. Functional requirements define what a software product must do: its features and functions. The non-functional requirements (NFR) define

DOI: 10.4018/978-1-6684-8582-8.ch001

the system attributes such as scalability, availability, maintainability, reliability, performance, and usability (Barrera et al.). The Non-Functional requirements are just as critical as functional Epic, Features, Story (Arseniev et al.). As a result, the system is made more usable and effective. In the absence of any one of these criteria, a system may fail to satisfy internal business, user, or market needs, or fail to meet regulatory or standard requirements. Figure 1 shows the Non-Functional Requirements. Let's discuss some of the Non-Functional Requirements:

THE KEY NON- FUNCTIONALREQUIREMENTS ARE

Figure 1. Non-functional requirements

Availability

Availability refers to the likelihood of a user being able to access a specific system at any given moment. It can be quantified by calculating the percentage of successful requests or the percentage of time the system is operational during a specified period. For instance, users may have access to the system for 99% of the time in a month. Availability is a crucial business requirement, but it also necessitates estimating the reliability and maintainability of the system to define it. Employing Multi Cluster or Blue Green deployment ensures that the application is available at all times (Behera et al. 2022). In order to achieve high availability, a system must be designed with redundancy and fault tolerance. Various methods can be used to achieve this, including redundant hardware, software, and network connections; failover mechanisms can also be used. Automated testing and system monitoring can be used to measure availability. In order to identify areas for improvement, it is important to regularly monitor and measure availability (Behera,. 2023) (Behera et al. 2022).

A software system's availability is influenced by many factors, including its architecture, its scalability, and its ability to handle peak usage periods. Creating highly available software systems makes them more reliable, resilient, and valuable to their users. The availability of a software system affects its overall effectiveness and reliability. High-availability software systems are more likely to be adopted and used by their users, and to deliver the desired benefits and outcomes (Aviv, I. T. Z. H. A. K. et al. 2023)). High availability can improve the effectiveness, value, and effectiveness of software systems (Bass et al. 2003).

Performance

It is very important to consider the app performance when developing a mobile application, since there is limited memory and capacity on mobile processors. A mobile widget turnaround time should be fast enough when compared to the turnaround time of .com applications as well. In the case of a mobile application, it is important to check the performance of every call made by the page (Robertson et al., 2006). There is a risk that the user will not like the mobile app if the response time of the app is high. It is important for the architecture to be designed in a way that meets the performance expectations of the customers in an efficient manner. The performance of the application will be enhanced if it has several cluster nodes to serve the request, as that way the request will not have to wait on the queue to get processed, which will improve performance. The load-balancer for the API Gateway needs to be configured to distribute load across the clusters, so that each request will be handled by a different node in the cluster. Horizontal and Vertical scaling will increase the performance of the application (Sachedina et al.).

A performance requirement can include metrics such as response time, throughput, and resource utilization. Throughput refers to the number of requests the system can handle per unit of time, while response time refers to how quickly a system responds to a user request. The utilization of resources, such as CPU, memory, and disk, measures the efficiency of the system. The performance requirements should be determined by the users' needs and the expected load on the system (Fowler et al., 1999) (Buschmann 1996). In order to ensure that users can complete their transactions quickly on an e-commerce website with many concurrent users, there may be a need for a high level of throughput. In order to achieve high performance, software systems optimize resource utilization, minimize bottlenecks, and utilize caching to reduce response times. Performing performance tests is another important aspect of software development, which involves simulating expected user loads ands analyzing system performance (Haw et al., 2023). The ability to achieve high performance can also come at a cost, since further hardware or software optimizations may be necessary in order to achieve that level of performance. The performance requirements for a software system should, therefore, be carefully analyzed based on both the needs of the users and the expected impact on the overall system as a whole (Habibullah et al. 2023).

Security

Security plays an important role in the design of the system; it helps to secure the data that is used in the application by ensuring the architecture of the system is secure. Mobile applications should have the ability to enable security mechanisms at the application level, such as passwords and other complex security mechanisms (Hooda, Susheela, et al. 2023). The protection of digital identity helps to guard against fraudulent attacks on the identity. Secure all internal and external rest service calls and connections. For enhancing the security all the data-source and external connection to be encrypted in the application level. Implemented the Auth0 SSO configuration for additional security in the system (Yasar et al.). Add message level and application-level security in the webservice calls. The following are some common security requirements:

1. Authentication: Users must be able to verify their identities and only authorized users can access the system.
2. Authorization: Users should be able to control access to specific resources and functions based on their roles and permissions.
3. Confidentiality: Data must be protected from unauthorized access.
4. Integrity: Data must not be tampered with or modified without authorization.

5. Availability: Authorized users must be able to access the system, and denial-of-service attacks must be prevented.

Various security mechanisms are included in software systems, including firewalls, intrusion detection and prevention systems, encryption, and access control (Kotusev et al., 2023). In addition, security testing is performed to identify vulnerabilities and ensure that the system will withstand attacks. Security is a continuous process that requires continuous monitoring and improvement. Based on the changing threat landscape and the evolving needs of the system, security requirements for software systems need to be carefully evaluated and updated regularly (Mukherjee et al. 2022).

Maintainability

The maintainability of a solution refers to the fact that it is designed to be able to be fixed quickly, changed to improve performance, or be adapted to changes in the environment in the future because of future changes (Nouman et al., 2023). There is a similar way in which the probability of a component being repaired over time can be expressed as a function of reliability. For a software system to be maintainable, the following requirements must be met:

1. Modularity: A modular system consists of well-defined and independent components that are easily modified or replaced.
2. Documentation: A detailed description of the architecture, design, and implementation of the system is necessary.
3. Code quality: Code must be well-structured, well-documented, and easy to understand and modify.
4. Testability: Components need to be easily testable, and the system must be designed with testing in mind.

Version control, code reviews, and automated testing are among the software engineering practices used to ensure maintainability of software systems. A regular maintenance program is also conducted for the purpose of keeping the system up-to-date and functional, including bug fixes, updates, and enhancements (Pang et al., 2022)(Rahy et al., 2022) (C. et al., 2007).

Scalability

Scalability measures the system's ability to scale up or down its performance, and the system should be able to handle the growing workload by adding more resources. A computer application or product (hardware or software) that is designed to meet the needs of a user must continue to function at its best despite any change in its size or volume. Make the application high scalable by implementing vertical scaling and horizontal scaling.A horizontal scaling occurs when more nodes (VMs) are added to the cluster in order to run in parallel with each other (Martin et al., 2017) (Serban et al., 2022). By doing so, it will be able to receive more resources not from a single node, but from several. A computer system is vertically scaled when there is an increase in resources. For instance, an element of your web application begins to consume more physical memory as it processes incoming requests (Trad et al., 2023). However, a single node is limited in capacity. Software systems can be scalable in the following ways:

- Vertical scalability: A single machine must be capable of handling increased traffic by adding additional CPUs, memory, or storage.
- Horizontal scalability: Adding more machines and distributing workload across them allows the system to handle increased traffic.
- Load balancing: To prevent any single machine from becoming a bottleneck, the workload must be distributed evenly across multiple machines.

A variety of techniques are used in software systems to ensure their scalability, such as caching, sharing, and distributed databases. The system's configuration is also optimized by performing performance tests. A software system's long-term viability depends on its scalability. It can be costly and risky for a system that is not able to handle increasing amounts of traffic or workload. In order to ensure the scalability of a software system based on the expected growth and evolution of the system, it is imperative that scalability requirements are carefully evaluated and prioritized.

Reusability

The reusability of components in a system improves the speed at which an application can be developed and reduces the cost of development, it also helps with the timely release of the product as reusability reduces effort. Reusability refers to the ability of the code to be used across a variety of platforms for a variety of functions. A programmer is not supposed to repeat himself, which is one of the basic principles of reusability of code. During the design of a system, identify common modules, such as exception handling, logging, security, connections, and database operation, during the process of system design. The common modules can then be used during the integration process of a system. Using this approach, you will be able to reduce the size of the application, increase its performance, and improve its maintainability. Software reusability requirements can include:

1. Modularity: Components must be well-defined and independent, so they can be reused across different parts of the system.
2. Standardization: Systems must adhere to industry standards and best practices to make them easier to integrate with other systems and reuse components.
3. Compatibility: It is important to design a system so that it can be easily integrated with other systems and reused.

Component-based development and service-oriented architecture are some of the software engineering practices used to ensure reusability. In addition, reusable components are stored in code repositories and libraries. By reducing the time and effort required for development and maintenance, reusability can result in significant cost savings and improved productivity. Based on the expected lifespan and evolution of the system, as well as the possibility of future integration, reusability requirements for a software system need to be carefully evaluated and prioritized.

Portability

Using the same software across different environments is called software portability. In the field of mobile application development, it is not uncommon for the mobile operating systems to be upgraded every year, and the application should be designed in such a way that it can seamlessly work with the new version of the OS, or a new model mobile device with a new set of hardware. When an application is designed with a good architecture, by keeping portability in mind, the impact of a change on the application will be very minimal, whenever any new hardware or new model is introduced to the market in the future (Velasco-Elizondo et al., 2022).

A software system's portability is determined by several factors. This includes the use of standard programming languages and interfaces, the avoidance of hardware-specific features, and the use of modular design principles. It is possible to create more portable software systems by adhering to these principles. Portability can sometimes require trade-offs with other NFRs in order to achieve. The performance of a system can be improved by optimizing it for a particular platform, but its portability may be compromised. It is therefore important for software engineers to achieve a balance between different NFRs.

Accessibility

A product or service's accessibility refers to whether it can be used by everyone, no matter how they meet it. The purpose of accessibility laws is to provide assistance to people with disabilities, such as visual impairments, hearing impairments, and motor impairments, but designers should attempt to accommodate all potential users in various contexts in which their products will be used. By implementing the following techniques can improve the accessibility of an application. These techniques are as follows. Create an appropriate structure and organize with headers. Use Alt tags for images. Select colors carefully. Use a keyboard to navigate the website.

There are number of factors that software engineers need to consider when designing for accessibility. For example, audio descriptions for visual content or text captions for audio content may be used to provide alternative access to information. People with visual impairments may also need software that has larger fonts, high contrast colors, or other features that make it easier for them to use.

As part of accessibility, software may also need to be designed to work with assistive technologies, such as screen readers and braille displays. It may be necessary to adhere to accessibility standards, such as the Web Content Accessibility Guidelines (WCAG) or the Americans with Disabilities Act (ADA).

Agility

Software agility refers to a system's ability to handle changing requirements and environments quickly and effectively. Adaptable software systems can be easily modified and adapted to meet new needs without requiring extensive redevelopment.

Modularity, flexibility, and scalability play a key role in achieving agility in software engineering. As a result of using these practices, it is possible to modify or extend software systems as needed, without having to introduce new errors or break existing functionality in the process. The use of agile development methodologies, such as Scrum and Kanban, is one of the most common ways to achieve agility within software engineering. As a result of these methodologies emphasizing iterative development and continuous feedback, software engineers are able to quickly adapt to changes in requirements or user needs in response to these changes. In general, agility plays an important role in NFRs because it allows software systems to remain relevant and effective over time, even as user needs, and requirements evolve. Software engineers can enhance the responsiveness, adaptability, and value of their systems by designing systems that are agile.

Usability

The usability of a software system refers to how easy it is to use and how easy it is to learn. An easily understandable, easily learned, and easily used software system is one that does not require extensive training or effort to use. Several design principles and practices are required to achieve usability in software engineering, including user-centered design, user testing, and iterative design. As a result of these practices, software engineers can design software systems that are intuitive, user-friendly, and efficient. The design of a software system's user interface, the clarity of its instructions and documentation, and the consistency of its behavior across contexts all contribute to its usability. A software engineer who adheres to these principles can create software systems that are more usable, effective, and provide a better user experience. It is important to consider usability as an NFR because it affects the overall effectiveness and adoption of a software system. It is more likely that users will adopt a software system that is easy to use and understand, resulting in the intended benefits and outcomes. Software engineers can design more effective, more valuable, and more widely used systems by designing usable ones.

Compatibility

A software system's compatibility refers to its ability to work seamlessly with other software, systems, or devices. In order for a software system to be compatible, it must be capable of integrating and communicating effectively with other systems or devices without causing errors or incompatibilities with them. A number of design principles and practices are required to ensure compatibility in software engineering, such as adherence to standards, use of open APIs, and careful testing. By following these practices, software engineers ensure that their software can integrate with other systems and provide the functionality and results they intended. The application compatibility should be checked across all the browsers and operating systems that an end user may use, including Chrome, Safari, Opera, Firefox, etc. Testing the code in different browsers, adding a doctype, and using CSS would help in achieving compatibility.

SOLID PRINCIPLES

Whenever there is an application development is concern, the developers should follow the SOLID principles. For Enterprise or the mobile application development, it's play an important part to develop the mobile application in a very effective manner. A set of five design principles referred to as SOLID principles is used in the development of object-oriented software to ensure that the system is more maintainable and scalable. These 5 design principles first letters make the name SOLID, and these named as:

S: Single Responsibility Principle.
O: Open-Closed Principle.
L: Liskov Substitution Principle.
I: Interface Segregation Principle.

D: Dependency Inversion Principle.

Single Responsibility Principle (SRP)

This Single Responsibility principles says that A module or a class should have one, and only one reason to change. The class should be responsible for one functionality, if there are more functionality came than that needs to split into separate class. The class functionality should not be change for any other functionality changes, it should be a independent one. This helps the loose coupling of the system. "A class should have only one reason to change." (Robert C. Martin, Agile Software Development). In the Figure 1 illustration of Single Responsibility Principle, how we can implement in application design. Software developers can write more modular and maintainable code by adhering to the SRP. There is a clear and focused responsibility for each class or module, which makes it easier to understand and modify. It is also possible to improve code reuse and testing through the use of the SRP, as individual classes or modules can be extracted and reused in other parts of the system as needed.

```
class AuthenticationManagerService {
public boolean authenticateUserDetails(String username, String
password) {
                    // Authentication logic for the user goes
here
          }
       }
class UserProfileManagerService {
          public UserProfile getUserProfile(String userId) {
                    // Get user profile logic goes here
          }
          public void updateUserProfile(UserProfile
userProfile) {
            // User profile update logic goes here
          }
}
```

This example uses the AuthenticationManagerService class to authenticate users based on their username and password. On the other hand, the UserProfileManagerService class manages user profile information, such as retrieving and updating it. The separation of these responsibilities into two classes allows us to make it easier to maintain and modify our code over time by ensuring that each class has only one reason to change. Additionally, separating these responsibilities allows us to reuse these classes in other parts of our system, which can reduce code duplication and save time.

Open-Closed Principle (OCP)

The mobile application should be open for extensions but closed for modification. This indicates that any application should be extendible and new more functionality can be added but the core functionality should be remain same. Any developer should be able to extend any class without changing its core functionality according to the Open/Close principle. In other words, it is open to expansion but closed to modification. This will avoid A cascade of changes to modules that are dependent on one another. In order to achieve this principle, it is generally recommended that you refer to Abstraction in Object-Oriented principles for dependencies, such as interfaces and abstract classes, rather than using concrete classes for dependency resolution. A new class that implements an interface can also provide additional functionality. This approach reduces the risk of adding new bugs to existing code, resulting in a more robust application. The following classes could be changed to the second one by applying this principle.

```
public class Triangle {
        public double base;
        public double height;
}
public class Square {
        public double side;
}
public class calculateArea {
        public double getAreaTraiangle(Triangle triangle) {
                return area = (triangle.base * triangle.
height) / 2;
        }
        public double getAreaSquare(Square square) {
                return square.side * square.side;
        }
}
```

And after applying the Open-Close Principle

```
public interface Shape {
        public double getArea();
```

```
}
public class Triangle implements Shape {
        double base;
        double height;
        public double getArea() {
                return (base * height) / 2;
        }
}
public class Square implements Shape {
        public double side;
        public double getArea() {
                return side * side;
        }
}
public class AreaCalculate {
        public double getShapeArea(Shape shape) {
                return shape.getArea();
        }
}
```

Liskov's Substitution Principle (LSP)

This principle is based on "substitutability", its states that an Object (as class) may be replaced by a sub-object (a class that extends the first class) without breaking the programs, in the Object-oriented programming. This principle is an extension of the open/close principles. In this Figure 2, the base class Vehicle standard Speed is 40 Km/h, whereas the derived class Car has superspeed of 120 Km/h; as this is derived from base class it can also have the access to base class property of standard Speed of 40 km/h. So when a new vehicle is designed than its not necessary to get extend from the Base class again, it would be ideal if it extends from the subclass as the subclass has inherited property and also his own new property. So with this approach, Liskovs' substitution principle satisfies and this will not break any existing functionality of the application or module. Figure 2 shows Liskovs' substitution principle example.

Figure 2. Liskovs substitution principal example

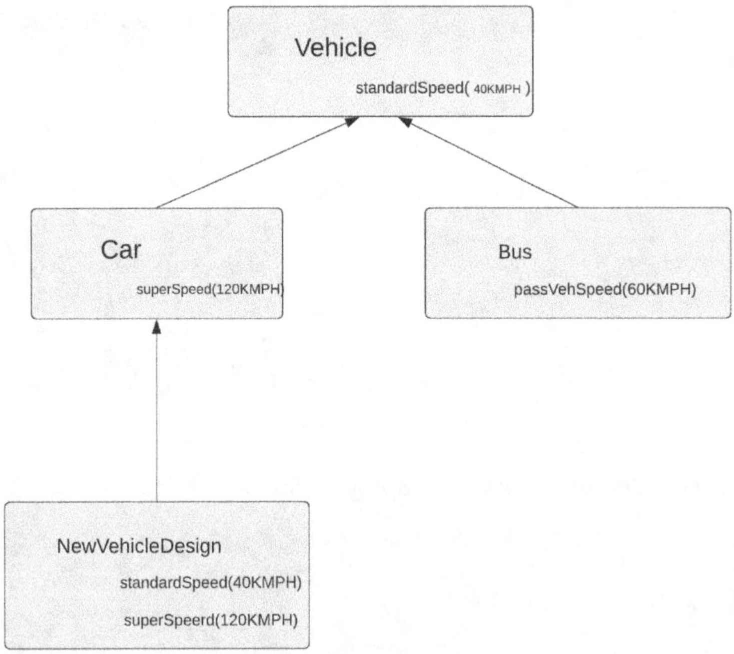

Interface Segregation Principle (ISP)

"Clients should not be forced to depend on methods that they do not use." (Martin, *Agile Software Development*). Interface Segregation Principle states it is better to have several client-specific interfaces than one general-purpose interface. Clients depending on a subset of interfaces need to change when other clients using a different subset change. In this below example, Figure 3, all the client operations in one Interface, whenever there is a new addition for any of the client interface the rest other client forced to implement and do the changes. Which violates the principle. So, after applying ISP- Interface segregation principle, in Figure-4, the major interface is splitted into client specific interfaces, this makes the system loosely coupled. And whenever any change happens to any client specific operations this will not force other clients to implement the changes. Figure 3 shows Client Operation design without ISP and Figure 4 shows Client Operation design by applying ISP.

Figure 3. Client Operation design without ISP

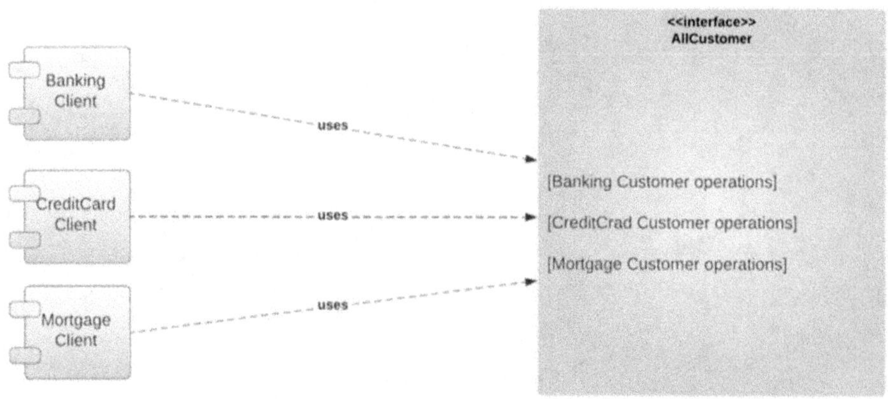

Figure 4. Client Operation design by applying ISP

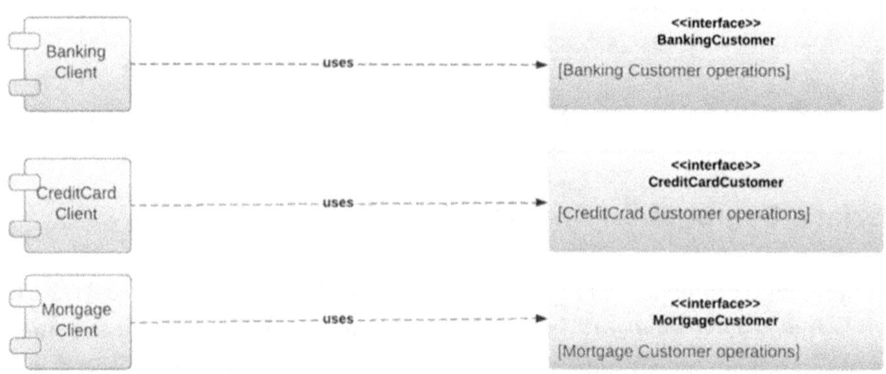

Dependency Inversion Principle (DIP)

The principle states that "*High-level modules should not depend on low-level modules. Both should depend on abstractions. Abstractions should not depend on details. Details should depend on abstractions.*"

This provides a loose coupling feature in software development.

a. Modules at the high level should not import anything from modules at the low level. Abstractions (e.g., interfaces) should underlie both.
b. Abstractions should not be dependent on details. Abstractions should be the basis for details (concrete implementations).

Without the Dependency Inversion Principle, Object A in Package A refers to Object B in Package B in the figure above. The Dependency Inversion Principle introduces an abstraction, Interface A, into Package A. Object A now refers to Interface A and Object B inherits from Interface A. As a result of the principle:

a. Interface A, the abstraction, now governs both Object A and Object B.
b. This inverted the dependency between Object A and Object B, making Object B dependent on the abstraction (Interface A).

Figure 5. Dependency Inversion Principle

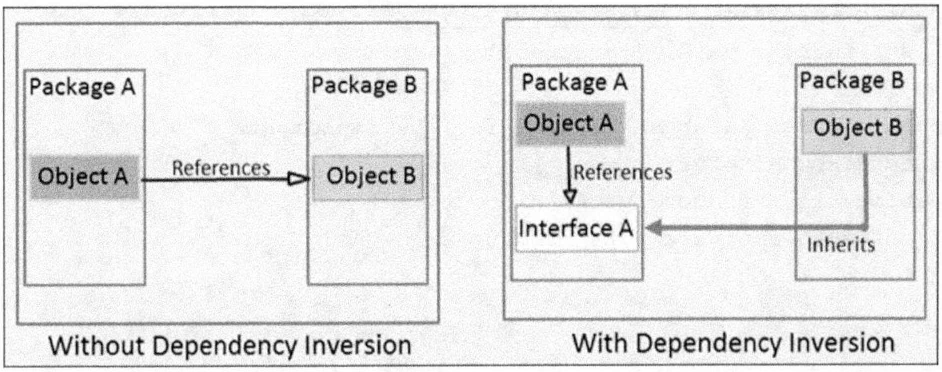

Example of Dependency Inversion Principle:

Suppose we have two classes, **DatabaseRetrievalService** and **WebpageDisplayService**, where **DatabaseRetrievalService** retrieves data from a database, and WebpageDisplayService displays the data on a webpage.

Without applying DIP, the **WebpageDisplayService** class would have a direct dependency on the **DatabaseRetrievalService** class:

```
package solid.pattern;
public class DatabaseRetrievalService {
public List<Product> getAllProducts() {
// code to retrieve data from database
}
}
public class WebpageDisplayService {
private DatabaseRetrievalService db = new
DatabaseRetrievalService();
```

```
public void displayProducts() {
List<Product> products = db.getAllProducts();
// code to display products on webpage
}
}
```

As you can see in the above example, the WebpageDisplayService class is tightly coupled with the DatabaseRetrievalService class.

In order to apply DIP, we create an interface, ProductRepositoryService, which will be implemented by the DatabaseRetrievalService class as follows:

```
public interface ProductRepository {
List<Product> getAllProducts();
}
public class DatabaseRetrievalService implements
ProductRepository {
public List<Product> getAllProducts() {
// code to retrieve data from database
}
}
```

In the next step, the WebpageDisplayService class will be modified to depend on the ProductRepositoryService interface rather than the DatabaseRetrievalService interface:

```
public class WebpageDisplayService {
private ProductRepository repository;
public WebpageDisplayService(ProductRepository  repository) {
this.repository = repository;
}
public void displayProducts() {
List<Product> products = repository.getAllProducts();
// code to display products on webpage
}
}
```

As shown in the example above, the WebpageDisplayService class now relies on the ProductRepositoryService interface, which, in turn, can be implemented by any class, not just the DatabaseRetrievalService class. This example illustrates the concept of loose coupling, which is one of the key benefits that the Dependency Inversion Principle has to offer.

BENEFITS OF SOLID PRINCIPLES

There are several benefits to using the SOLID principles when it comes to the development of object-oriented programs. The following are a few key advantages:

1. Maintainability: Code that adheres to SOLID principles is more maintainable over time. Modularity encourages isolation and bug fixing, as well as adding new features without affecting other systems.
2. Testability: Code can also be tested more easily using SOLID principles. It is easier to create unit tests for each module when code is organized into smaller, more focused modules. Bugs can be identified and fixed earlier in the development process as a result.
3. Flexibility: Code design should be flexible according to the SOLID principles. It is easier to adapt code to changing requirements or technologies when abstractions are used rather than concrete implementations.
4. Scalability: SOLID principles encourage modularity and reusability, making it easier to scale a codebase as a project grows.
5. The use of agreed-upon principles is an effective way for developers to work together as a team. Coding principles help reduce conflicts and misunderstandings by providing a common vocabulary and expectations for code design.

In general, the SOLID principles can lead to a better code design and provide the framework for a more maintainable, testable, and flexible software system. Developers can create code that is easier to work with and more adaptable to changes over time by adhering to these principles.

CONCLUSION

In the fast-growing software world, application development is a key component of a successful digital transformation. In addition, they will directly benefit people in their daily lives. It becomes very important to develop applications that adhere to the Architecture Standards and follow the principles. Having a good architectural design will ensure that the software is user-friendly, resilient, of high quality, and available. The non-functional requirements should be given the same importance as the functional requirements in any software development project. Architects and developers can make a well-structured project and reduce the pitfalls associated with application development by using SOLID principles.

REFERENCES

(2023). *Antoine. "Integrating a Holistic Enterprise Architecture Pattern: A Proof of Concept." Handbook of Research on Digitalization Solutions for Social and Economic Needs*. IGI Global.

Arseniev, D. G. (2023). Software Engineering Principles Apply to Artificial Intelligence Systems. In *Cyber-Physical Systems and Control II* (pp. 151–158). Springer International Publishing. doi:10.1007/978-3-031-20875-1_14

Aviv, I. T. Z. H. A. K., Barger, A., Kofman, A., & Weisfeld, R. (2023). Reference Architecture for Blockchain-Native Distributed Information System. *IEEE Access: Practical Innovations, Open Solutions*, *11*, 4838–4851. doi:10.1109/ACCESS.2023.3235838

BarreraJ. M. (2023). An Extension of Istar for Machine Learning Requirements by Following the PRISE Methodology. SSRN 4358075. doi:10.2139/ssrn.4358075

Bass, L., Clements, P., & Kazman, R. (2003). *Software architecture in practice*. Addison-Wesley Professional.

BeheraT. (2022). How Blockchain Solves the Supply Chain Problems Using RFID Techniques. SSRN: https://ssrn.com/abstract=4287240 or doi:10.2139/ssrn.4287240

Behera, T., & Panda, B. S. (2023). *Master Data Management using Machine Learning Techniques: MDM Bot*. TechRxiv., doi:10.36227/techrxiv.21818040.v1

Behera, T., & Tripathi, K. (2022). *Root Cause Analysis Bot using Machine Learning Techniques*. TechRxiv., doi:10.36227/techrxiv.21588159.v2

Buschmann, F., Meunier, R., Rohnert, H., Sommerlad, P., & Stal, M. (1996). A System of Patterns. *Pattern-Oriented Software Architecture*, (*volume 1*).

Enterprise Architecture As Strategy. (1999). *Creating a Foundation for Business Execution, Fahim Sachedina, Everything There Is To Know About Automated Testing Fowler, Martin. Refactoring: Improving the Design of Existing Code*. Addison-Wesley.

Habibullah, K. M., Gay, G., & Horkoff, J. (2023). Non-functional requirements for machine learning: Understanding current use and challenges among practitioners. *Requirements Engineering*, 1–34. doi:10.100700766-022-00395-3

Haw, M. A., MacDonald, M. E., & Colom, S. V. (2023). Big-data Efficient and Automated Science Transfer (BEAST): an open-source software architecture for arc jet data management, modeling, and automation. AIAA SCITECH 2023 Forum. ARC. 10.2514/6.2023-2712

Hohpe, G. (2020). *The Software Architect Elevator.* O'Reilly Media, Inc.

Hooda, S. (Ed.). (2023). *Agile Software Development: Trends, Challenges and Applications*. Kinza Yasar, Software testing. https://www.techtarget.com/whatis/definition/software-testing

Kotusev, S., Kurnia, S., & Dilnutt, R. (2023). Enterprise architecture artifacts as boundary objects: An empirical analysis. *Information and Software Technology*, *155*, 107108. doi:10.1016/j.infsof.2022.107108

Martin, R. C. (2017). *Clean Architecture: A Craftsman's Guide to Software Structure and Design* (1st ed.). Prentice Hall Press.

Martin, R. C., & Code, C. (2007). *A Handbook of Agile Software Craftsmanship*.

Mukherjee, P. K. (2022). Artificial Intelligence Based Smart Government Enterprise Architecture (AI-SGEA) Framework. *Artificial Intelligence: First International Symposium, ISAI 2022*. Cham: Springer Nature Switzerland.

Nouman, M., Azam, M., Saleh, A. M., Alsaeedi, A., & Abuaddous, H. Y. (2023). A systematic review of non-functional requirements mapping into architectural styles. *Bulletin of Electrical Engineering and Informatics*, *12*(2), 1226–1236. doi:10.11591/eei.v12i2.4081

Pang, C.-Y. (2022). Applying Software Engineering Design Principles to Agile Architecture. In *Research Anthology on Agile Software, Software Development, and Testing, edited by Information Resources Management Association* (pp. 330–355). IGI Global. doi:10.4018/978-1-6684-3702-5.ch016

Rahy, S., & Bass, J. M. (2022). Managing non-functional requirements in agile software development. *IET Software*, *16*(1), 60–72. doi:10.1049fw2.12037

Robertson, D., Ross, J. W., & Weill, P. (2006). *Everything There is to Know About Automated Testing*. Global App Testing. https://www.globalapptesting.com/blog/everything-there-is-to-know-about-automated-testing

Serban, A., & Visser, J. (2022). Adapting software architectures to machine learning challenges. *2022 IEEE International Conference on Software Analysis, Evolution and Reengineering (SANER)*. IEEE. 10.1109/SANER53432.2022.00029

Velasco-Elizondo, P. (2022). Software Architecture: Developing Knowledge, Skills, and Experiences. In *Research Anthology on Agile Software, Software Development, and Testing, edited by Information Resources Management Association* (pp. 17–39). IGI Global. doi:10.4018/978-1-6684-3702-5.ch002

Chapter 2
A Wellness Mobile Application for Smart Health

B. Narendra Kumar Rao
School of Computing, Mohan Babu University, India

ABSTRACT

The suggested wellness mobile application uses many processes, both automatic and manual, to record various parameters on a daily basis. The wellness app also calculates and monitors the appropriate indicators monitoring nutrition, physical activity, and sleep quality to motivate the user to alter their lifestyle. An inference reasoning module analyses these signs to identify the types of recommendations the subject needs to adopt the desired behaviours. These recommendations might include, for example, maintaining the Mediterranean diet correctly, which is acknowledged as the suitable diet for preventing cardiovascular disease, and/or improving the activity plan. Guidance, exercise, sleep, and mindfulness are among linked aspects that Flutter, a wellness app for Android and iOS, takes care of.

INTRODUCTION

Promoting health and fitness through a balanced diet, frequent exercise, and healthy sleep habits is a major goal in many countries. The Wor1d Health Organization estimates that 31% of adults undertake inadequate physical activity, which results in about 3.2 million deaths per year. Additionally, a study found that certain conditions are to blame for a whopping 86% of fatalities, 77% of years spent bedridden, and 75% of healthcare costs in Europe. These problems include illnesses of the heart, malignancies, diabetes, the chronic respiratory system, the mind, and the

DOI: 10.4018/978-1-6684-8582-8.ch002

musculoskeletal system. Common risk factors for these disorders include smoking, being overweight or obese, abusing alcohol, eating too few fruits and vegetables, and not exercising (Abraham et al. 2008).

LITERATURE SURVEY

Obtaining a diet plan through the current dietitian system is entirely manual. To learn about the recommended diet, people must physically visit their neighbourhood dietitian. The user has to wait for their dietitian appointment. The user occasionally may have to wait for many hours. Receiving their diet plan as a result is quite inconvenient for the individuals. Even merely learning what they should eat requires them to wait. When seen from the perspective of the end user, this is ineffective (Alberti-Fidanza, et al. 2017). Under the previous system, conditioning algorithms and data mining were frequently used to construct diet charts, which encouraged database use and only relied on the database, necessitating repetitive data entry. Current systems consider a user's height, weight, and diet while ignoring daily activities, medical issues, and dietary preferences, which is a severe problem. Given that the domains are different, the AI domain has an advantage in producing a healthy diet plan, which other systems lack. The following are some drawbacks of prior systems: It doesn't consider the user's health state, such as diabetes or cardiac patients. A group of dietitians, personal trainers, and behavioural psychologists support the Noom programme, which emphasises on changing behaviours rather than restricting one's diet or following a rigorous exercise routine. It is based on how our behaviours connect to food and the psychology of eating (Alshawmar, M et al. 2021). The app also allows you to monitor your exercise, weight, blood sugar, and blood pressure. It also provides articles to motivate you, health quizzes, and one-on-one health counselling (Alderfer, C.P. et al. 2019).

Middle-Earth About a thousand guided meditations on subjects including stress, anxiety, acceptance, happiness, inspiration, focus, and breath are available in Meditopia's library. The fitness app by Lucy Mecklenburgh. It results in Wellness Lifestyle takes pride in being able to house all of your health apps in one place, with more than 1,000 workouts, 600+ meals, four mindset courses, 11 organised programmes, and monthly live workout sessions. We enjoy the "pain clinic" portion, which focuses on managing back discomfort. Delectably Ella's app With her de l icious plant-based recipes delivered directly to your phone, Deliciously Ella's popular app is like carrying a recipe book around with you. There are about 300 nutritious options to choose from, and the upgrade for 2021 will add even more for you to savour.

TOP NUTRITION APPS

Information quality: We searched for apps that provide trustworthy and beneficial dietary advice, especially from an authority figure like a qualified dietitian. User interface: We searched for apps with simple user interfaces and clear navigation (Brewer, L.C.et al. 2015). Nutritional requirements: We included apps that meet a range of nutritional requirements, from pregnancy and weight loss to meal planning and food tracking. Customer reviews: Friends, family, and coworkers were asked for ideas, read through many user reviews, and searched the web for highly rated apps. Apps have all been reviewed to make sure they adhere to Healthline's business and medical standards. Read more about how we screen candidates (Buysse et al., 1989).

TOP PICK FOR THE BEST NUTRITION APPS

MyPlate Calorie Counter: The app keeps track of your dietary consumption and physical activity and offers a variety of recipes, meal plans, and workouts. Additionally, included is a community element that provides motivation, peer support, and member-submitted guidance. Additionally, the app gives users access to premium services like personalised nutrition labels, daily averages, and detailed information on their nutrient intake and general progress (Rao et al., 2019).

Best Meal Planning App: Before creating an individual meal plan and shopping list for the family, it employs a questionnaire to collect information about your lifestyle, culinary preferences, and nutrition. You can choose certain ingredients you want to stay away from, how much diversity you prefer, and the portion sizes you want in addition to choosing your meal plan (Rao et al. 2019).

Best Food Tracking App: A large food database, a barcode scanner, a recipe importer, a restaurant logger, food insights, a calorie counter, and countless other functions are all included in this useful tool. Using MyFitnessPal to keep track of your nutrition makes it simple to shed pounds, alter eating patterns, or simply hold oneself responsible.

Best Healthy Recipe App: Yummly is my go-to app for healthy cooking, offering more than 2 million recipes as well as a variety of alternatives to meet the tastes, needs, and preferences. Because the recipes are organised by cuisine, course, diet, and necessary components, it is simple to choose one that suits you. Plus, you may link smart appliances to the app, add items to your shopping list, and construct unique collections of your favourite recipes.

Best Weight Loss App: The software Lifesum offers features including a recipe library, food journal, and calorie counter to make losing weight easier. You are asked questions about your height, weight, age, and specific goals when you sign up for the app in order to create a plan that is tailored to your requirements. Also, it has a health test that is done once a week and provides insights into your habits as well as prospective areas for improvement (Rao et al., 2019).

Best Mindful Eating App: Ate Food Journal, which is marketed as mindful food journaling software that doesn't track calories, aims to teach you how to analyse your eating habits and how certain meals affect your mood. With the app, you can capture images of your meals and activities to create a visual diary of the day. You can also take part in studies where you eat with your non-dominant hand while describing the companions, method of cooking, and flavour of each meal.

Best Macro Counting App: The MyNet Diary Calorie Counter is a terrific tool for meal journaling while you're on the road, whether you're attempting to boost your protein intake, decrease your carb intake, or understand the macronutrient composition of your diet. Depending on the particular diet plan you choose, the app assists you in setting objectives, monitoring your consumption, and observing changes in your weight. Also, it offers comprehensive nutritional information for every item in your meal record along with a daily analysis to help you stay on track (Eysenbach et al. 2011).

Best Nutrition Education App: Because it offers individualised meal plans, weekly challenges, and access to a virtual coaching team, Noom is well-liked by dieters all around the world. For instance, according to my mother, the succinct, uncomplicated lesson plans made it easier for her to understand how her food decisions affected her health. Noom not only offers instructional material but also features tools for tracking your development and exercise schedules to increase your everyday activity (Forastiere et al., 2016).

Best Food Allergy App: It gives evidence on whether food goods are safe and allergen-free and enables you to choose the precise substances you need to avoid. To identify allergy-safe dining and lodging options, you can also read reviews left by previous customers for establishments all over the world.

Best Pregnancy Nutrition App: When you're expecting, the many changes in nutritional requirements that occur during pregnancy can be overwhelming. You may now be more confident than ever that you're getting the nourishment you need for a healthy pregnancy thanks to Ovia. The app offers a food safety guide, a meal and activity record, a recipe collection, and prescription reminders. Also, it offers useful tools, advice, and thorough details regarding the development of your pregnancy (Fortino et al., 2015).

OBJECTIVE AND SCOPE

Objective

The project's main purpose is to create a Wellness App utilising the Flutter Framework that aids the user in reaching their wellness objectives. The apps also encourage behaviour changes, including self-monitoring, in relation to things like increasing physical activity, measuring health and fitness parameters, calculating Body Mass Index (BMI), setting weight goals, getting nutrition and diet recommendations, tracking sleep, and getting exercise advice. New exercise and health goals can be set with these applications.

Scope

This app is a user-friendly application provides information about weight loss, weight gain and also diet plans, daily health tips.

ADDING UP CALORIES BY NUTRIENT

Typically, nutritional information refers to the nutrient contents of foods. To find out which foods are good for them, or which foods fit into a particular diet plan they are following, people frequently check for nutritional information. Based on a chemical analysis, the nutritional value of food might include details like calories, fat, carbs, sugars, and protein as well as vitamins, minerals, and other nutrients. Foods that are packaged frequently have nutrition labels with detailed nutritional information. Finding nutritional information by just reading labels is frequently simple. When you buy whole foods, including fruits and vegetables, and goods that don't come in a container with nutritional information, you can often find the calorie and nutrient content of single ingredients and particular foods in the USDA's National Nutrition Database (Ghasemzadeh et al., 2014). Find the nutrition information on the product's packaging. Food manufacturers are required by law in several regions of the world to list nutritional information on packaged food products. On the back or side of the container, normally, you can find a chart outlining this information. You should start by reading the nutrition facts label if you're curious about what you're eating. The nutrition facts on a food can provide you with all the information you require about what is in it, including a detailed list of the ingredients and an explanation of each of the main macronutrients (Fortino et al., 2013).

Take note of the item's protein, carbohydrate, and fat content. Protein, carbohydrates, and fat are the three factors you should consider when evaluating a food's nutritional value. All of the item's calories are accounted for by these macronutrients (aside from calories from alcohol). Add the calorie equivalent of each macronutrient (Gravina et al., 2017). About 4 calories are thought to be present per gramme of protein. A gramme of fat contains a whopping 9 calories, while a gramme of carbohydrates only has 4. If a food item comprises xg of protein, yg of carbohydrates, and zg of fat, you would multiply that amount by Xx4, Yx4, and Zx9 to get the number of calories provided by each macronutrient—4X, 4Y, and 9Z. Grams are the standard unit of measurement for nutrients. Be sure you use the right criterion when calculating the calories in food on your own.

For each macronutrient, add up the calories. Now that you are aware of how the calories are distributed, add together each individual count to obtain the total calories in one serving of the food. Using the previous illustration as a guide, 4X+ 4Y+9Z = 355 calories. This figure must match the estimate printed on the product's box. You can discover how to incorporate a particular food type into a balanced diet by breaking down the calorie count by macronutrient as opposed to simply reading it off the box. 355 calories might not sound like a lot, but if you're attempting to consume less fat, you might be alarmed to realise that fat grammes account for nearly half of the total. Take serving size into account. Be aware that the nutritional data only show a single suggested portion when it comes to calories and macronutrients. The overall number of calories in the package will actually be significantly higher if there are many servings. If you track calories as part of a diet or exercise plan, this can be a crucial detail to keep in mind (Higgins, J.P. et al. 2016).

According to their recommended daily values, compare the caloric content of various nutrients. You should take between 46 and 65 percent of your daily calories of carbohydrates, 10 to 35% from protein, and 25 to 30 percent from fats, according to dietitians and other food experts. You can determine how much of those proportions you're receiving from an item by looking at the recommended daily value (DV) column in the nutrition information (Jenkins et al., 2015).

IMPLEMENTATION AND ALGORITHM

The aim of the coding phase, often known as programming, is to convert the system design created during the design phase into code in a particular programming language that can be run by a computer and performs the computation indicated by the design. The goal of the coding phase is not to make the programmer's job easier. Instead, the task of the tester and maintenance should be made simpler (Siri, W.E. et al. 1956). The following properties, such as ease of translating designs into code,

code efficiency, memory efficiency, response time, maintainability, and security, are taken into consideration when writing code. simple, easily comprehensible code, Effective and reliable reasoning (Metzgar et al., 2015).

Design

Design shows how the system functionality is designed for the wellness app. The design of the system can be shown using UML diagrams. For example, use case diagrams model the functionality of the system. It consists of actors, use cases and relationships. Fig.1 shows Use case diagram and Fig.2 shows Sequence diagram.

1) Use case diagram

Figure 1. Use case diagram

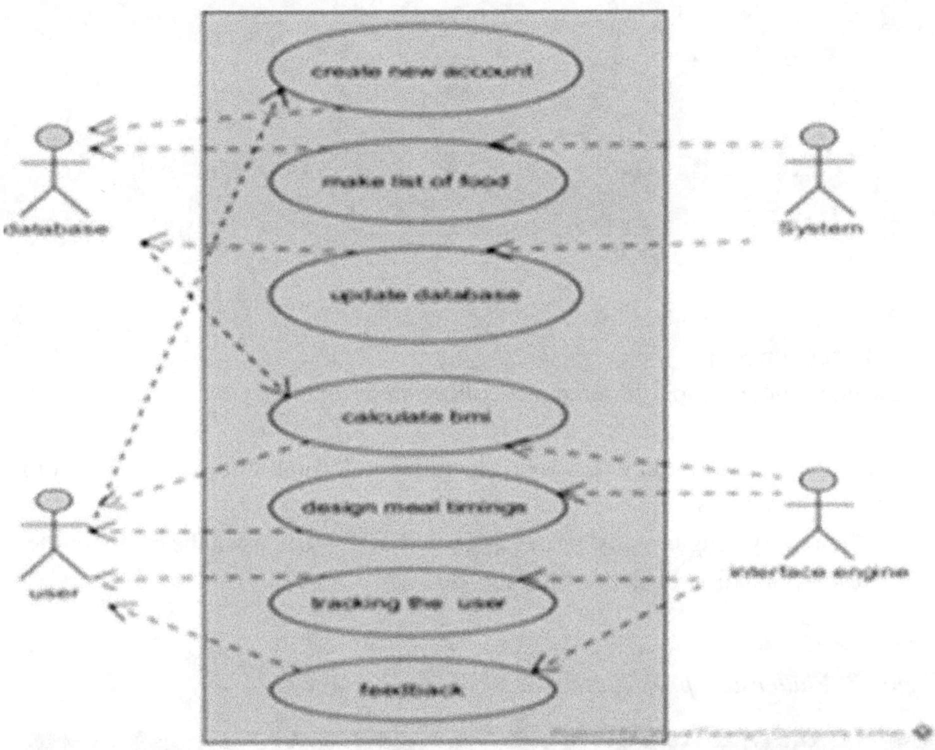

Sequence diagram will depict the interaction between objects in a sequence order.

2) Sequence Diagram

Figure 2. Sequence diagram

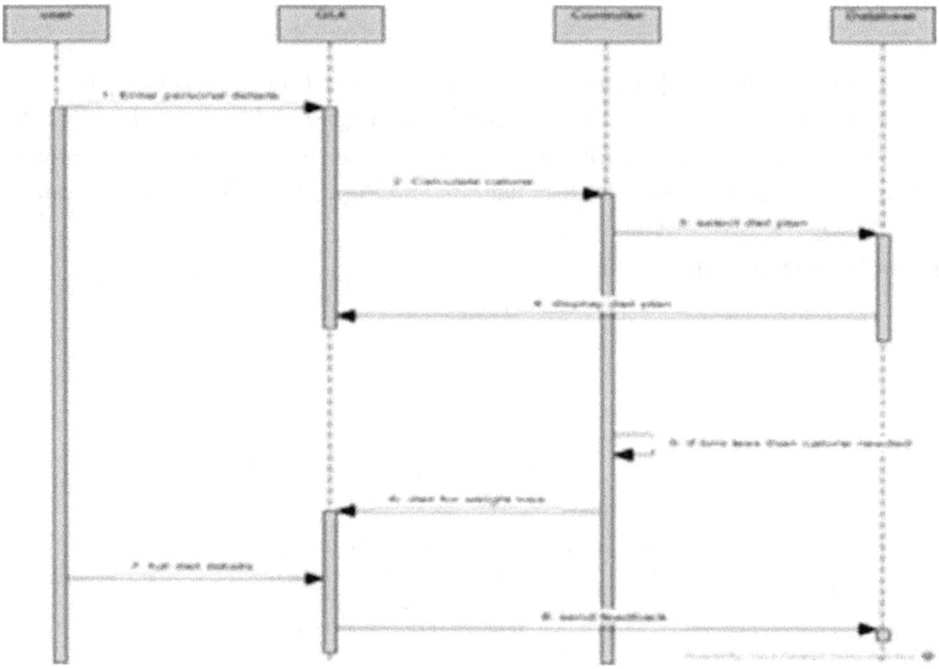

B. Execution Steps

Create a Flutter App. Flutter app is collection of widgets.

1. Make sure the Flutter extension is installed in Visual Studio Core. We enjoy using Android Studio or Visual Studio Core.
2. When everything is ready, press CMD + SHIFT + P to open the create Flutter: New Application Project window.

Figure 3. Flutter new project creation

3. Provide our project a name. We made the simple decision to use the word "fitness." It is the name we will use throughout our work on the app, much like the Firebase setup. Now choose the "Projects" folder; this is where the programme will be kept (Sannino, G. et al. 2016). Fig. 3 shows Flutter new project creation and Fig. 4 shows Pubspec.yaml.
4. Write the main code in the main.dart file, which is where the execution begins.
5. Create a folder inside the assets folder and add the Font Family to our pubspec. yaml file. You probably have guessed that this folder will include all the typefaces we require for our programme. The Font Family of our choosing is available for download online.

Figure 4. Pubspec.yaml

```
fonts:
  - family: NotoSansKR
    fonts:
      - asset: assets/fonts/NotoSansKR/NotoSansKR-Regular.otf
      - asset: assets/fonts/NotoSansKR/NotoSansKR-Bold.otf
        weight: 700
```

The project's dependencies, such as certain packages (and their versions), fonts, or picture files, are specified in the pub-spec file. Other requirements are often stated, such as specific Flutter SDK version restrictions or developer package dependencies (such as testing or mocking packages) (Middelweerd et al., 2014).

C. Inference Method

This module implements routines to give the user feedback regarding all the observed parameters and inputs. We employed the T-test inference technique. The T-test is a statistical technique for contrasting the averages or ratios of two samples taken from either the same group or from distinct groups. It aims to assess a population-specific hypothesis using the hypothesis testing method (Phillips, A.C. et al. 2010)(Robertson, J. et al. 2000).

Formula of T-test is $t = (\bar{x} - \mu) / (s / \sqrt{n})$

Where $\bar{x}$ is observed mean, μ is theoretical mean, s is standard deviation and n is sample size.

D. Algorithm

```
import "package:flutter/material.dart";
class MyApp extends StatelessWidget
Begin:
const MyApp({Key key}): super(key: key);
@override
Widget build(BuildContext context)
Begin:
return MaterialApp(
debugShowCheckedModeBanner:false,
 home:First());
End.
End.
Main()
Begin:
runApp(MyApp());
End:
```

Flutter has a built-in class called MaterialApp. It is probably the primary or central element of flutter. We can use this widget to create visually appealing apps.

RunApp widget executes the code. The simplest Flutter app just uses a widget to invoke the runApp() function: The runApp () function uses the supplied Widget to establish it as the widget tree's primary node

RESULTS

Figure 5 and Figure 6 represents Wellness page, Figure 7 shows Exercise page, Figure 8 projects My Diet image, Figure 9 shows Food page and Figure 10 shows Food article page.

Figure 5. Wellness page

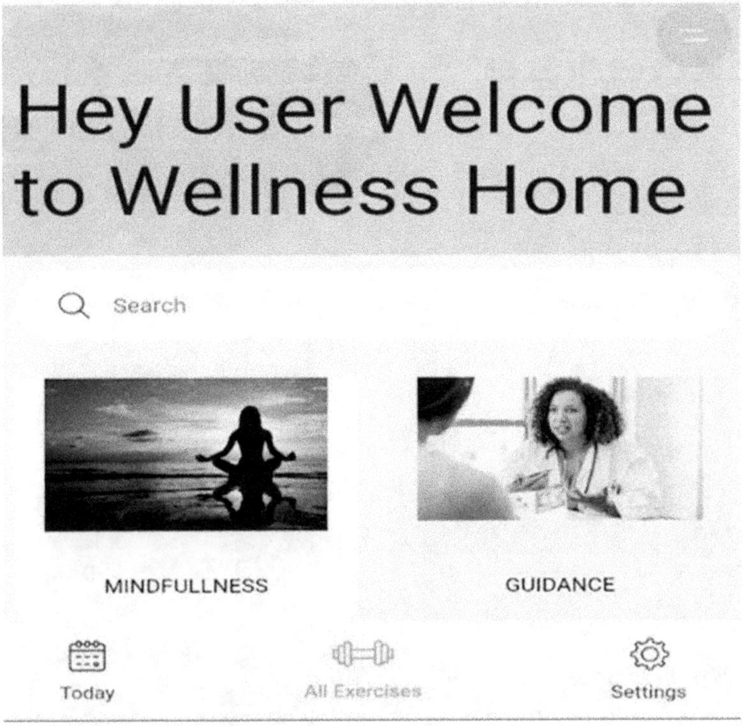

Figure 6. Wellness page

Figure 7. Exercise page

Figure 8. My Diet image

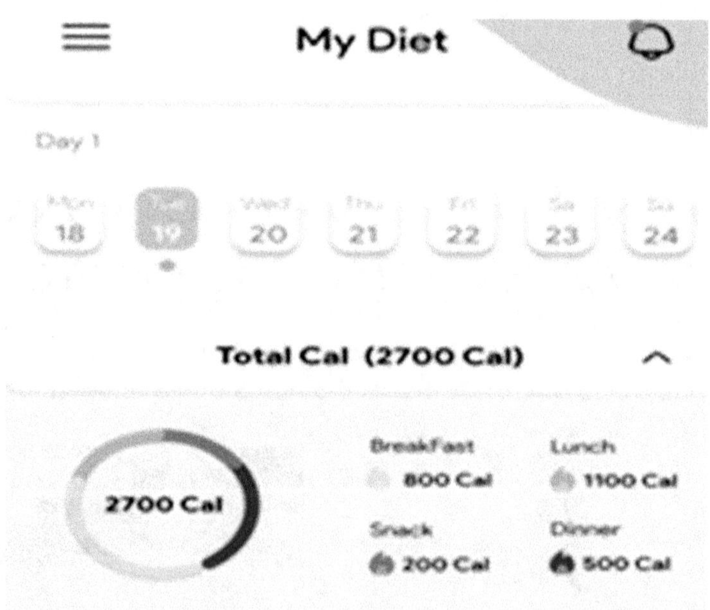

Figure 9. Food page

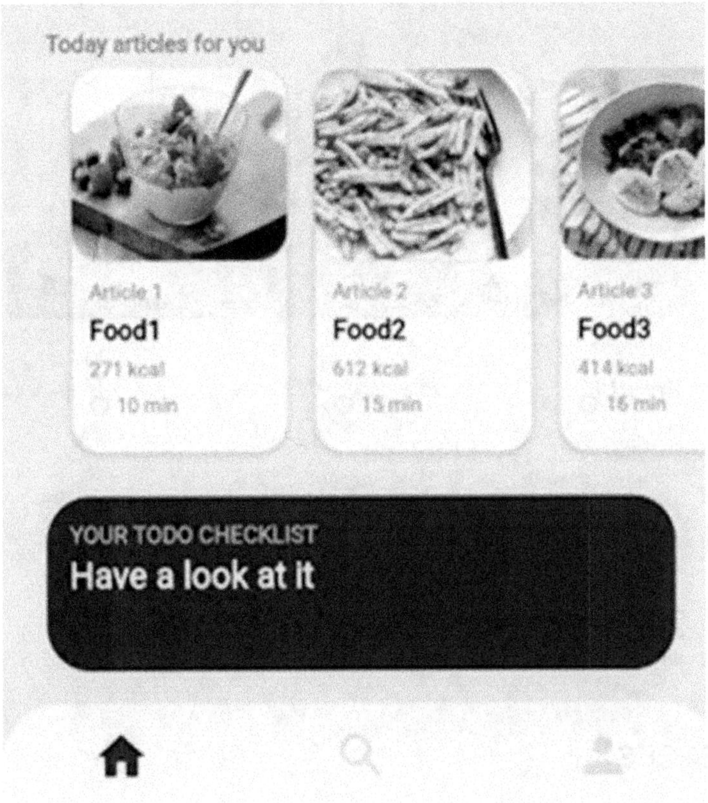

CONCLUSION AND FUTURE WORK

This project's objective was to design and build an online platform for managing wellbeing. The usage of this software makes it easy for users to accomplish their goals when dieting, which is a crucial aspect of life. The diet regimen promotes a wide range of foods, including beans, lean meat, chicken, fish, and vegetables, as well as whole grains, fruits, and non- or low-fat dairy products. Yet, the system creates a meal plan individually for each circumstance because every person has a unique dietary pattern and a range of health issues. The project is designed so that users can simply access their diet plan via an app that is available to both Android and iOS users without having to visit a dietician's office.

The future enhancements for the website can be: Create two separate apps for dietician & patient, Sharing the medical test reports of patients to dieticians through this app, Tracking the user health through AI based scanner

Figure 10.Food article page

REFERENCES

Abraham, C., & Michie, S. (2008). A taxonomy of behavior change techniques used in interventions. *Health Psychology*, *27*(3), 379–387. doi:10.1037/0278-6133.27.3.379 PMID:18624603

Alberti-Fidanza, A.; Fidanza, F.; Chiuchiu, M.; Verducci, G.; Fruttini, D. (1999). Dietary studies on two rural Italian population groups of the Seven Countries Study, Trend of food and nutrient intake from 1960 to 1991. *Eur. J. Clin. Nutr.*, *53*, 854–860.

Alderfer, C.P. (1972). *Existence, Relatedness, and Growth: Human Needs in Organizational Settings*. APA.

Alnasser, A., Kyle, J., Aloumi, N., Al-Khalifa, A., & Marais, D. (2019). The Twazon Arabic Weight Loss App: App-Based Intervention for Saudi Women with Obesity. *JMIR mHealth and uHealth*, *7*(5), e10923.

Alshawmar, M. (2021). A Review of the Applications of Affordance Theory in Mhealth App Research. *Proceedings of the 54th Hawaii International Conference on System Sciences*, (pp. 3595). Scholar Space. 10.24251/HICSS.2021.437

Brewer, L. C., Kaihoi, B., Zarling, K. K., Squires, R. W., Thomas, R., & Kopecky, S. (2015). The use of virtual world-based cardiac rehabilitation to encourage healthy lifestyle choices among cardiac patients: Intervention development and pilot study protocol. *JMIR Research Protocols*, *4*(2), e39. doi:10.2196/resprot.4285 PMID:25857331

Buysse, D. J., Reynolds, C. F. III, Monk, T. H., Berman, S. R., & Kupfer, D. J. (1989). The Pittsburgh Sleep Quality Index:A new instrument for psychiatric practice and research. *Psychiatry Research*, *28*(2), 193–213. doi:10.1016/0165-1781(89)90047-4 PMID:2748771

Eysenbach, G. (2011). Consort-EHEALTH Group. CONSORT-EHEALTH: Improving and standardizing evaluation reports of Web-based and mobile health interventions. *Journal of Medical Internet Research*, *13*(4), e126. doi:10.2196/jmir.1923 PMID:22209829

Forastiere, M., de Pietro, G., & Sannino, G. (2016). An mHealth application for a personalized monitoring of one's own wellness: Design and development. In *Innovation in Medicine and Healthcare 2016* (pp. 269–278). Springer. doi:10.1007/978-3-319-39687-3_26

Fortino, G., Galzarano, S., Gravina, R., & Li, W. (2015). A framework for collaborative computing and multi-sensor data fusion in body sensor networks. *Information Fusion*, *22*, 50–70. doi:10.1016/j.inffus.2014.03.005

Fortino, G., Giannantonio, R., Gravina, R., Kuryloski, P., & Jafari, R. (2013). Enabling effective programming and ñexible management of efðcient body sensor network applications. *IEEE Transactions on Human-Machine Systems*, *43*(1), 115–133. doi:10.1109/TSMCC.2012.2215852

Ghasemzadeh, H., Panuccio, P., Trovato, S., Fortino, G., & Jafari, R. (2014). Power-aware activity monitoring using distributed wearable sensors. *IEEE Transactions on Human-Machine Systems*, *44*(4), 537–544. doi:10.1109/THMS.2014.2320277

Gravina, R., Alinia, P., Ghasemzadeh, H., & Fortino, G. (2017). Multi-sensor fusion in body sensor networks:State-of-the-art and research challenges. *Information Fusion*, *35*, 68–80. doi:10.1016/j.inffus.2016.09.005

Higgins, J. P. (2016). Smartphone applications for patients' health and ðtness. *The American Journal of Medicine*, *129*(1), 11–19. doi:10.1016/j.amjmed.2015.05.038 PMID:26091764

Jenkins, C. D. (2015). Building Better Health: A Handbook of Behavioral Change. Pan American Health Organization. doi:10.2196/resprot.4008

Metzgar, C. J., & Nickols-Richardson, S. M. (2015). Determinants of weight gain prevention in young adult and midlife women: Study design and protocol of a randomized controlled trial. *JMIR Research Protocols*, *4*(1), e36. doi:10.2196/resprot.4008

Middelweerd, A., Mollee, J. S., van der Wal, C. N., Brug, J., & te Velde, S. J. (2014). Apps to promote physical activity among adults: A review and content analysis. [Google Scholar] [CrossRef] [PubMed]. *The International Journal of Behavioral Nutrition and Physical Activity*, *11*(1), 97. doi:10.118612966-014-0097-9 PMID:25059981

Phillips, A. C., Der, G., & Carroll, D. (2010). Self-reported health, self-reported ðtness, and all-cause mortality:Prospective cohort study. *British Journal of Health Psychology*, *15*(2), 337–346. doi:10.1348/135910709X466180 PMID:19619405

Rao, B. N. K. & Rao, B. K. (2019). Block chain Based Implementation of Electronic Medical Health Record. *International Journal of Innovative Technology and Exploring Engineering, 8*(8).

Rao, B. N. K. & Rao, B. K. (2019). Clustering based Test Suite Selection for Ranking of Program Execution Sequence using Improved Precision in Regression testing. *International Journal of Innovative Technology and Exploring Engineering*, *8*(7).

Rao, B. N. K. & Rao, B. K. (2019). Defect Detection in Printed Board Circuit using Image Processing. *International Journal of Innovative Technology and Exploring Engineering*, *9*(2).

Robertson, J., & Robertson, S. (2000). Volere. [Google Scholar]. *Requir. Specif. Templates*, *16*, 1–80.

Sannino, G., Graziani, A., de Pietro, G., & Pratola, R. (2016). *Wellness & LifeStyle Server: A platform for anthropometric and LifeStyle data analysis.* In Proceedings of the International Conference on P2P, Parallel, Grid, Cloud and Internet Computing, Asan, Korea.

Siri, W. E. (1956). The gross composition of the body. [Google Scholar]. *Advances in Biological and Medical Physics*, *4*, 513. doi:10.1016/B978-1-4832-3110-5.50011-X PMID:13354513

Smart Health 2.0. (2014). *Home.* Smart Health 2.0. http://www.smarthealth2.com/

World Health Organization. (2008). *2008–2013 Action Plan for the Global Strategy for the Prevention and Control and Diabetes.* World Health Organization.

Zephyr Techonology. (2016). *New Zephyr BioPatch Monitoring Device for Human Performance*. Zephyr Techonology. [Google Scholar]

Chapter 3
Mobile Architecture for Version Control Systems

Blendi Rrustemi
Rochester Institute of Technology, Kosovo

Deti Baholli
Rochester Institute of Technology, Kosovo

Herolind Balaj
Rochester Institute of Technology, Kosovo

ABSTRACT

Mobile architecture is an essential but often overlooked aspect of mobile app development. It determines the structure and organization of an app, and it plays a crucial role in managing the complexity of code base and enabling collaboration among developers. In today's rapidly evolving mobile environment, the importance of mobile architecture cannot be overstated. But what exactly is mobile architecture, and how can it be used to create successful mobile apps? One of the key components of mobile architecture is the use of version control systems, which are powerful tools that help developers track and manage changes to the source code of an app. In this chapter, the authors will explore the role of version control systems in mobile architecture, and discuss how these systems can help developers create high-quality and efficient mobile applications that are able to handle the challenges of the mobile environment. So, any developers looking to create the next big mobile app, read on to learn more about the importance of mobile architecture and the role of version control systems.

DOI: 10.4018/978-1-6684-8582-8.ch003

INTRODUCTION

Mobile architecture is a term used to describe the design and layout of mobile devices, such as smartphones and tablets. This type of architecture is necessary to ensure that mobile devices are able to perform their functions efficiently and effectively. One of the key challenges of mobile architecture is the need to design devices that are both compact and powerful. Because mobile devices are typically much smaller than desktop computers, the components that make up a mobile device must be carefully arranged to maximize space efficiency (Kolassa et al., 2013). This often involves using compact, low-power components that are able to deliver a high level of performance despite their small size. Another important aspect of mobile architecture is the need to optimize the device for portability. Mobile devices are designed to be used on the go, so they need to be lightweight and easy to carry around (A. Alali et al., 2008). This means that the materials used in their construction must be chosen carefully to ensure that the device is as light as possible without sacrificing durability. In addition to these physical design considerations, mobile architecture must also consider the software that runs on the device (Hindle et al., 2008) (Kolassa et al., 2013). Because mobile devices have limited processing power and storage compared to desktop computers, the operating system and other software must be designed to run efficiently on the device. This often involves using specialized algorithms and data structures that are optimized for the mobile environment. One of the key trends in mobile architecture in recent years has been the development of multi-core processors (Hou et al., 2014) (Bird et al., 2006). These processors contain multiple CPU cores, which can work together to perform complex tasks more quickly and efficiently. This has allowed mobile devices to become more powerful and capable of running demanding applications, such as games and productivity software. Another trend in mobile architecture is the use of graphics processing units (GPUs) to improve the performance of graphics-intensive applications (Gote et al., 2019). GPUs are specialized chips that are designed to handle the complex calculations required to render graphics on a screen. By using a GPU, a mobile device can display high-quality graphics without putting too much strain on its CPU (Zingg et al., 2020)(Newman et al., 2004). In addition to these technical advances, mobile architecture must also consider the needs of users. This means designing devices that are easy to use and intuitive, with user-friendly interfaces and controls. It also means ensuring that the device is capable of connecting to the internet and other devices, so that users can access the information and services they need.

KEY FACTORS THAT CONTRIBUTE TO THE SUCCESS OF MOBILE APP ARCHITECTURE

Overall, mobile architecture plays a crucial role in the design and performance of mobile devices. By carefully arranging the physical components of the device and optimizing the software that runs on it, mobile architects are able to create powerful, portable devices that are capable of meeting the needs of users. As mobile technology continues to evolve, the field of mobile architecture will continue to be an important area of research and development. Good mobile app architecture is one that is well-designed, efficient, and scalable (Winkler et al., 1990) (Nanumyan et al., 2020). There are several key factors that contribute to the success of mobile app architecture. One of the most important factors is the ability to handle a high volume of users and transactions. Good mobile app architecture must be able to support a large number of users and handle a high volume of data and transactions without experiencing performance issues or downtime. This requires a robust and scalable design that can easily accommodate growth and changes in user behaviour (Wiese et al., 2016) (E. Kouters et al., 2012). Another key factor is the ability to integrate with other systems and services. Good mobile app architecture should be able to easily integrate with other systems and services, such as databases, cloud services, and APIs. This allows the app to access and share data and functionality with other systems, which can greatly enhance its capabilities and usefulness. A third important factor is security. Good mobile app architecture should include robust security measures to protect user data and prevent unauthorized access (Landis et al., 1977). This may include encryption, authentication, and other security technologies to ensure that user data is safe and secure. A fourth key factor is usability. Good mobile app architecture should be easy to use and intuitive, with a user-friendly interface and controls. This makes it more likely that users will be able to quickly and easily understand how to use the app, which can improve their overall experience and increase the app's popularity.

Finally, good mobile app architecture should be flexible and adaptable. As user needs and preferences change, the app architecture should be able to evolve and adapt to accommodate these changes. This allows the app to remain relevant and useful to users over time, which can help to ensure its long-term success. When dealing with mobile architecture we will have to construct diagrams. When designing a mobile architecture diagram, there are several important things to consider (Ventura et al., 2015). First and foremost, the diagram should be clear and easy to understand, even for those who are not familiar with mobile architecture. This means using clear, concise labels and explanatory text to help users understand the various components and their relationships (Sarigol et al., 2014). Another important consideration is the level of detail included in the diagram. While it is important to provide enough information to be useful, the diagram should not be overly complex or cluttered. This can make it difficult to understand and may cause confusion. Instead, the diagram should strike a balance between providing useful information and being easy to understand. It is also important to consider the intended audience for the diagram. Different audiences may have different levels of knowledge and expertise, so the diagram should be tailored to the intended audience. For example, a diagram intended for experts in the field of mobile architecture may include more technical details and jargon, while a diagram intended for a general audience may use more simplified language and explanations. Another important thing to consider when designing a mobile architecture diagram is the layout and visual design. The diagram should be well-organized and easy to read, with clear lines and labels that help users understand the relationships between the various components. Additionally, the diagram should use colors and other visual elements to help differentiate between different types of components and highlight important information.

The mobile architecture diagram should be kept up-to-date. As mobile technology evolves and new technologies and approaches are developed, the diagram should be updated to reflect these changes. This will help ensure that the diagram remains accurate and useful for those who rely on it. Designing a mobile architecture diagram requires careful consideration of a variety of factors, including clarity, level of detail, intended audience, visual design, and timeliness. By taking these factors into account, it is possible to create a diagram that is both informative and easy to understand, and that accurately reflects the current state of mobile architecture.

TYPES OF MOBILE APPLICATION ARCHITECTURE

There are three main types of mobile application architecture: Android, iOS, and hybrid. Android mobile application architecture is based on the Android operating system, which is developed by Google. Android apps are typically written in the Java programming language, and they are designed to run on devices that use the Android operating system. Android apps are typically built using the Android Studio development environment, which includes tools and libraries for creating and testing Android apps (E. Jimenez-Ruiz et al., 2009) (J. Kim et al., 2018).

iOS mobile application architecture is based on the iOS operating system, which is developed by Apple. iOS apps are typically written in the Swift or Objective-C programming languages, and they are designed to run on devices that use the iOS operating system, such as iPhones and iPads. iOS apps are typically built using the Xcode development environment, which includes tools and libraries for creating and testing iOS apps (Ma et al., 2014).

Hybrid mobile application architecture is a combination of both Android and iOS architectures. Hybrid apps are built using web technologies, such as HTML, CSS, and JavaScript, and they are designed to run on both Android and iOS devices (Goeminne et al., 2013). Hybrid apps are typically created using tools like Apache Cordova, which allow developers to create a single app that can be deployed on both Android and iOS devices.

Each of these mobile application architectures has its own strengths and weaknesses. Android apps have the advantage of being able to run on a wider range of devices, including those made by different manufacturers. iOS apps, on the other hand, have the advantage of being able to access the full range of features and capabilities available on Apple devices (Arafat & Riehle, 2009). Hybrid apps, meanwhile, have the advantage of being able to run on both Android and iOS devices, but they may not be able to take full advantage of the capabilities of each platform.

METHODOLOGIES

Methodologies for version control vary from simple to the most complex, which are created by using different hardware and software to manage and store multiple versions of the same document. Version control is a procedure to store and manage different versions of documents created. Every day we store and manage different documents on our computers, but the way we do that differs from the version control (Vinay Singh et al., 2019). If in a computer we have several files which have almost the same content but each of them is slightly different from each other, we can see dates in which we have updated the files, but this doesn't provide any information on how the document has been changed from each of the versions, also it doesn't show why it has been changed, or how are these files associated to each other. Also, this file system managed by humans is more prone to errors, than software which is made specifically for version controls. Below is a screenshot of a human managed file system, which we can see the different versions of the project, but not the progress or the changes made on it.

Storing content, tracking changes, and distributing or committing the changes with other collaborators is the main three functionalities of a Version Control System. Version control systems date back when the software development rose and the developers felt the need to work together and at the same time with each other. Also, tracking progress and revising their changes was another need for the programmers at that time. Thus, nowadays we have different systems which help us solve these problems. These systems are like git, subversion, visual source safe, etc. but, the most used version control system is git. Theoretically, a list of file-based changes is how most other version control systems store information except git (Vinay Singh et

Figure 1. Screenshot of a human managed file system

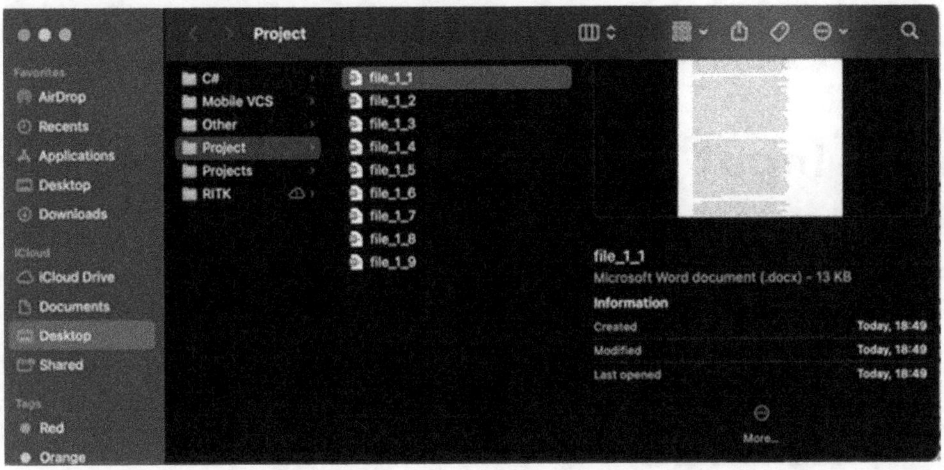

al., 2020). Delta based version control system, is storing information as sets of files and tracking the changes that are made to each of the files over a period of time, and this is how most other systems think about their data. Git uses a sequence of snapshots for every tiny file system. This means that, every time a project is saved, git takes a snapshot of all the files at that moment, and saves a reference to that photo. Also, git is very space efficient, as it doesn't save the files multiple times if anything has changed to the project, rather it saves only a link to that same file which is stored, and with this we can understand that git uses the stream of snapshots as a way of thinking about data. Also, for git to work, usually all the files are locally operated, this means that most of the information is not needed to be allocated from other computers in the network. This works because the history of the files we are working on the project are saved locally on the disk, that is why the processes are pretty quick. As an instance, if we want to browse the history of the project we are working on, Git has no need to search in the servers outside the computer and show it to us, but it reads directly locally from our database. Git also uses a checksum, which is a feature built in the low level so git cannot lose files in transfer, or corrupt them in any way. This feature is called hashing, and it is a string consisting of 40 characters from 0-9 and a-f. This hash looks like:

Git is composed of modified, staged, and committed states. Making a change in the file, but not committing it to the database, is called the modified state. Modifying a file in that current version and marking it to go to the commit snapshot means that file is staged. Finally, committing the file means that file is locally stored in the database. Also, git consists of sections like, working tree, staging area, and directory (Luczak-Rösch et al., 2010). A single checkout of the project we are working on

Figure 2. Hash

24b9da6552252987aa493b52f8696cd6d3b00373

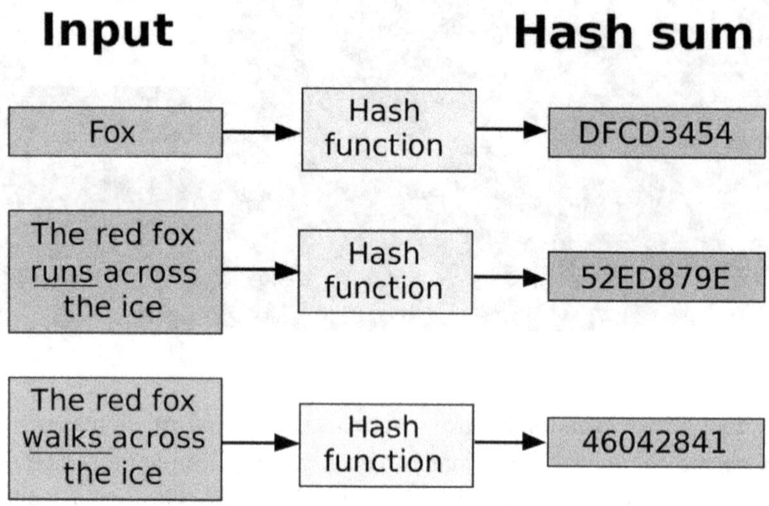

is called the working tree, where files are located in the disk from the database, so we can modify them. The place that stores the files which are going to go into the commit is called the staging area. Storing the metadata and the object database for the project is called the git directory; this means that this part is responsible for cloning the repository from other computers.

VERSION CONTROL SYSTEM

A version control system (VCS) is a valuable tool for teams working on the same project. Without a VCS, team members are likely to be working on the same set of files in a shared folder. This can be problematic because it is easy for one person to overwrite the changes made by another person. This can lead to errors and confusion, and it is not an acceptable way of working. With a VCS, everyone on the team has complete freedom to work on any file at any time. This makes it much easier for team members to collaborate and make changes to a project without running into conflicts. The VCS maintains a central, shared location for storing and managing different versions of a project. This means that there is no doubt about where the most recent version of a file or the entire project is. It is always in a shared, central location: the version control system. Saving a version of your project after making changes is an important habit to develop. Without a VCS, this quickly becomes tiresome and confusing. It can be difficult to keep track of which files have been modified and which files are still the same. It is also not always clear how much data to save. Should you save only the modified files, or the entire project? If you save only the modified files, you won't be able to see the entire project at any point in time (Sarigolet al., 2017). If you save the entire project, you will end up with a lot of unneeded data on your hard disk. Another problem with working without a VCS is that it can be difficult to keep track of different variations of a project. For example, you might need to produce one version with a header area and another version without a header area. Without a VCS, it is easy to lose track of these variations and end up with a confusing mess of files.

The most significant advantage of using a VCS is that it allows you to know exactly what is different between different versions of a project. Without a VCS, it is unlikely that you will take the time to carefully record each significant change and add it to a README file in the project folder. With a VCS, you are asked to provide a brief summary of what was changed each time you save a new version of your project. You can also check what was modified in the file's content (if it is a code or text file). This allows you to see how your project changed between versions. Another benefit of using a distributed VCS like Git is that it can serve as a backup. Each team member has a full copy of the project on their own disk, complete with the project's history. If the central server fails (and your backup drives fail), you can recover the project from a local Git repository on one of your team members' computers (Cohen et al., 1960).

Overall, using a VCS is a smart decision for any team working on a project. It makes it easier for team members to collaborate and ensures that changes to a project are well-organized and safe. It also provides the ability to easily undo changes and restore previous versions of a project. And in the case of a distributed VCS like Git, it provides an additional backup of the project, making it possible to recover from a catastrophic failure of the central server. Version control system has a lot of benefits, some of them are: Increases project development pace by facilitating effective collaboration. Improves staff productivity, product delivery, and skills through improved communication and help. Reduce the possibility of errors and disputes during project development by tracking every tiny change. Employees or collaborators to the project can contribute from anywhere, regardless of geographical location, using this VCS.

There are three types of Version Control Systems, these are Local, Distributed, and Centralized Version Control Systems. Files are simply copied into a separate directory locally in a local version control system. Versions of the same file are saved so that any particular version at any point in time can be easily retrieved. This approach is widely used for tiny personal projects or files since it allows you to easily version your project locally. The local VCS is easy to set up and very cheap to run, but has disadvantages like, as files are stored locally this means that it is unsafe. Also, it is more prone to errors and it is not appropriate for projects working on teams. If you're part of a team and two or more members are working on the same project, you'll probably want the most recent modifications made to the project by other team members to be reflected in your version of the project anytime you want.

Figure 3. Local computer

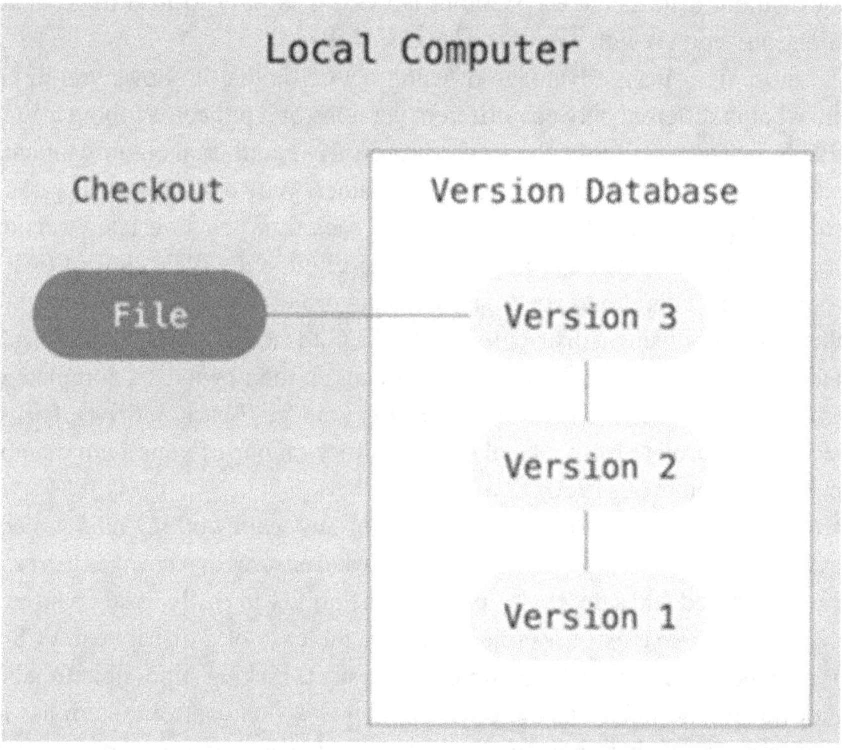

CENTRALIZED VERSION CONTROL SYSTEM

A centralized version control system allows the main version of the project to be stored in a single centralized location from which each team member can 'check-out' any specific version of the project. After making the necessary modifications to the project, each team member can 'check-in' the changes to the centralized repository, making the new updates available to the rest of the team (Purushothaman et al., 2005). Likewise, centralized VCS has its own advantages and disadvantages. It is very easy to set up, has different options to choose from, like open source and proprietary, between team members it allows to share the files working on, and the project is stored in servers which are more reliable than storing the project in local storage. Disadvantages include that if the server storing the project fails, then all the changes made to it won't be obtainable. Also, if a lot of people are updating the project, means that it could lead to file conflicts between the updates of the files, and all the files could be mixed up and will not work properly (Svajlenko et al., 2016).

Figure 4. Centralized version control systems

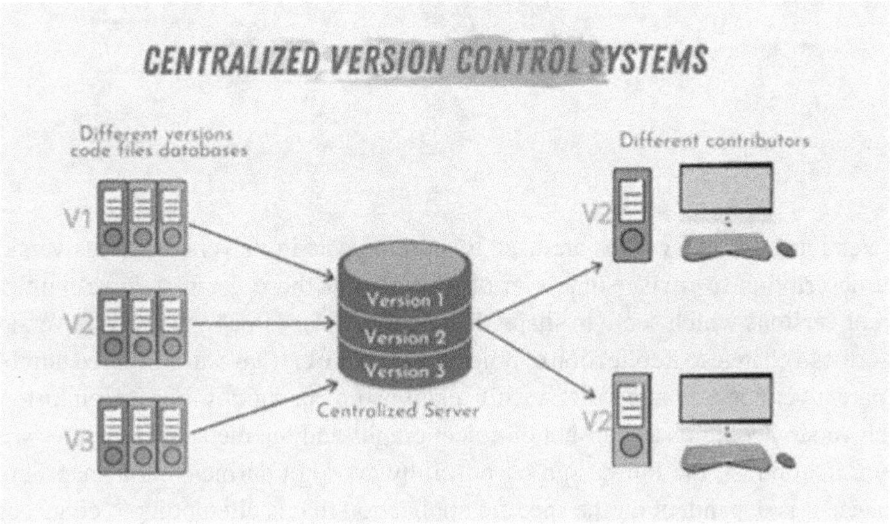

Distributed Version Control Systems seek to address one of the shortcomings of centralized VCS systems by allowing each 'check-out' version to serve as a repository source for all other members working on the same project (Pedregosa et al., 2011). While this is a more difficult procedure to organize, it has the advantage of being more stable and less reliant on a single central repository. In a distributed VCS, users don't just check out a version of the file; they fully mirror the repository and make a backup of all available versions for all files. The advantages of distributed version control systems are that it is very reliable because all the team members working on the project have a copy of all the changes on it. It also allows sharing the files between the members of that project and like centralized VCS, it has a lot of options to choose from. Considering the disadvantages, the distributed VCS is the most complicated to set up, and it takes a lot of storage in the local computers.

Figure 5. Distributed version control systems

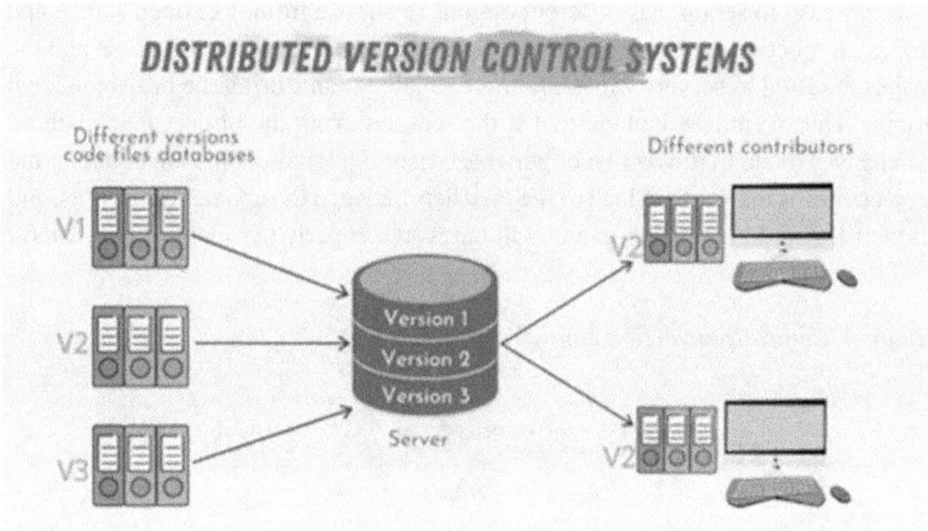

Version Control Systems are used to store objects in its versions. This version is a description to revise some sets of objects, but these versions have multiple parent versions which serve to shape the version history in a repository. A version describes a change to a collection of objects. A version can have an unlimited number of parent versions. In a repository, this relationship constitutes the version history. Each version contains a snapshot of object graphs and the metadata that goes with them. In contrast, the Pur design purposefully does not define what a snapshot is because it is dependent on the specific application that is attempting to connect to a version control system (G. Robles et al., 2005). Pur, on the other hand, defines a store entity that can create and restore snapshots, updating the items in the store (S. Amreen et al., 2020). The meaning of "restore" varies depending on the type of business. The example Pur for Newspeak implementation presented in defines two stores: an image store for capturing classes and methods from the Newspeak environment into a snapshot and loading them back from a snapshot, and a file store for writing snapshots to files and reading such files back into a snapshot.

Now that we have discussed the basics of mobile architecture and version control systems, let's explore the connections between the two. As we have seen, mobile architecture is an important aspect of mobile app development, and it determines the structure and organization of an app. Similarly, version control systems are essential tools that help developers track and manage changes to the source code of an app. But how do these two concepts fit together? In fact, the use of version control systems is a crucial part of mobile architecture. By using these systems, developers can manage the complexity of the codebase and ensure that the different components of the app work together smoothly. In short, the connection between mobile architecture and version control systems is essential for creating successful mobile apps.

One of the key components of mobile architecture is the use of version control systems. Version control systems, as mentioned earlier, are tools that allow developers to track and manage changes to the source code of an app. These systems help developers to collaborate on a project, and they enable them to keep track of different versions of the code and easily revert back to earlier versions if necessary. When it comes to mobile architecture, version control systems are essential because they allow developers to manage the complexity of the codebase. With mobile apps, there are often many different components, such as the user interface, the data storage, and the network communication, all of which need to be managed and coordinated. Using a version control system, developers can easily track and manage changes to each of these components, ensuring that the app continues to function properly. In addition to helping with code management, version control systems also enable developers to work together on a project more effectively. With these systems, multiple developers can work on the same codebase at the same time, and the version control system will automatically merge their changes together and resolve any conflicts. This allows developers to focus on their individual tasks without having to worry about coordinating their work with each other.

One of the key benefits of using version control systems in mobile architecture is their ability to help developers manage the complexity of the codebase. As mobile apps continue to become more powerful and feature-rich, their codebases often become larger and more complex. This can make it difficult for developers to keep track of all the different components and ensure that they are working together properly. Using a version control system, however, developers can easily track and manage changes to the code, making it easier to maintain the app and ensure that it continues to function properly. Additionally, version control systems allow developers to collaborate more effectively, enabling them to work on the same codebase simultaneously and easily merge their changes together. This can help to speed up the development process and ensure that the app is delivered on time (Hattori et al., 2008).

In addition to these benefits, using version control systems in mobile architecture can also help to improve the quality of the code. By using these systems, developers can easily track and review changes to the code, making it easier to identify and fix bugs and other issues (Hofmann et al., 2009). For example, imagine that a developer is working on a mobile app that allows users to book flights. The app has a feature that allows users to select their departure and arrival airports, and it calculates the flight price based on the route and the date. However, a bug has been introduced that causes the app to crash when the user tries to select certain airports. Using a version control system, the developer can easily track down the source of the bug and fix it. For instance, they might use the system to identify the specific commit where the bug was introduced, and then they can revert back to an earlier version of the code and fix the issue. This can help to improve the overall stability and reliability of the app, ensuring that it provides users with a seamless and enjoyable experience.

Furthermore, using version control systems can help to ensure that the app is maintainable and scalable, making it easier to add new features and capabilities in the future (Arafat et al., 2009). For example, the developer might use a version control system to create a separate branch for a new feature, allowing them to work on the feature without disrupting the main codebase. This allows the developer to experiment and iterate on the feature without having to worry about breaking the app (Sarawagi et al., 2002). In short, the use of version control systems in mobile architecture can help developers create high-quality and efficient mobile apps that are able to handle the challenges of the mobile environment (Sinatra et al., 2016). By using these systems, developers can easily track and manage changes to the code, ensuring that the app continues to function properly and that all of its components work together smoothly (Singh et al., 2021).

On the other hand, one of the main challenges with using version control systems in mobile architecture is the learning curve. These systems can be complex and difficult to understand, especially for developers who are new to the field. For example, developers may need to learn how to use specialized commands and tools to interact with the version control system, and they may need to understand how to manage branches and merge changes. This can be a steep learning curve, and it can take time for developers to become proficient with these systems. Additionally, using version control systems can add overhead and complexity to the development process. These systems often require developers to follow strict rules and procedures, such as committing changes frequently and using descriptive commit messages (Levenshtein et al., 1966). This can make it more difficult for developers to work quickly and efficiently, and it can lead to delays and other problems. Additionally, using version control systems can require extra coordination and communication among developers, which can be challenging in larger teams or distributed environments. In short, while using version control systems in mobile architecture can be beneficial in many ways, there are also some potential drawbacks to consider.

In conclusion, the use of version control systems in mobile architecture is an essential part of the development process. These systems enable developers to track and manage changes to the codebase, ensuring that the app continues to function properly and that all of its components work together smoothly. Additionally, version control systems allow multiple developers to work on the same codebase simultaneously, enabling them to collaborate effectively and focus on their individual tasks without having to worry about coordinating their work. By using these systems, developers can create high-quality mobile apps that provide users with a seamless and enjoyable experience, and they can ensure that their apps are able to handle the challenges of the mobile environment. In short, the use of version control systems is an important part of creating successful mobile applications. As a final thought, it's worth considering the potential future developments in mobile architecture and version control systems. For example, as mobile apps continue to become more complex and powerful, it's likely that we will see more advanced version control systems that are able to handle even larger and more complex codebases. Additionally, we may see the rise of new approaches to mobile architecture that are better able to take advantage of the unique capabilities of mobile devices. Whatever the future holds, it's clear that the use of version control systems will continue to play a crucial role in the development of mobile applications.

CONCLUSION

Mobile architecture is a crucial but frequently disregarded component in developing mobile applications. It establishes an app's structure and organization, and it is essential for controlling the complexity of the code base and fostering developer cooperation. It is impossible to exaggerate the significance of mobile architecture in the fast-changing mobile environment of today. But what precisely is mobile architecture, and how can it be used to the development of effective mobile apps? The usage of version control systems, which are potent instruments that assist developers in tracking and managing changes to an app's source code, is one of the essential elements of mobile architecture. The function of version control systems in mobile architecture will be examined in this paper, and it will be discussed how these systems can aid developers in producing effective and high-quality mobile apps that can withstand the demands of the mobile environment. As a result, read on to learn more about the significance of mobile architecture and the function of version control systems if you're a developer trying to build the next great mobile app.

REFERENCES

Alali, A., Kagdi, H., & Maletic, J. (2008). What's a Typical Commit? A Characterization of Open Source Software Repositories. *Proc. the 16th IEEE Int'l Conf. Program Comprehension (ICPC'08)*, (pp. 182-191). IEEE. 10.1109/ICPC.2008.24

Amreen, S., Mockus, A., Zaretzki, R., Bogart, C., & Zhang, Y. (2020). ALFAA: Active learning fingerprint based anti-aliasing for correcting developer identity errors in version control systems. *Empirical Software Engineering*, *25*(2), 1–32. doi:10.100710664-019-09786-7

Arafat, O., & Riehle, D. (2009). The Commit Size Distribution of Open Source Software", Proc. the 42nd Hawaii Int'l Conf. Syst. Sci. (HICSS'09), pp. 1-8, 2009.

Arafat, O., & Riehle, D. "The Commit Size Distribution of Open Source Software. *Proceedings of the 42nd Hawaii International Conference on Systems Science (HICSS'09)*, (pp. 1-8). IEEE.

Bird, C., Gourley, A., Devanbu, P., Gertz, M., & Swaminathan, A. (2006). Mining email social networks. *Proceedings of the 2006 International Workshop on Mining Software Repositories*, (pp. 137-143). ACM. 10.1145/1137983.1138016

Cohen, J. (1960). A coefficient of agreement for nominal scales. *Educational and Psychological Measurement*, *20*(1), 37–46. doi:10.1177/001316446002000104

Goeminne, M., & Mens, T. (2013). A comparison of identity merge algorithms for software repositories. *Science of Computer Programming*, *78*(8), 971–986. doi:10.1016/j.scico.2011.11.004

Gote, C., Scholtes, I., & Schweitzer, F. (2019). git2net - Mining time-stamped co-editing networks from large git repositories. *2019 IEEE/ACM 16th International Conference on Mining Software Repositories (MSR)*, (pp. 433-444). ACM, IEEE. 10.1109/MSR.2019.00070

Hattori, L., & Lanza, M. (2008). On the nature of commits. Proc. the 4th Int'l ERCIM Wksp. Softw. Evol. and Evolvability (EVOL'08), (pp. 63-71). IEEE. doi:10.1109/ASEW.2008.4686322

Hindle, A., Germán, D., & Holt, R. (2008). What do large commits tell us?: a taxonomical study of large commits. *Proc. the 5th Int'l Working Conf. Mining Softw. Repos. (MSR'08)*, (pp. 99-108). ACM. 10.1145/1370750.1370773

Hofmann, P., & Riehle, D. (2009). Estimating Commit Sizes Efficiently. *Proc. the 5th IFIP WG 2.13 Int'l Conf. Open Source Systems (OSS'09)*, (pp. 105-115). Springer. 10.1007/978-3-642-02032-2_11

Hou, Q., Ma, Y., Chen, J., & Xu, Y. (2014). An Empirical Study on Inter-Commit Times in SVN. *Int. Conf. on Software Eng. and Knowledge Eng*, (pp. 132-137). Springer.

Jim'enez-Ruiz, E., Grau, B. C., Horrocks, I., & . Llavori, R. B. (2009). *Contentcvs: A cvs-based collaborative ontology engineering tool.* SWAT4LS.

Kim, J. (2018). Evaluating author name disambiguation for digital libraries: A case of DBLP. *Scientometrics, 116*(3), 1867–1886. doi:10.100711192-018-2824-5

Kolassa, C., Riehle, D., & Salim, M. (2013). A Model of the Commit Size Distribution of Open Source. *Proc. the 39th Int'l Conf. Current Trends in Theory and Practice of Comput. Sci. (SOFSEM'13),* (pp. 52-66). IEEE. 10.1007/978-3-642-35843-2_6

Kolassa, C., Riehle, D., & Salim, M. (2013). A Model of the Commit Size Distribution of Open Source. *Proceedings of the 39th International Conference on Current Trends in Theory and Practice of Computer Science (SOFSEM'13),* (pp. 5266). ACM. 10.1007/978-3-642-35843-2_6

Kouters, E., Vasilescu, B., Serebrenik, A., & Van Den Brand, M. G. (2012). Who's who in Gnome: Using LSA to merge software repository identities. *2012 28th IEEE International Conference on Software Maintenance (ICSM)*, (pp. 592-595). IEEE. 10.1109/ICSM.2012.6405329

Landis, J. R., & Koch, G. G. (1977). An application of hierarchical kappa-type statistics in the assessment of majority agreement among multiple observers. *Biometrics, 33*(2), 363–374. doi:10.2307/2529786 PMID:884196

Levenshtein, V. I. (1966). Binary codes capable of correcting deletions insertions and reversals. *Soviet Physics, Doklady*, *10*(8), 707–710.

Luczak-Rösch, M., Coskun, G., Paschke, A., Rothe, M., & Tolksdorf, R. (2010). Svont-version control of owl ontologies on the concept level. *GI Jahrestagung, 176*(2), 79–84.

Ma, Y., Wu, Y., & Xu, Y. (2014). Dynamics of Open-Source Software Developer's Commit Behavior: An Empirical Investigation of Subversion. Proceedings of the 29th Annual ACM Symposium on Applied Computing (SAC'14), (pp. 1171-1173). ACM. 10.1145/2554850.2555079

Nanumyan, V., Gote, C., & Schweitzer, F. (2020). Multilayer network approach to modeling authorship influence on citation dynamics in physics journals. *Physical Review E, 102*, (pp. 032303). https://link.aps.org/doi/10.1103/PhysRevE.102.032303

Newman, M. E. (2004). *Who is the best connected scientist? A study of scientific coauthorship networks in Complex Networks*. Springer.

Pedregosa, F., Varoquaux, G., Gramfort, A., Michel, V., Thirion, B., & Grisel, O. (2011). Scikit-learn: Machine learning in Python. *Journal of Machine Learning Research, 12*, 2825–2830.

Purushothaman, R., & Perry, D. E. (2005). Toward Understanding the Rhetoric of Small Source Code Changes. *IEEE Transactions on Software Engineering, 31*(6), 511–526. doi:10.1109/TSE.2005.74

Robles, G., & Gonzalez-Barahona, J. M. (2005). Developer identification methods for integrated data from various sources. *Software Engineering Notes, 30*(4), 1–5. doi:10.1145/1082983.1083162

Sarawagi, S., & Bhamidipaty, A. (2002). Interactive deduplication using active learning. *Proceedings of the Eighth ACM SIGKDD International Conference on Knowledge Discovery and Data Mining*, (pp. 269-278). ACM. 10.1145/775047.775087

Sarigol, E., Garcia, D., Scholtes, I., & Schweitzer, F. (2017). Quantifying the effect of editor-author relations on manuscript handling times. *Scientometrics, 113*(1), pp. 609-631. https://link.springer.com/article/10.1007/s11192-017-2309-y

Sarigol, E., Pfitzner, R., Scholtes, I., Garas, A., & Schweitzer, F. (2014). Predicting scientific success based on coauthorship networks. *EPJ Data Science, 3*(9). http://www.epjdatascience.com/content/3/1/9

Sinatra, R., Wang, D., Deville, P., Song, C., & Barabási, A.-L. (2016). Quantifying the evolution of individual scientific impact. *Science, 354*(6312), aaf5239. doi:10.1126cience.aaf5239 PMID:27811240

Singh, V., Aggarwal, A., Kumar, N., & Saini, A. K. (2020). A Novel Approach for Pre-Validation Auto Resiliency & Alert Notification for SVN To Git Migration Using Iot Devices. *PalArch's Journal of Arch. of Egypt/Egyptology, 17*(9), 7131-7145.

Singh, V., Aggarwal, A., Kumar, A., & Sanwal, S. (2019). The Transition from Centralized (Subversion) VCS to Decentralized (Git) VCS: A Holistic Approach. *Journal of Electrical and Electronics Engineering (Oradea), 12*(1), 7–15.

Singh, V., Alshehri, M., Aggarwal, A., Alfarraj, O., & Sharma, P. (2021). A holistic proactive and novel approach for pre during and post migration validation from subversion to git. *Computers Materials & Continua, 66*(3), 2359–2371. doi:10.32604/cmc.2021.013272

Svajlenko, J., & Roy, C. K. (2016). A machine learning based approach for evaluating clone detection tools for a generalized and accurate precision. *International Journal of Software Engineering and Knowledge Engineering, 26*, pp. 1399-1429.

Ventura, S. L., Nugent, R., & Fuchs, E. R. (2015). Seeing the non-stars: (Some) sources of bias in past disambiguation approaches and a new public tool leveraging labeled records. *Research Policy, 44*(9), 1672–1701. doi:10.1016/j.respol.2014.12.010

Wiese, I. S., Da Silva, J. T., Steinmacher, I., Treude, C., & Gerosa, M. A. (2016). *"Who is who in the mailing list? Comparing six disambiguation heuristics to identify multiple addresses of a participant", 2016 IEEE international conference on software maintenance and evolution.* IC-SME.

Winkler, W. E. (1990). String comparator metrics and enhanced decision rules in the Fellegi-Sunter model of record linkage. *Proceedings of the Section on Survey Research Methods*, (pp. 433-444). ACM.

Zingg, C. Nanumyan, V., & Schweitzer, F., (2020).Citations driven by social connections? A multilayer representation of coauthorship networks. *Quantitative Science Studies, 1*(4), pp. 1493-1509. doi: 1. doi:0.1162/qssa00092

Chapter 4
Machine Learning-Based Prediction Model for the Measurement of Mobile Addiction

Alma Beluli
Rochester Institute of Technology, Kosovo

ABSTRACT

Mobile phones are now one of the most important elements of our lives. We know that they are helpful and enable us to perform different services for our requirements online. But there are limits to everything. It is excessive that we cannot even eat or go to sleep without our mobile phones. According to some statistics, a person uses their mobile phone on average 3 hours and 15 minutes a day. The part of the population that is affected most by this problem is teenagers. Most of them suffer from "nomophobia," which is defined as "the fear of not having a mobile phone." Teenagers are the most impacted by mobile addiction because they cannot manage their screen time and they still have not developed the ability to self-control. According to some research, it is ascertained that 27% of the population who own a mobile phone are people around 11-13 years old that don't turn off their phone, not even to sleep.

DOI: 10.4018/978-1-6684-8582-8.ch004

INTRODUCTION

The mobile phone is now one of the most important elements of our lives. They are useful and helpful, enabling us to perform different services for our requirements online. But excessive usage of smartphones can be seriously harmful. There are a lot of research and studies based on mobile addiction and the measurement of the people that are prey to it. One of these studies is the "Machine-Learning Predictive Models for Dependency on Smartphones Based on Risk Factors" research by Giraldo-Jiménez et al. They performed an analytical study on 1228 students from a private university. "The tests were 1) smartphone dependency assessment and 2) the Nordic musculoskeletal symptoms questionnaire." (Giraldo-Jiménez et al). One of the best uses of mobile phones is as educational, learning tools and this use of them has been very common in these past years. But, according to Giraldo-Jiménez et al, online learning via phones has a major impact on student's academic performance, different from traditional learning, which could be a better approach. This study has shown that the rate of excessive use of mobile phones is 38.9%, with young people being a large percentage of this number (G. Kumar et al. 2022). This behavior can result in poor decision-making, inhibition, insomnia, and memory efficiency. The excessive use of mobile phones by this percentage of people can be related to one of the factors such as family, environmental risks, marital status, socioeconomic status, and so on. Regarding socioeconomic status, this study shows that of 385 students that were active phone users, 51.7% were employed whereas the other 48.3% didn't have a job. Xie and Molassiotis concluded from a group of 1500 individuals that "44.8% presented high income, 29.2% medium income and 26% low income, therefore establishing a relationship in the use of the smartphone with a prevalence of 78.9% over the target population" (R. Das Gupta et al. 2022). (Xie & Molassiotis) (Hennig et al., 2014) According to Freitas & Cols, young people that have high family incomes are most likely to develop mobile phone dependency. In relation to the marital status being a factor in mobile phone dependency, from 402 university students, 91.1% were single, 7.9% married, and 1.0% were divorced, leading us to the conclusion that for the percentage of unmarried and divorced students, the loneliness and not having a partner to share their life with, leads to excessive use of mobile phone (D. Singh et al. 2022). The academic program of students can also be a factor in the dependency on mobile devices (Guha et al. 2022). The results conclude that 253 (84.3%) students part of the engineering program show phone dependency, 259 (77.8%) from health school, 150 (50.0%) from economic sciences, and 201 (68.1%) from law school (Chinnery, G et al. 2006) (Burch, Z. A. et al. 2019). When concerned about musculoskeletal discomforts, the statistics show that the most risked part of the body from the excessive use of smartphone is the wrist, with a 95% range. As for other body parts such as the neck, people with smartphone dependency have 1.42

times more risk to have neck pains than people without smartphone dependency (D. Samanta et al. 2022) (Kukulska-Hulme et al. 2016).

In 2018, Nahas and Hlais did a research based on the usage of smartphones by adults of age 18-65 (Dutta et al., 2022) (Canuel, et al., 2015) (Galanek, J. D. et al. 2019) (Chen, et al., 2011). The results indicated that smartphones are mostly used for chatting, mostly by young adults and unmarried people, also people who use mobile data. Apart from chatting, these past few years, smartphones have been used for a lot of other purposes and services, which makes them even more useful and raises the possibility of people becoming more dependent on them. For this reason, in 2021, Swati Aggarwal et al conducted a study on smartphone addiction through different Android applications and obtained a correlation for every type of application used by the people part of this study (Althar et al., 2022) (Compton et al., 2018) (EDUCAUSE et al., 2019) (Fosnacht et al., 2017). This study was operated via an application called Activity Tracker, which collected every smallest activity on the phone such as calls, messages, screen activity, and so on. The authors of the study separated the applications into 5 classes such as gaming, utility, entertainment, social, and shopping. One of the most influential factors in the usage of mobile phones is screen time (Carlos et al., 2012) (Chen et al., 2013) (Oz, et al., 2013) (Galanek et al., 2018). The higher the screen time duration and the number of messages, the higher the possibility to get phone addicted. "The maximum effect can be seen through the usage of apps present in the "social" bucket such as Facebook, WhatsApp, and Instagram, with a positive correlation of 0.208."(Aggarwal et al) When talking about smartphone addiction, shopping and gaming apps have also a positive correlation, whereas utility apps show a negative correlation which indicates that they do not impact smartphone addiction. This study was also gender-based where a group of 96 people, male and female were part of the experiment (Canuel et al., 2016) (Nagavelli et al., 2022) (Davidson et al., 2016) (Carter et al., 2015) (Gierdowski et al., 2020) (Garcia et al., 2016). Of 47 males in the group, 35 of them are smartphone addicted and the remaining 12 are not, whereas of 49 females, 26 of them have smartphone dependency and 23 others do not. The results of the analysis showed that males are more phone dependent than females. Those smartphone dependent people mostly use social and gaming applications. The final results of the study show that the correlation for social applications is a positive number of 0.208, 0.18 for gaming applications, and 0.201 for shopping applications/food and beverages. Meanwhile, the utility applications such as Google Maps, Healthify Me, Video Recorder, used for the study, have a negative correlation which tells us that they do not affect at all smartphone addiction (Chen et al., 2015) (Hanbridge et al., 2018) (Hinze et al., 2017).

The usage of smartphones has significantly increased recently. Its impact is indeed beneficial but harmful at the same time. A large number of people have started to develop addiction and phone dependency to the point that a lot of them suffer from "nomophobia", which is the fear of being without a smartphone. This is a really serious concern that people have shared in the last few years. This is the main factor that made us start this project. We wanted to research and find statistics about the age and number of people with phone addiction. According to those findings, teenagers are the most impacted by mobile addiction. The main reason for this is that they cannot manage their screen time and they still have not developed the ability to self-control. One other objective of the research is the negative effects of smartphone dependency on our health. Excessive phone usage can result in insomnia, memory damage, and poor decision-making. Also, the statistics have shown that the most risked body part from phone usage is the wrist, leaving the neck the second. With our research project, we wanted to find statistics and results like this and maybe propose a small partial solution for smartphone dependency.

RESULT AND EXPERIMENT

Table 1. 'Social' category with purpose of uses

Serial Number	Name	Purpose
1	Instagram	Social Media used to contact and interact with our friends form all over the world.
2	Snapchat	Snapchat is a social media who allows us to share photos and texts with our friends from the whole world. The photos can be edited in different interesting filters.
3	Whatssapp	Whatsapp is a free messaging and calling app.
4	Viber	Viber is same as Whatsapp. It gives us the possibility to call and chat with our friends for free.
5	Facebook	A social media where you can share posts, see other's posts, chat or call your friends and also play games with them.
6	BeReal	BeReal is a new application. It allows you to post only once a day, when you get the notification from the app. You post whatever you are doing in the moment and there is no filter you can use because you have to "BeReal".
7	LinkedIn	LinkedIn is a social media but used for professional reasons. You can share all your academic information and also look for jobs.
Utility' Category with Purpose of Uses		
1	Reminder	We can set reminders for any work we have to do which we might forget.
2	Calculator	Enables us to do any type of mathematical calculation.
3	Weather	It tells us the weather forecast every day, maximum for a week.
4	Samsung Notes	Samsung Notes allows us to write notes or any other thing that we may forget. It also allows us to add photos and make our notes more colorful.
5	Bixby	Bixby is a virtual assistant for Samsung which makes it easier to use our phone.

continues on following page

Table 1. Continued

Serial Number	Name	Purpose
6	Calculator	Enables us to do any type of mathematical calculation.
7	Clock	Clock is an application that provides us the possibility to set alarms, add timers and run a stopwatch.
8	Chrome	A browser that provides access to any webpage that is on the internet.
9	Contacts	Contacts is like a little phone book that you keep in your pocket, which contains all of our contacts and gives us the possibility to call them.
10	Calendar	As we understand from its name, it gives us access to the year calendar and allows us to add reminders for any specific day.
11	mKlik NLB Kosovo	This is a bank app. It allows me to check my balance or use any other bank service. This application is unique for every bank.
'Games' Category with Purpose of Uses		
1	WoW	Words of Wonder is a crossword game where you can improve your vocabulary. You can play it in any language you want.
2	Ludo King	Ludo King is a very famous boarding game. You can play with your friends or even strangers from all over the world.
3	Traffic Rider	Traffic Rider is a racing game. With your motorcycle, you have to arrive at the finish line in time otherwise you have to play again.
4	Happy Color	It is like a coloring book. You can color pictures of any kind by coloring the numbers with their specific colors.
'Educational' Category with Purpose of Uses		
1	Pulse	Pulse is a student's application. It gives you access to your courses and it does work for any university in the world. You get notification when your grade is updated or when you have a new assignment.
2	Zoom	Through Zoom, you can hold conference meetings or classes in distance.
3	Socratic	Socratic is also a student's application. Whenever you have any problem in any subject, you just scan it and Socratic will find articles on the web related to that problem.
4	Duo	Duo lets users make and receive one-to-one video calls with high definition.
5	Student Calendar	You can add dates of your exams or your assignments and Student Calendar will remind you of your exam dates.
Photography' category with purpose of uses		
1	Photos	Photos is an app similar to Gallery. It gives us access on our photos and edit them.
2	Camera	Camera is a photo application where you can take photos and edit them with a majority of filters.
3	Gallery	Gives us access to our photos where we can also edit them.
4	Snapchat	Snapchat is a social media who allows us to share photos and texts with our friends from the whole world. The photos can be edited in different interesting filters.
5	BeReal	BeReal is a new application. It allows you to post only once a day, when you get the notification from the app. You post whatever you are doing in the moment and there is no filter you can use because you have to "BeReal".

continues on following page

Table 1. Continued

Serial Number	Name	Purpose
6	Instasize	Instasize is a picture editing application. You can change your picture's background, and size, edit it with different filters and colors, etc.
'Communication' Category with Purpose of Uses		
1	Messages	Every message we receive in our phone number goes directly to Messages, where we can read and delete them if we want.
2	Gmail	Enables us to have access on our gmail account and read and send emails.
3	Outlook	Gives us the possibility to receive and send emails.
4	Contacts	Contacts is like a little phone book that you keep in your pocket, which contains all of our contacts and gives us the possibility to call them.
5	Phone	Allows us to call any phone number or use any service provided by our provider.
'Entertainment' Category with Purpose of Uses		
1	Spotify	A music app that requires payment, where you can listen to any music you want anytime.
2	Netflix	Netflix gives us access to different movies and TV shows and we can download them to save for later. Netflix also requires payment.
3	Samsung Music	Gives us the possibility to listen to the music we download on our phone. We can also make playlists based on our preferences.
4	Game Launcher	Every game that you download in your Samsung will be stored there and you have access to them anytime you want to play.
5	AR Zone	With AR Zone we can create bitmoji with our features and it allows us to experience augmented reality.
6	YouTube	Allows us to watch, share, like, and comment on videos.

Table 2. Category Vs Number of Apps

SL No	Category	Number of apps
1	Social	7
2	Utility	11
3	Games	4
4	Educational	5
5	Photography	6
6	Communication	5
7	Entertainment	6

Figure 1. Graphical representation of category vs number of apps

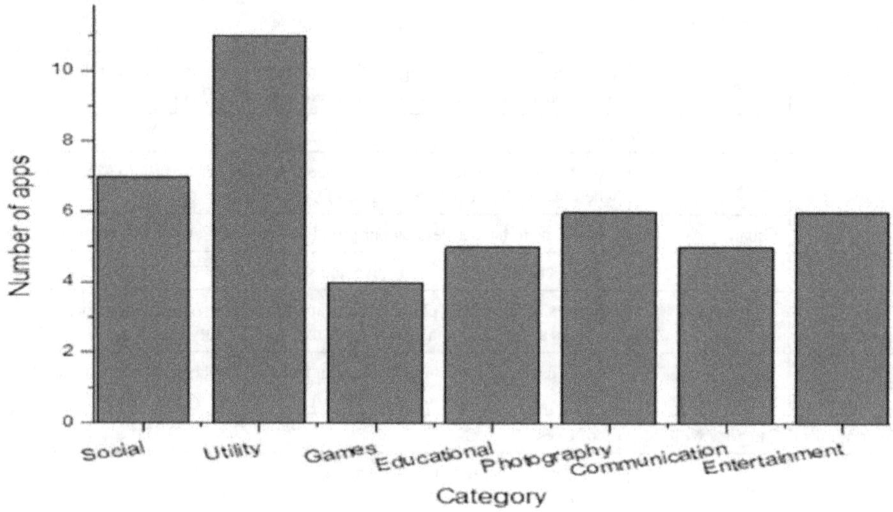

CONCLUSION

Today, a mobile phone is among the most crucial aspects of our life. We are aware that they are beneficial and give us the ability to execute various services for our needs online. But everything has its limits. It is outrageous that we need our phones so much that we cannot even eat or sleep without them. A person uses their cell phone on average for 3 hours and 15 minutes per day, according to some statistics. Teenagers are the group of people who are most impacted by this issue. Numerophobia, or "the fear of not having a cell phone," is a condition that affects the majority of people. Due to their inability to control their screen time and lack of self-discipline, teenagers are the group most affected by mobile addiction. According to some studies, people between the ages of 11 and 13 who own a mobile phone make up 27% of the population and never turn their phone off, not even to go to sleep. Phone addiction is influenced by real-world issues like family issues, strained relationships at school, psychological issues, and more. There are countless detrimental effects of a smartphone addiction. Real physical and mental health issues may result from it. Without a doubt, we are aware of how essential mobile phones are to our daily lives. Time management is therefore the best approach to overcoming addiction. Although using a phone is a need for daily life, we can at least control and limit the amount of time we spend on it.

REFERENCES

Althar, R. R., & Samanta, D. (2022). BERT-Based Secure and Smart Management System for Processing Software Development Requirements from Security Perspective. In *Machine Intelligence and Data Science Applications: Proceedings of MIDAS 2021*, , (pp. 427–440). Springer Nature. 10.1007/978-981-19-2347-0_34

Burch, Z. A., & Mohammed, S. (2019). Exploring Faculty Perceptions about Classroom Technology Integration and Acceptance: A literature Review. The Richard W. Riley College of Education and Leadership Publications.

Canuel, R., & Chrichton, C. (2015). Leveraging apps for research and learning: A survey of Canadian academic libraries. *Library Hi Tech*, *33*(1), 2–14. doi:10.1108/LHT-12-2014-0115

Canuel, R., MacKenzie, E., Senior, A., & Torabi, N. (2016). Apps for academic success: Developing digital literacy and awareness to increase usage. *Education for Information*, *33*(3), 1–18. doi:10.3233/EFI-160085

Carlos, A. (2012). Research on the go: Mobile tools for conducting research. *The Reference Librarian*, *53*(4), 433–440. doi:10.1080/02763877.2012.707462

Carter, A., Liddle, J., Hall, W., & Chenery, H. (2015). Mobile phones in research and treatment: Ethical guidelines and future directions. *JMIR mHealth and uHealth*, *3*(4), 1. doi:10.2196/mhealth.4538 PMID:26474545

Chen, B., Seilhamer, R., Bennet, L., & Bauer, S. (2015). *Students' mobile learning practices in higher education: A multi-year study*. Educause Review. https://er.educause.edu/articles/2015/6/students-mobile-learning-practices-in-higher-education-a-multiyear-study

Chen, B., Sivo, S., Seilhamer, R., Sugar, A., & Mao, J. (2013). User acceptance of mobile technology: A campus-wide implementation of blackboard's MobileTM learn application. *Journal of Educational Computing Research*, *49*(3), 327–343. doi:10.2190/EC.49.3.c

Chen, G. (2011). Mobile research: Benefits, applications and outlooks. Internationalization. *Design and Global Development*, *1*, 11–16.

Chinnery, G. (2006). Going to the MALL: Mobile assisted language learning. *Language Learning & Technology*, *10*, 9–16.

Compton, H., & Burke, D. (2018). The use of mobile learning in higher education: A systematic review. *Computers & Education*, *123*, 53–64. doi:10.1016/j.compedu.2018.04.007

Das Gupta, R., Samanta, D., & Joseph, N. P. (2022). Rendering View of Kitchen Design Using Autodesk 3Ds Max. In *Micro-Electronics and Telecommunication Engineering: Proceedings of 5th ICMETE 2021.* Springer Nature. 10.1007/978-981-16-8721-1_33

Dutta, S., Das, A. K., Ghosh, S., & Samanta, D. (2022). Data Analytics for Social Microblogging Platforms. Elsevier.

Davidson, J., Paulus, T., & Jackson, K. (2016). Speculating on the future of digital tools for qualitative research. *Qualitative Inquiry*, *22*(7), 606–610. doi:10.1177/1077800415622505

EDUCAUSE. (2019). Horizon Report Preview: 2019 Higher Education Edition. EDUCAUSE. https://library.educause.edu/-/media/files/library/2019/2/2019horizonreportpreview.pdf

Fosnacht, K., Sarraf, S., Howe, E., & Peck, L. K. (2017). How important are high response rates for college surveys? *The Review of Higher Education*, *40*(2), 245–265. doi:10.1353/rhe.2017.0003

Galanek, J. D., & Gierdowski, D. C. (2019). *ECAR Study of Faculty and Information Technology, 2019. Research report.* ECAR. https://www.educause.edu/-/media/files/library/2019/12/facultystudy2019.pdf?la=en&hash=F0F8AF83BD870239A7E5CD285FBDBE157AA51135

Galanek, J. D., Gierdowski, D. C., & Brooks, D. C. (2018). *ECAR study of undergraduate students and information technology, 2018. Research report.* ECAR. https://library.educause.edu/-/media/files/library/2018/10/studentitstudy2018.pdf?la=en&hash=C590C1F6C62B77792711BFAC1F642254A5618590

Garcia, B., Welford, J., & Smith, B. (2016). Using a smartphone app in qualitative research: The good, the bad and the ugly. *Qualitative Research*, *16*(5), 508–525. doi:10.1177/1468794115593335

Gierdowski, D. C., Brooks, D. C., & Galanek, J. D. (2020). *EDUCAUSE 2020 Student Technology Report: Supporting the Whole Student, Research report.* CO: ECAR. https://www.educause.edu/ecar/research-publications/student-technology-report-supporting-the-whole-student/2020/technology-use-and-environmental-preferences

Gikas, J., & Grant, M. M. (2013). Mobile computing devices in higher education: Student perspectives on learning with cellphones, smartphones & social media. *The Internet and Higher Education*, *19*, 18–26. doi:10.1016/j.iheduc.2013.06.002

Guha, A., Obaidat, M. S., Samanta, D., & Islam, S. K. H. (2022). Clustering-based Optimal Resource Allocation Strategy in Title Insurance Underwriting. In *2022 International Conference on Computer, Information and Telecommunication Systems* (CITS), (pp. 1–5). IEEE. 10.1109/CITS55221.2022.9832993

Hahn, J. (2014). Undergraduate research support with optical character recognition apps. *RSR. Reference Services Review*, *42*(2), 336–350. doi:10.1108/RSR-09-2013-0045

Hanbridge, A. S., Tin, T., & Sanderson, N. (2018). Information literacy skills on the go: Mobile learning innovation. *Journal of Information Literacy*, *12*(1), 118–136. doi:10.11645/12.1.2322

Hennig, N. (2014). *Apps for librarians: Using the best mobile technology to educate, create and engage.* Libraries Unlimited. https://ebookcentral-proquest-com.ezproxy.waikato.ac.nz/lib/waikato/detail.action?docID=1813321

Hinze, A., Vanderschantz, N., Timpany, C., Cunningham, S. J., Saravani, S. J., & Wilkinson, C. (2017a). *Use of mobile apps for teaching and research. (Working Paper Series 01/2017a).* Department of Computer Science, The University of Waikato.

Kukulska-Hulme, A. (2016). *Personalization of Language Learning through Mobile Technologies*. Cambridge University Press.

Kukulska-Hulme, A., & Shield, L. (2008). An overview of mobile assisted language learning: From content delivery to supported collaboration and interaction. *ReCALL, 20*(3), 271–289. doi:10.1017/S0958344008000335

Kumar, G., & Samanta, D. (2022). Effective View of Swimming Pool Using Autodesk 3ds Max: 3D Modelling and Rendering. In *Internet of Things and Its Applications: Select Proceedings of ICIA 2020,* (pp. 47–53). Springer Nature. 10.1007/978-981-16-7637-6_5

Nagavelli, U., Samanta, D., & Thomas, B. (2022). ML-Based Prediction Model for Cardiovascular Disease. In *Emerging Technologies in Data Mining and Information Security: Proceedings of IEMIS 2022*, (volume 2, pp. 91–98). Springer Nature.

Oz, H. (2013). Prospective English teachers' ownership and usage of mobile device as m-learning tools. *Procedia: Social and Behavioral Sciences*, *141*, 1031–1041. doi:10.1016/j.sbspro.2014.05.173

Samanta, D. (2022). *3D Modeling Using Autodesk 3ds Max With Rendering View*. IGI Global.

Singh, D. (2022). Search Engine Optimization for Digital Marketing to Raise the Rank, Traffic, and Usability of the Website. In *Human-Centric Smart Computing: Proceedings of ICHCSC 2022,* (pp. 217–226). Springer Nature Singapore Singapore.

Singh, D., Shivamurthaiah, M., Samanta, D., Bhattacharya, A., & Dutta, S. (2022). Applicability of Search Engine Optimization for WordPress (WP) Website. In *Human-Centric Smart Computing: Proceedings of ICHCSC 2022,* , (pp. 227–237). Springer Nature.

Chapter 5
An Approach With Iterative and Incremental Development (IID) for Mobile Applications

Jona Ndrecaj
Rochester Institute of Technology, Kosovo

Shkurte Berisha
Rochester Institute of Technology, Kosovo

Erita Çunaku
Rochester Institute of Technology, Kosovo

ABSTRACT

There are several studies on the software development life cycle (SDLC) and various approaches that can be utilized in planning and designing a software project. The contemporary norm for the software development process across industries is to build products using iterative or incremental design. Iterative and incremental development (IID) is useful for different types of software. The authors have provided a literature review of the current definitions of iterative and incremental design, and also discuss the most common pitfalls for each approach. The authors also analysed which approach is better for different stages of the development of mobile applications. At the end of this research, it was concluded that the iterative approach is the best approach for front-end development, while the incremental approach is a better fit for back-end development. Finally, the authors provided an analysis of utilizing IID for applications in the banking sector.

DOI: 10.4018/978-1-6684-8582-8.ch005

INTRODUCTION

Every software development has a process or a life cycle that can be broken down in different approaches. These processes define the software development flow and what steps must be followed in every software development phase (Tyagi et al., 2017). These different approaches to the software development process further clarify what must be performed by the development team, what deliverables should be provided, when and how to get user requirements etc. (Flora et al., 2014). The two development processes we are going to further discuss and explore their real-life application is iterative and incremental development (Parmar et al., 2021). By defining what iterative and incremental development are, their mobile application and taking the banking sector as a real-life application, we can understand their impact on software design and progress.

ITERATIVE SOFTWARE DEVELOPMENT

The first software development approach is iterative development. The term iteration describes a repeating process that aims to achieve a certain target or goal. The iterative development takes the large software development process and breaks it down to smaller portions or iterations (Cao et al., 2012). The outcome of one iteration is the start of the next one. Each iteration builds upon each other to create the full software development cycle or process. Different from other development processes, in iterative development each iteration contains the plan, design, development and test. This way when one iteration is completed the team evaluates that particular iteration, suggests any changes that need to be made and then uses those ideas to start the next iteration which also contains the planning, designing, development and testing (Potasso-Justino 2019). The first stage in an iterative development model is the planning phase. In this phase, the development team decides what hardware or software requirements we need so that those requirements are used for the other upcoming phases. The second phase in an iteration is designing where the business logic, database models and any technical requirement is defined. The next phase is development or implementation where the planning and design requirements are implemented, and coding takes place (Carlucho 2017). The last phase is testing, where the development team identifies any possible bugs or issues that show up during the implementation phase. Iterative development is more cyclical rather than the traditional step-by-step process that other models use. Figure 1 visualizes how iterative development works (Kaleel et al., 2013).

Figure 1. Iterative Development Cycle

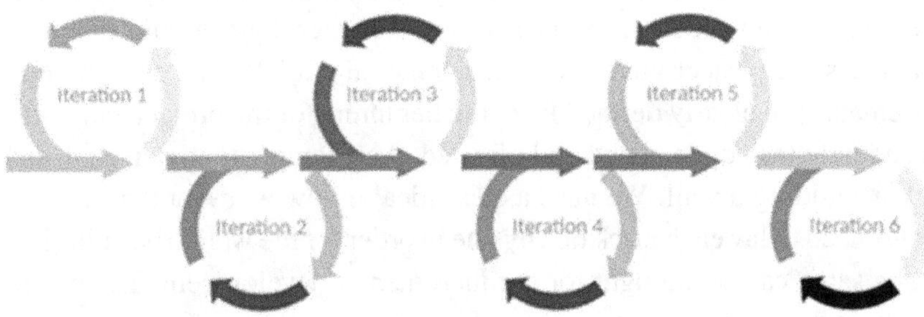

Using iterative development can be attractive to different software companies. One of the reasons is because of the iterations created (van Kessel et al., 2021). Iterations means building a software step by step, which allows the development team to identify defects at an early stage. Detecting defects as early as possible saves money, time and most importantly makes sure the development process is completed successfully without dragging previous errors into the next stage of development (Li et al., 2019). This also means that changing the project scope is easier to implement and less costly as well. Using iterations also helps in receiving feedback from customers. At the end of each iteration, feedback is given about what must be changed for the next iteration. Another advantage of using iterative development is that it saves time on documentation, as this process is more focused on design rather than documenting the process and development (Zein et al., 2016). Lastly, using iterations make it easier to measure the development process, which can be essential in the success of the project.

Even though iterative development does have advantages, there are some drawbacks to using such a process. Having to go through the cycle of planning, designing, development and testing for each iteration means that you will need resources and resources are costs for development (Abhilasha 2013). The most dangerous disadvantage of iterative development is that it is difficult to know when the project is fully developed. It is hard to pin down the end date for the development of that product, since each iteration may result in different improvements that the next iteration must meet (Cao et al., 2009). Lastly, the process itself may be a constraint since requirements for the overall project are not truly defined but rather are specific for each iteration.

INCREMENTAL SOFTWARE DEVELOPMENT

The second software development approach is incremental development. In this software engineering process, the project is taken step-by-step and incremented until it reaches the project goal (Hosseinzadeh et al., 2022). The end product and its requirements are clearly defined from the beginning of the project and each step is independently tested, designed and planned. An example to understand such an approach is building a wall. We have a clear idea of how we want the wall to look like, but we need to lay each brick one by one in order for the wall to be built (Parmar 2021). The same can be thought for the incremental development. The benefit of using such an approach is that it allows for the users to get a faster hand on the software. This is because you are able to release whatever you have built so far and show it to the users. Another benefit is the quality. Because you concentrate only on a small portion or system at a time, you can easily ensure that there is quality in each sub-system (Barbareschi 2022). However, just like any other methodology, it does have its pitfalls.

Because the software development process is divided into subsystems, it can occur that the cost of the overall system can exceed the costs of the initial project costs. Without a solid planning design, the software will not be successful (Riesener et al., 2021). The overall idea of incremental development takes a lot of time for a product to be developed. Another disadvantage of this approach is its rigid incremental phases, which can be challenging in overlapping the phases with one another and changing a requirement from the beginning (M. Halpern et al., 2016). Even though incremental development does seem very similar to iterative, we have to understand that the main difference between the two is that incremental development process is focused on every possible detail of a final product, while iterative improvements are made at any moment and all team members work at the same time (Lee 2022). Figure 2 illustrates the way that the incremental development process is performed.

MOBILE APPLICATION DEVELOPMENT

Now that we have a better understanding about the similarities and differences between the incremental and iterative models, we can focus on the utilization of these models for mobile application development (Gu et al., 2022).

The process for creating software applications that run on mobile devices is called mobile application development. The steps for that process are shown in Figure 3.

Figure 2. Incremental Development Process

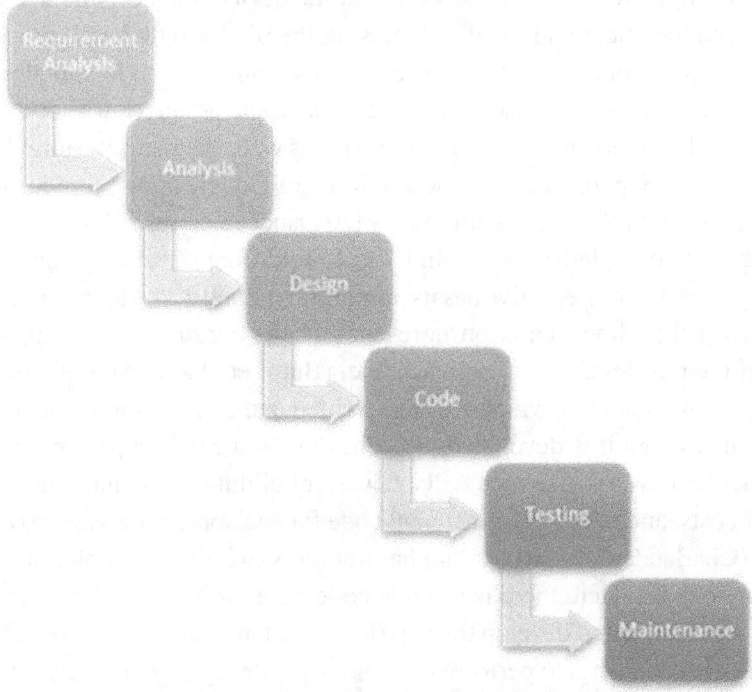

Figure 3. Mobile Development Process

Every development of a mobile application starts with an idea. In the first phase, the clients communicate that idea to the mobile developers regarding the applications, its functionality, design, expectations, and other information regarding the app (Francis 2017). Once the requirements are gathered, the requirements are analyzed to see what can be done and what might need to change if a requirement cannot be accomplished (Nguyen et al., 2015). Since mobile applications are usually built for the specific platform and are native to the device, meaning that the needed software packages are installed and stored on the device, the decision needs to be made between the client and the developers on the platform they will proceed the development on (Kemper et al., 2002). The main operating systems for mobile devices are iOS and Android. iOS is made exclusively for Apple products, while Android is made by Google but is used by other companies as well, such as Samsung, Xiaomi, Huawei, Amazon, Oppo, etc., with the condition of Google apps being installed in the device, and other specific requirements (Ashishdeep et al., 2016).

Having those two platforms in mind, there are different ways we can go about developing apps. Each alternative has its pros and cons, and it's up to the developer to decide which path to choose based on the resources, time, constraints of the application, strategy of the business, user experience, etc. (Butor et al., 2004). Let's start with native mobile applications, which are apps written in the programming language of the operating system it is developed for. It has the best runtime performance, and you get direct access to the device APIs, however building and maintaining the app has higher costs, and needs to be different code for each operating system (iOS and Android) (Charland et al., 2011). Another option is to build cross-platform native mobile applications. Here there is a single code base for both platforms, and it's a lot easier to maintain and develop the app (Robson et al., 2016). However, there are limitations when it comes to performance and dependency on the libraries that join the two platforms. Hybrid mobile applications are developed using web technologies and are packed into installation packages (The Modern Static Analysis Platform). The code base can be shared with the app and the web platform, but the performance is lower than having a native installation bundle, as well as there are limitations to support of native device features. Lastly, progressive web applications are the same app for the web app as well as the mobile device app (Arnold et al., 2001). In this case, no installation is needed, and the mobile app can be accessed through a URL. This alternative also has limitations, including limited support for the features of the device, and all apps depend on the browser the user is using. Which platform you choose to develop depends on the skills of the developing team, the existing infrastructure, and the target mobile (Chae et al., 2001).

Regardless of the alternative mobile development approach, all mobile apps are composed of two parts: the front-end development process and back-end development process (Lucia-Palacios 2016). The front-end development focuses on user interface design. An important task of mobile applications is making the interface of the application user-friendly. A long line of research shows that the success of the application is heavily linked with user satisfaction and application performance (Jeong et al., 2008). Constraints for making the application interface are the changing screen size of the device, contexts of the users, limited attention of the user, and other constraints. The functionality of the user interface relies on the back-end development (Arzenšek et al., 2014). The back-end development focuses on data integration, connection to APIs, security, authentication, and other services. Without the functionality, the application interface is just a skeleton (Abbas et al., 2008). The front-end and back-end developers work together to create the application curated to the needs of the client.

FRONT-END APPLICATION DEVELOPMENT

The design process for mobile app development usually starts with designing the interface of the application (Sillitto et al., 2019). When designing the interface, it is crucial we test the design and its usability with the end users. Based on the feedback of those users, we make needed modifications to our design, until the review process is satisfied with the results of the testing, and we can move on to deployment (Fazal Rehman Shamil). Here, when designing the front-end portion of the application, it is best to use the iterative model for software development. Due to the iterative nature of the user experience and interface design, it is best to create a version, test it, and then add or change features as required, until the best version is reached (Francis, 2017). This allows for maximum user input, we can detect problems at a much earlier stage, allows for parallel development, testing and debugging is a lot easier as well.

BACK-END APPLICATION DEVELOPMENT

When developing the back-end portion of the application, it is best to take on an incremental approach (Roberto et al., 2019). Here, based on the design of the application, we can divide the functionality into different parts, such as Login and authenticity functionality, sharing or commenting on a post, sending a friend request, inputting a grade for a specific class, etc. For each increment we gather the requirements and start working on the design and development of it (Fazal Rehman Shamil). This way we can test each component of the application and make sure the customer is satisfied with the outcome. One by one each component of the application is developer and finally all are integrated together into one complete application (Wang et al., 2022). Incremental development is very common in the software industry. Basically, every application that maintains its system and offers updates from time to time relies on incremental development to do so. The model allows for user feedback, faster delivery of functionality, more flexible and also easier to test.

BANKING SECTOR

Up until now, we have established the grounds for defining iterative and incremental development. We also discussed the ways we can employ these models in mobile applications and what implications our choice of either model will have in the process of software development (Liu et al., 2010). Our focus will become even more narrow now as we analyze all that we have laid down, but this time from the POV of the banking sector.

Financial services are becoming more and more digitized, so we must take careful consideration in our SW design approach whether that is iterative or incremental. Banking software is the connective tissue between customers and their bank branches (Friedenthal, 2015). Mobile applications can make the life of banking staff and clients a lot easier by providing the following services: handling transactions, managing financial documents, communicating with customers, etc. (Kaur et al., 2022). It is no surprise, then, that there is great demand for mobile applications that manage banking services.

The modern market already provides us with various examples of successful banking software (Wang et al., 2022). The entrepreneur, Ashish Parmar, mentions the following software solutions for the banking sector: "Temenos, Avaloq, Cleartouch, Mambu, Finastra, Oracle FLEXCUBE, Coreplus, FIS Profile, EBANQ, and BankWare"(2021).

The aforementioned list of banking applications is not exhaustive, as there are many more solutions already out there or perhaps in the process of development (Legner et al., 2016). We can certainly take inspiration from existing and proven software to discuss IID for mobile applications in the Banking Sector. But is that really the best way to discuss the value of the IID models?

Figure 4. Software Solutions for Banking Services

We suggest that IID discussion should come not from popular examples, but rather by analyzing "the challenges that arise during mobile banking software development" (Parmar, 2021). Each model can bring their own flavor of solution to the following software development "problems" in banking: authentication, adaptability, legal compliance, data migration, security, and training. Let's see what solutions we can provide for these problems (Dennison et al., 2013).

Authentication is a "problem" that does not necessarily rely on the development model, but we must take great consideration in implementing proper identity verification. Whether we employ an incremental or iterative approach, development teams should deal with the changes that the banking sector has experienced in authentication processes (Francino, 2011). Before the digitization of banking services, authentication was done via passwords and identifying documents. For quite some time, this process was simply carried over to online banking systems. Very soon banks started to realize that while applications had many benefits, they also provided a platform for identity theft. Robert D. Lee, a Senior Technology Specialist writing for the Federal Deposit Insurance Corporation (FDIC), provides the following solutions to the authentication challenge: secret information shared between the bank and the customer, device identification, geolocation, Internet Protocol (IP) identification, placing encrypted cookies, and out-of-band communication via cell phone calls or messages to provide verification (Oulasvirta et al., 2011). When we develop a mobile application for the banking sector, we should use two or three of the aforementioned solutions along with traditional authentication methods such as passwords and IDs.

Adaptability is another problem that arises with banking apps (Sakshi et al., 2022). Financial tech is an ever-changing industry, so unexpected changes in requirements are a staple in banking SW development. For this reason, we might want to pick an iterative approach for our software lifecycle. The iterative development model allows us to be more flexible with changes.

Legal compliance is a problem that refers to our software's adherence toward regulations such as PCI DSS, CCPA, GDPR, and PSD2.

Figure 5. Legal Regulations

Making our software compatible with legal requirements is a process that requires due diligence (Montalvo et al., 2022). This task is difficult because legal regulations get constantly updated and changed. To develop a successful mobile application, we must stay updated on the legal requirements and adapt a model that allows for flexibility. Iterative design once again gives us the space for drastic changes.

Data migration is a rather costly problem in mobile application development. This "problem" refers to the task of transforming data from the physical system to another technology. Data migration is a big investment, and it also requires a systematic approach (Kienitz, 2021). For this purpose, we might want to consider the incremental model when we develop mobile applications for banking services.

When it comes to banking services, there is simply no such thing as negotiating with security requirements. Whatever development model we use, our application should be averse to digital security risks, such as: "phishing, pharming, spyware, malware, worms, midas, viruses, buffer overflows, and spam" (Lee, 2022).

As for the final problem, we will be pointing out a factor that is often overlooked by developers: the human element. This is especially important because, ultimately, the banking sector is about customer service and the staff should be well acquainted with banking mobile applications, so they can provide proper support for their clients. This is generally a very costly process which does not depend on the development model, but rather on a good feedback loop and effective communication between departments.

We have noticed that incremental and iterative models are useful for different aspects of mobile software development in banking. This happens because software projects relating to banking software are huge undertakings that can branch out in several smaller projects. The banking sector requires a flexible approach, so we can adapt to changes but also a systematic approach that allows us to deal with non-negotiable requirements, such as security. Dividing the development lifecycle into smaller and more focused projects allows us to employ both models and utilize the strengths of iterative and incremental design in different areas (Cruz et al., 2022). As a final example we can suggest a hypothetical arrangement where Data Migration, for instance, is a separate project that uses the incremental model, while Backend Development is a separate project that uses the iterative model, so we can adapt to changes in legal and security requirements.

CONCLUSION

We have described in depth both iterative and incremental development. Our exploration of the IID models allowed us to see the implications of utilizing different development methods for mobile applications. Finally, we discussed the unique challenges of IID models for mobile apps pertaining to the banking sector.

REFERENCES

DeepSource. (n.d.). *The Modern Static Analysis Platform.* DeepSource, https://deepsource.io/glossary/incremental-development/

Abbas, N., Gravell, A. M., & Wills, G. B. (2008). Historical roots of agile methods: Where did "agile thinking" come from? *In International conference on agile processes and extreme programming in software engineering,* (pp. 94-103). Springer.

Abhilasha, S. A. (2013). Test effort estimation in regression testing, in: Innovation and Technology in Education (MITE). *IEEE International Conference in MOOC*, (pp. 343–348). IEEE.

Arnold, U., Essig, M., & Kemper, H.-G. (2001). ìTechnologische Entwicklungen im mobilen Internet und ihre R¸ckwirkungen auf die Unternehmensstrategie,î. In A. T. Nicolai, T. Petersmann (eds.) Strategien im M-Commerce. Stuttgart.

Ashishdeep, A., Bhatia, J., & Varma, K. (2016). Software process models for mobile application development: A review. *Computer Science and Electronic Journal*, *7*(1), 150–153.

Arzenšek, B., & Heričko, M. (2014). Criteria for selecting mobile application testing tools. *CEUR Workshop Proceedings*, 1–8.

Barbareschi, M. (2022, December). Scrum for Safety: An Agile Methodology for Safety-Critical Software Systems. *Software Quality Journal*, *30*(4), 1067–1088.

Butor, M., & Howard, R. (2004). *Mobile*. Amazon, Dalkey Archive Press. https://aws.amazon.com/mobile/mobile-application-development/

Cao, G., Yang, J., Zhou, Q., & Chen, W. (2012). *Software Testing Strategy for Mobile Phone, Advances and Applications in Mobile Computing*. InTech.

Cao, L., Mohan, K., Xu, P., & Ramesh, B. (2009). A framework for adapting agile development methodologies. *European Journal of Information Systems*, *18*(4), 332–343.

Carlucho, I. (2017). Incremental Q-Learning Strategy for Adaptive PID Control of Mobile Robots. [ScienceDirect, https://doi.org/10.1016/j.eswa.2017.03.002.]. *Expert Systems with Applications*, *80*(Sept), 183–199.

Chae, M., & Kim, J. (2001). Information Quality for Mobile Internet Services: A Theoretical Model with Empirical Validation. In V. Storey, S. Sarkar, and J. I. DeGross (eds.), *Proceedings of the Twenty-Second International Conference on Information Systems*, (pp. 43-54). IEEE.

Charland, A., & Leroux, B. (2011). mobile application Development : Web vs. native. *Communications of the ACM*.

Cruz, M. (2022). *A Model-Based Approach for Specifying Changes in Replications of Empirical Studies in Computer Science*. Springer. doi:10.1007/s00607-022-01133-x

Dennison, L., Morrison, L., Conway, G., & Yardly, L. (2013). Opportunities and challenges for smartphone applications in supporting health behavior change: Qualitative study. *Journal of Medical Internet Research*, *15*(4), e86. doi:10.2196/jmir.2583

Flora, H. K., Wang, X., & Chande, S. V. (2014). An investigation into mobile application development processes: Challenges and best practices. *International Journal of Modern Education and Computer Science*, *6*(6), 1.

Francino, Y. (2011). *"What Is Iterative Development? - Definition from Whatis.com." SearchSoftwareQuality*. TechTarget. https://www.techtarget.com/searchsoftwarequality/definition/iterative-development

Francis, P. (2017). *Mobile App Development Process - an in-Depth Overview*. The BHW Group. https://thebhwgroup.com/blog/mobile-app-development-process

Friedenthal, S. (2015). *A practical guide to SysML: The systems modeling language*. Morgan Kaufman.

Gu, C. (2022, October). What Do Users Care about? Research on User Behavior of Mobile Interactive Video Advertising. [ScienceDirect, https://doi.org/10.1016/j.heliyon.2022.e10910.]. *Heliyon*, *8*(10), e10910.

Hosseinzadeh, M. (2022, December). Clustering for Smart Cities in the Internet of Things: A Review. *Cluster Computing*, *25*(6), 4097–4127.

Jeong, Y.-J., Lee, J.-H., & Shin, G.-S. (2008). Development process of mobile application sw based on agile methodology. In *Advanced communication technology, 2008. icact 2008. 10th international conference on* (*Vol. 1*, pp. 362-366). ICACT.

Li, K., Tao, M., & Chen, Z. (2019). A computation-communication tradeoff study for mobile edge computing networks. *2019 IEEE International Symposium on Information Theory, (ISIT)* (pp. 2639-2643). doi:10.1109/ISIT.2019.8849601

Kaleel, S. B., & Harishankar, S. (2013). *Applying agile methodology in mobile software engineering: Android application development and its challenges*. Computer Science Technical Reports.

Kaur, A., & Kaur, K. (2022). Systematic Literature Review of Mobile Application Development and Testing Effort Estimation. *Journal of King Saud University - Computer and Information Sciences, 34*(2), pp. 1–15. ScienceDirect. doi:10.1016/j.jksuci.2018.11.002

Kemper, H. G. & Wolf, E. (2002). Iterative Process Models for Mobile Application Systems: A Framework. *ICIS 2002 Proceedings, 37*. PSU.

Kienitz, P. (2021). *The Pros and Cons of Iterative Software Development.* One Beyond. https://www.one-beyond.com/pros-cons-iterative-software-development/

Lee, D. R. (2022). *FDIC: Supervisory Insights - Authentication in Internet Banking: A Lesson in Risk Management.* Fdic.gov. www.fdic.gov/regulations/examinations/supervisory/insights/siwin07/article05_authentication.html

Legner, C. (2016, September). Mobile Business Application for Service and Maintenance Processes: Using Ex Post Evaluation by End-Users as Input for Iterative Design. [ScienceDirect, https://doi.org/10.1016/j.im.2016.03.001.]. *Information & Management, 53*(6), 817–831.

Liu, Z., Gao, X., & Long, X. (2010). Adaptive random testing of mobile application. In *Computer engineering and technology (iccet), 2010 2nd international conference on* (*Vol. 2*, pp. V2-297).

Lucia-Palacios, L. (2016, July). Complementary IT Resources for Enabling Technological Opportunism. [ScienceDirect, https://doi.org/10.1016/j.im.2016.02.002.]. *Information & Management, 53*(5), 654–667.

Halpern, M. (2016). Mobile CPU's rise to power: quantifying the impact of generational mobile CPU design trends on performance, energy, and user satisfaction. *Proceedings of the 2016 IEEE international symposium on high performance computer architecture (HPCA)*. IEEE.

Montalvo, A. (2022, December). Model-Driven System-Level Validation and Verification on the Space Software Domain. [Springer Link, https://doi.org/10.1007/s10270-021-00940-8.]. *Software & Systems Modeling, 21*(6), 2367–2394.

Oulasvirta, A., Wahlström, M., & Ericsson, K. A. (2011). What does it mean to be good at using a mobile device? an investigation of three levels of experience and skill. *International Journal of Human-Computer Studies*, *69*(3), 155–169.

Parmar, A. (2021). *A Comprehensive Guide on Banking Software Development.* Prismetric. www.prismetric.com/guide-to-banking-software-development/

Potasso-Justino. Lucas. "What Is Iterative Development and How Does It Work? " Planbox, 16 Apr. 2019, https://www.planbox.com/what-is-iterative-development-and-how-does-it-work/

Riesener, M. (2021). Methodology for Iterative System Modeling in Agile Product Development. [ScienceDirect, https://doi.org/10.1016/j.procir.2021.05.101.]. *Procedia CIRP*, *100*(Jan), 439–444.

Roberto, R. (2019). Geometrical and Statistical Incremental Semantic Modeling on Mobile Devices. [ScienceDirect, https://doi.org/10.1016/j.cag.2019.09.003.]. *Computers & Graphics*, *84*(Nov), 199–211.

Robson, C., & McCartan, K. (2016). *Real world research*. John Wiley & Sons.

Sakshi & Kukreja, V. (2022). A Dive in White and Grey Shades of ML and Non-ML Literature: A Multivocal Analysis of Mathematical Expressions. *Artificial Intelligence Review*. Springer. doi:10.1007/s10462-022-10330-1

Shamil, F. R. (n.d.). Incremental Model, Advantages, Examples, Case Study. *T4Tutorials.Com*. https://t4tutorials.com/incremental-model-in-software-engineering-advantages-disadvantages-examples-case-study-of-incremental-model/

Shamil, F. R. (n.d.). *Iterative Model in Software Engineering Advantages Disadvantages*. T4Tutorials.Com. https://t4tutorials.com/what-is-iterative-model-software-engineering/

Sillitto H., Martin J., McKinney D., Griego R., Dori D., Krob D., Godfrey P., Arnold E., Jackson S. (2019). Systems Engineering and Systems Definitions.

T. Nguyen, T. (2015). Structural modeling from depth images. *IEEE Trans Visual Comput Gr.* IEEE.

Tyagi, V., & Kumar, A. (2017). Internet of things and social networks: a survey. *International Conference on Computing, Communication and Automation, ICCCA*, (pp. 1268-1270). IEEE. doi:10.1109/CCAA.2017.8230013 van Kessel, K. (2021). Design Considerations for a Multiple Sclerosis Fatigue Mobile App MS Energize: A Pragmatic Iterative Approach Using Usability Testing and Resonance Checks. *Internet Interventions: the Application of Information Technology in Mental and Behavioural Health*, *24*(Apr), 100371.

Wang, J. (2022, December). A Review on Extreme Learning Machine. *Multimedia Tools and Applications*, *81*(29), 41611–41660.

Wang, M. (2022). Users' Experience Matter: Delay Sensitivity-Aware Computation Offloading in Mobile Edge Computing. *Digital Communications and Networks*. ScienceDirect. doi:10.1016/j.dcan.2022.08.005

Zein, S., Salleh, N., & Grundy, J. (2016). A systematic mapping study of mobile application testing techniques. *Journal of Systems and Software*, *117*, 334–356.

Chapter 6
Uses of Mobile Apps for Youth in Education Purposes

Yllka Totaj
Rochester Institute of Technology, Kosovo

ABSTRACT

People nowadays use their phones constantly throughout the day, for a variety of reasons. It seems reasonable that they may be used for educational purposes as well. The process of education has become much easier with mobile applications. Through these mobile apps, students can access all kinds of books and information. They can save a lot of time without the need to go to the library to acquire it. Thanks to these mobile apps, classes and lectures are not confined to physical classrooms only. Students no longer have to wait until class to ask questions; instead, they may get in touch with their professors whenever the need appears. When schools were closed and lessons were done online during Covid-19, mobile applications were also incredibly helpful.

INTRODUCTION

The main goal of this research is to determine the extent of smartphone addiction and how male and female users respond differently when engaging in various smartphone activities. Mobile technology has brought the globe closer together by enabling contact between individuals in various locations and developing new capabilities like internet access, gaming, email sending and receiving, music listening, and book reading (Chinnery et al., 2006) (Burch et al. 2019). The second-largest telecom market worldwide is reportedly found in India. Phones are considered essential for sustaining social connections and carrying out other everyday chores. Despite the

DOI: 10.4018/978-1-6684-8582-8.ch006

numerous benefits they offer, they also create a great deal of issues (Canuel et al. 2015). Face-to-face interaction is decreasing as a result of people being glued to their mobile devices. Machine learning is an aspect of artificial intelligence that focuses on creating software that can access data and utilize it for its own purposes. Unsupervised learning (when the model is trained in an unlabeled dataset) and supervised learning (where the model is trained using a labeled dataset) are both possible (Compton et al. 2018). Every piece of information used in this particular study is unlabeled and unclassified. According to this survey, men access books and e-books on cell phones more frequently than women do. Males had the highest percentage of phone possessions lasting fewer than six hours, while females have the highest percentage of phone possessions lasting more than 12 hours (Canuel et al. 2016). The findings also revealed that most guys had phones with batteries that last for a day, although the percentage of ladies who answered "yes" and "no" was about equal. The entire population is divided into three clusters: "extremely addicted," "moderately addicted," and "non-addicted" (Carlos et al. 2012).

One of the key educational topic areas where technology has played a significant role over the past few decades is foreign languages. E-readers are becoming widely used, particularly among young people (Kukulska-Hulme et al. 2008). Mobile applications allow students in contemporary schools to take a break from the traditional classroom setting while continuing to learn. But as technology grew in popularity, students' spans of concentration grew shorter (Davidson, J. et al. 2016) (Carter, A. et al. 2015). This study examines how Moodle, a mobile application, affects the instruction of ESL students (Chen, B. et al. 2015). In 2005, the initial version of this mobile application was released. It is renowned for offering social interaction, uniqueness, and engagement to language learners. The effectiveness and use of the Mobile Aided Language Learning Portal (MALLP), whose population is the tenth grade of Laboratory High School (LHS), served as the foundation for this study (EDUCAUSE et al. 2019) (Fosnacht, K. et al. 2017). Investigating how the MALLP assisted ELLs become more fluent in the English language in comparison to the use of standard resources in teaching English Language was one of the goals of this study (Chen, B. et al. 2013) (Oz, H et al. 2013).

Fifty LHS students participated in the experiment for three to four weeks. The control group and the experimental group were each given a portion of the respondents. The experimental group was taught utilizing mobile applications while the control group received instruction using the normal method of education (Galanek et al. 2019) (Chen, G. et al. 2011). According to the study, the experimental group performed better on the test, with a passing percentage of 75%. The control group, on the other hand, received lower marks, with a 60% passing rate (Galanek et al. 2018). The experimental group of students considered the mobile app for learning the English language to be beneficial because it was downloaded into their phones and had all the necessary tools and information (Gierdowski et al. 2020) (Garcia et al. 2016). The researchers found it to be quite useful because the learner had access to the app at all times and in any location. The pupils claimed that using the program was very simple and pleasant (Gikas et al. 2013).

The use of it did have some consequences, though (Hahn, J. et al. 2014). The students in the experimental group indicated that a few of their peers chose not to take part in the mobile app testing (Hanbridge et al. 2018) (Hinze et al. 2017). Although they were actually utilizing other mobile applications, the pupils were claiming to utilize their smartphones for educational purposes (Kukulska-Hulme et al. 2016). In general, during this study, some students enjoyed using the app, while others first found it to be quite challenging. Data connections were another area where they struggled a lot (Hennig et al. 2014)).

The reason why I chose this research is because of the interest and enthusiasm I have for using this form of technology in classrooms. The usage of smartphones and tablets in the classroom has increased rapidly over the past few years. There has also been a dramatic increase in the number of apps available to students. This study set demonstrates how people use their phones throughout the day and how they are used to enhance learning outcomes. To make classrooms more engaging, it is crucial to strive to apply moderate approaches in classrooms nowadays. Moderate approaches emphasize the importance of encouraging parents and teachers to develop methods that are appropriate for children's ages, backgrounds, and cultural values. That's why mobile apps need to be used in the classroom. A variety of apps are used to help students concentrate on their studies, assist with problem-solving and be accessible at any time and location. Some apps can assist students with time management issues in creating and following a timetable. As a result, this study helps me know what is happening in various classrooms across the world; where they are using mobile phones, how often they use them, and how many students are in those classrooms. I think it is important for everyone to know about the usage of mobile apps in classrooms. And Im sure that this study will help come up with innovative ways to improve education for people across the globe.

RESULT AND EXPERIMENT

We experimented with 53 different mobile apps. The results represented in Table1 shows Mobile Apps' Icon with Details, Table 2 projects 'Finance' Category with Purpose of Uses, Table 3 represents 'Educational' Category with Purpose of Uses, Table 4 shows 'Photo & Camera' Category with Purpose of Uses, Table 5 projects 'Social' Category with Purpose of Uses, Table 6 informs 'Entertainment' Category with Purpose of Uses, Table 7 shows 'Utilities' Category with Purpose of Uses, Table

8 projects Category Vs Number of Apps, Figure 1 shows Graphical representation of Category Vs Number of Apps.

Table 1. Mobile Apps' Icon with Details

SL	App Name	Purpose	Icon
1	App Store	An Apple app to buy and download applications.	
2	Mail	It has the sole purpose of sending emails.	
3	Contacts	It is an app where you can create contacts and view the existing ones.	
4	Calendar	In this app you can view the calendar, schedule meetings and make reminders.	
5	Notes	In this app you can store different notes, including images, drawings, checklists etc.	
6	Reminders	You can schedule a reminder and get a notification when its time. You can also make to-do lists.	
7	Voice Memos	An app for recording audios.	
8	Phone	An app to make calls and check you call list.	
9	Messages	An app to send messages to the numbers on your contact list.	
10	Face Time	An app to make video calls.	
11	Safari	It is a browsing database, with the same purpose as Google Chrome.	
12	Stocks	An app to keep track of your financial circumstances.	
13	Weather	An app to view the weather in every place in the world.	
14	Translate	An app to translate words from every language.	
15	Maps	A GPA app that offers satellite imagery, street maps, traffic conditions, route planning for traveling by car, bus, bicycle or foot	
16	Compass	It shows the direction your phone is pointing at the moment and elevation.	
17	Measure	It's an app to turn the device into a tape measure	

Table 2. Mobile Apps' Icon with Details

SL	App Name	Purpose	Icon
18	Shortcuts	It provides shortcuts of every app to add in the home screen and you can change the name or icon of the app	
19	Health	It keeps track of daily steps, burned calories, headphone audio levels, cycle tracking, sleep tracking, medication etc.	
20	TV	Its an app to watch TV shows, channels, buy or rent movies, etc.	
21	Photos	An app to store, view, edit and share pictures on your phone.	
22	Camera	App to take pictures or videos.	
23	Books	An app where you can read books, pdf or epub files.	
24	Podcasts	An app where you can listen to downloaded audio or video files.	
25	Game Center	App to invite friends to play games, track achievements and compare scores	
26	Air Brush	App for editing pictures	
27	BeReal	Is a social media app, where you share a picture once a day with all your friends.	
28	Dazz Cam	App for taking and editing pictures.	
29	Drive	App to store pictures and videos and access them from different devices.	
30	Edge	A browsing app that helps to protect data, save time and money.	
31	Facebook	Allows users who sign up for free to connect with people they know and those they don't	
32	Gmail	An email service that provides users with a gigabyte of storage	
33	Goodreads	App to keep track of the books you've read, the books you want to read, and news from authors you follow.	
34	HERE WeGo	A GPA app where you can download the map of every city in the world and access it without internet connection.	
35	Instagram	A social media app where you can share photos and videos and connect with people from every part of the world.	
36	Lens Buddy	An app to take pictures without having to press buttons or hold the device, but by setting a time interval.	
37	Linkedln	A professional app to find a job or internship and learn new skills for your professional career.	

Table 3. Mobile Apps' Icon with Details

SL	App Name	Purpose
38	Netflix	A paid streaming service to watch movies and TV shows.
39	Outlook	App to send and receive emails, manage your calendar etc.
40	Photomath	An app where you can take pictures of math problems and it gives you a step-by-step solution.
41	Pinterest	A social network where you can share pictures and videos and also make boards where you pin your favorite pictures and videos.
42	Reports+	An app that keeps track of your followers and likes on Instagram.
43	Snapchat	A social media app where you can talk and send snaps to people all around the world.
44	TebMobile	Banking service app
45	TikTok	A social media app where you can create and share videos from your everyday life.
46	Township	A game where you create a whole city with a farm, helicopter, mine and a zoo.
47	Tubidy FM	An app to listen to music for free.
48	Turbo VPN	Allows users to access features that are only available in other countries.
49	Viber	An app that allows users to make audio and video calls, send messages and stickers and also keep notes.
50	VSCO	App for editing pictures
51	WhatsApp	Allows users to make audio and video calls, send messages and share you status with your friends.
52	Zombie Catchers	A game where you catch zombies and make juices out of them
53	Zoom	Allows users to make video conferences and meetings.

Figure 1. Graphical representation of Category Vs Number of Apps

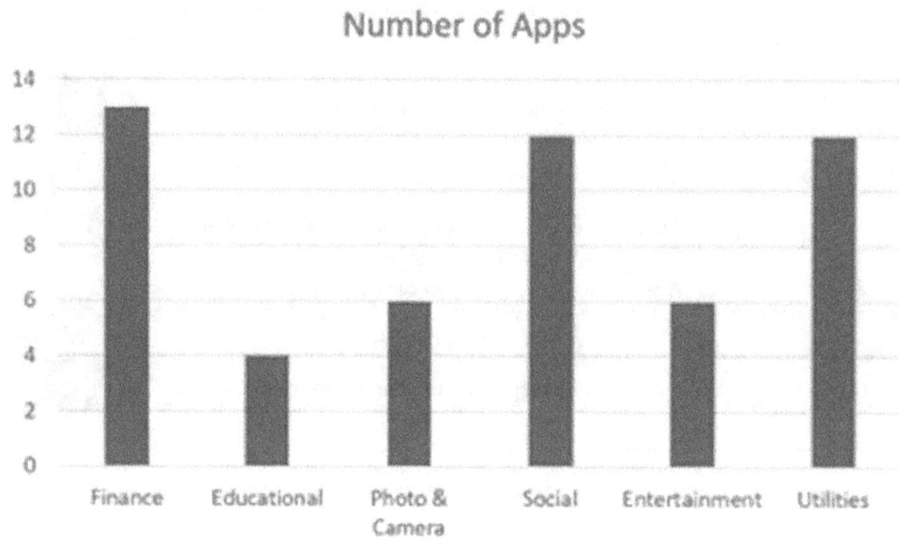

Table 4. 'Finance' Category with Purpose of Uses

SL	App Name	App Purpose
1	Calendar	In this app you can view the calendar, schedule meetings and make reminders.
2	Contacts	It is an app where you can create contacts and view the existing ones.
3	Drive	App to store pictures and videos and access them from different devices.
4	Files	App where all the downloaded files are stored.
5	Gmail	An email service that provides users with a gigabyte of storage
6	Mail	It has the sole purpose of sending emails.
7	Notes	In this app you can store different notes, including images, drawings, checklists etc.
8	Outlook	App to send and receive emails, manage your calendar etc.
9	Reminders	You can schedule a reminder and get a notification when its time. You can also make to-do lists.
10	Shortcuts	It provides shortcuts of every app to add in the home screen and you can change the name or icon of the app
11	TebMobile	Banking service app
12	Turbo VPN	Allows users to access features that are only available in other countries.
13	Zoom	Allows users to make video conferences and meetings.

Table 5. 'Educational' Category with Purpose of Uses

SL	App Name	App Purpose
1	Books	An app where you can read books, pdf or epub files.
2	Goodreads	App to keep track of the books you've read, the books you want to read, and news from authors you follow.
3	Translate	An app to translate words from every language.
4	Photomath	An app where you can take pictures of math problems and it gives you a step-by-step solution.

Table 6. 'Photo & Camera' Category with Purpose of Uses

	App Name	App Purpose
1	Camera	App to take pictures or videos.
2	Photos	An app to store, view, edit and share pictures on your phone.
3	AirBrush	App for editing pictures.
4	Dazz Cam	App for taking and editing pictures.
5	Lens Buddy	An app to take pictures without having to press buttons or hold the device, but by setting a time interval.
6	VSCO	App for editing pictures and posting them to your profile.

Table 7. 'Social' Category with Purpose of Uses

SL	App Name	App Purpose
1	BeReal	Is a social media app, where you share a picture once a day with all your friends.
2	Facebook	Allows users who sign up for free to connect with people they know and those they don't.
3	Instagram	A social media app where you can share photos and videos and connect with people from every part of the world.
4	LinkedIn	A professional app to find a job or internship and learn new skills for your professional career.
5	Messages	An app to send messages to the numbers on your contact list.
6	Phone	An app to make calls and check you call list.
7	FaceTime	An app to make video calls.
8	Reports+	An app that keeps track of your followers and likes on Instagram.
9	Snapchat	A social media app where you can talk and send snaps to people all around the world.
10	TikTok	A social media app where you can create and share videos from your everyday life.
11	Viber	An app that allows users to make audio and video calls, send messages and stickers and also keep notes.
12	WhatsApp	Allows users to make audio and video calls, send messages and share you status with your friends.

Table 8. 'Entertainment' Category with Purpose of Uses

SL	App Name	App Purpose
1	Pinterest	A social network where you can share pictures and videos and also make boards where you pin your favorite pictures and videos.
2	Netflix	A paid streaming service to watch movies and TV shows.
3	Tubidy FM	An app to listen to music for free.
4	Township	A game where you create a whole city with a farm, helicopter, mine and a zoo.
5	Zombie Catchers	A game where you catch zombies and make juices out of them
6	Game Center	App to invite friends to play games, track achievements and compare scores

Table 9. 'Utilities' Category with Purpose of Uses

SL	App Name	App Purpose
1	App Store	An Apple app to buy and download applications.
2	Calculator	An app to solve mathematical problems.
3	Clock	An app to make alarms and tell you the time in every place in the world.
4	Compass	It shows the direction your phone is pointing at the moment and elevation.
5	Edge	A browsing app that helps to protect data, save time and money.
6	Measure	It's an app to turn the device into a tape measure
7	Safari	It is a browsing database, with the same purpose as Google Chrome.
8	Voice Memos	An app for recording audios.
9	HERE WeGo	A GPA app where you can download the map of every city in the world and access it without internet connection.
10	Weather	An app to view the weather in every place in the world.
11	Maps	A GPA app that offers satellite imagery, street maps, traffic conditions, route planning for traveling by car, bus, bicycle or foot
12	Health	It keeps track of daily steps, burned calories, headphone audio levels, cycle tracking, sleep tracking, medication etc.

Table 10. Category Vs Number of Apps

SL	Category	Number of Apps
1	Finance	13
2	Educational	4
3	Photo & Camera	6
4	Social	12
5	Entertainment	6
6	Utilities	12

CONCLUSION

Nowadays, people frequently use their phones throughout the day for a number of purposes. It makes sense that they may also be applied in classroom settings. Mobile applications have greatly simplified the educational process. Students can access all of these books and resources using these mobile apps. Without having to visit the library to get it, they may save a lot of time. These smartphone apps have made it possible for lectures and lessons to take place anywhere. Students can now contact their professors whenever they need to, rather than waiting until class to pose a question. Mobile applications were also very useful during Covid-19 when schools were closed, and lessons were conducted online. Back then, the two most widely used online learning tools were "Zoom" and "Google Classroom." Today, a range of apps are utilized to aid in problem-solving, help students focus on their studies, and be available at any time and place. Some apps can help students who have trouble managing their time create and adhere to a schedule. Due to instructional mobile applications, learning has become more enjoyable and engaging and has improved student performance.

REFERENCES

Burch, Z. A., & Mohammed, S. (2019). Exploring Faculty Perceptions about Classroom Technology Integration and Acceptance: A literature Review. The Richard W. Riley College of Education and Leadership Publications.

Canuel, R., & Chrichton, C. (2015). Leveraging apps for research and learning: A survey of Canadian academic libraries. *Library Hi Tech*, *33*(1), 2–14. doi:10.1108/LHT-12-2014-0115

Canuel, R., MacKenzie, E., Senior, A., & Torabi, N. (2016). Apps for academic success: Developing digital literacy and awareness to increase usage. *Education for Information*, *33*(3), 1–18. doi:10.3233/EFI-160085

Carlos, A. (2012). Research on the go: Mobile tools for conducting research. *The Reference Librarian*, *53*(4), 433–440. doi:10.1080/02763877.2012.707462

Carter, A., Liddle, J., Hall, W., & Chenery, H. (2015). Mobile phones in research and treatment: Ethical guidelines and future directions. *JMIR mHealth and uHealth*, *3*(4), 1. doi:10.2196/mhealth.4538 PMID:26474545

Chen, B., Seilhamer, R., Bennet, L., & Bauer, S. (2015). Students' mobile learning practices in higher education: A multi-year study. *Educause Review*. https://er.educause.edu/articles/2015/6/students-mobile-learning-practices-in-higher-education-a-multiyear-study

Chen, B., Sivo, S., Seilhamer, R., Sugar, A., & Mao, J. (2013). User acceptance of mobile technology: A campus-wide implementation of blackboard's MobileTM learn application. *Journal of Educational Computing Research*, *49*(3), 327–343. doi:10.2190/EC.49.3.c

Chen, G. (2011). Mobile research: Benefits, applications and outlooks. Internationalization. *Design and Global Development*, *1*, 11–16.

Chinnery, G. (2006). Going to the MALL: Mobile assisted language learning. *Language Learning & Technology*, *10*, 9–16.

Compton, H., & Burke, D. (2018). The use of mobile learning in higher education: A systematic review. *Computers & Education*, *123*, 53–64. doi:10.1016/j.compedu.2018.04.007

Davidson, J., Paulus, T., & Jackson, K. (2016). Speculating on the future of digital tools for qualitative research. *Qualitative Inquiry*, *22*(7), 606–610. doi:10.1177/1077800415622505

EDUCAUSE. (2019). Horizon Report Preview: 2019 Higher Education Edition. *Educause*. https://library.educause.edu/-/media/files/library/2019/2/2019horizonreportpreview.pdf

Fosnacht, K., Sarraf, S., Howe, E., & Peck, L. K. (2017). How important are high response rates for college surveys? *The Review of Higher Education*, *40*(2), 245–265. doi:10.1353/rhe.2017.0003

Galanek, J. D., & Gierdowski, D. C. (2019). *ECAR Study of Faculty and Information Technology, 2019. Research report*. ECAR. https://www.educause.edu/-/media/files/library/2019/12/facultystudy2019.pdf?la=en&hash=F0F8AF83BD870239A7E5CD285FBDBE157AA51135

Galanek, J. D., Gierdowski, D. C., & Brooks, D. C. (2018). *ECAR study of undergraduate students and information technology, 2018. Research report*. ECAR. https://library.educause.edu/-/media/files/library/2018/10/studentitstudy2018.pdf?la=en&hash=C590C1F6C62B77792711BFAC1F642254A5618590

Garcia, B., Welford, J., & Smith, B. (2016). Using a smartphone app in qualitative research: The good, the bad and the ugly. *Qualitative Research*, *16*(5), 508–525. doi:10.1177/1468794115593335

Gierdowski, D. C., Brooks, D. C., & Galanek, J. D. (2020) *EDUCAUSE 2020: Student Technology Report: Supporting the Whole Student.* CO: ECAR. https://www.educause.edu/ecar/research-publications/student-technology-report-supporting-the-whole-student/2020/technology-use-and-environmental-preferences

Gikas, J., & Grant, M. M. (2013). Mobile computing devices in higher education: Student perspectives on learning with cellphones, smartphones & social media. *The Internet and Higher Education*, *19*, 18–26. doi:10.1016/j.iheduc.2013.06.002

Hahn, J. (2014). Undergraduate research support with optical character recognition apps. *RSR. Reference Services Review*, *42*(2), 336–350. doi:10.1108/RSR-09-2013-0045

Hanbridge, A. S., Tin, T., & Sanderson, N. (2018). Information literacy skills on the go: Mobile learning innovation. *Journal of Information Literacy*, *12*(1), 118–136. doi:10.11645/12.1.2322

Hennig, N. (2014). *Apps for librarians: Using the best mobile technology to educate, create and engage*. Libraries Unlimited. https://ebookcentral-proquest-com.ezproxy.waikato.ac.nz/lib/waikato/detail.action?docID=1813321

Hinze, A., Vanderschantz, N., Timpany, C., Cunningham, S. J., Saravani, S. J., & Wilkinson, C. (2017a). *Use of mobile apps for teaching and research. (Working Paper Series 01/2017a).* Department of Computer Science, The University of Waikato.

Kukulska-Hulme, A. (2016). *Personalization of Language Learning through Mobile Technologies*. Cambridge University Press.

Kukulska-Hulme, A., & Shield, L. (2008). An overview of mobile assisted language learning: From content delivery to supported collaboration and interaction. *ReCALL*, *20*(3), 271–289. doi:10.1017/S0958344008000335

Oz, H. (2013). Prospective English teachers' ownership and usage of mobile device as m-learning tools. *Procedia: Social and Behavioral Sciences*, *141*, 1031–1041. doi:10.1016/j.sbspro.2014.05.173

Chapter 7
Voice-Guided Mobile Robot System for Disabled People

Hla Myo Tun
Yangon Technical University, Myanmar

Devasis Pradhan
https://orcid.org/0000-0002-8201-2210
Acharya Institute of Technology, India

ABSTRACT

Mobility is one of the most important problems for totally and partially disabled people in daily life. Disabled people need some assistive stick or device to avoid obstacles before an accident. This research work proposes to assist disabled people in their daily routines without human guidance. There are four main sections in this paper. Firstly, in the obstacle detection system, five ultrasonic sensors sense the obstacles in the front side, left side, and right side of the user's way on the ground and another ultrasonic sensor senses the overhanging obstacles near the user's head level. Secondly, in an obstacle avoidance system, the servo steering motor is equipped with the front wheel of the stick's base. The wheel with servo motor will turn easily to the left direction or right direction depending on the detecting or sensing of the ultrasonic sensors. Thirdly, in the voice-guided system, there are six recorded human voice messages in the Myanmar language for disabled users in the author's country, Myanmar.

DOI: 10.4018/978-1-6684-8582-8.ch007

INTRODUCTION

Vision is a precious gift to human beings. Vision allows people to see and understand the surrounding of the world. Nowadays, there are many different types of devices to assist or help the totally or partially blind people because of the improvement of the modern technologies and research. Those devices are smart phone, laser, ultrasonic sensors, GPS, RFID, and so on. Smart phone allows the visually impaired people to listen and hear to the voice mails, and, even write and send mails. Laser cane transmits invisible laser beam to detect the objects. Ultrasonic sensor is used to sense the distance of the obstacle. GPS system is used to provide the information regarding the location of the blind user to his family andindicate the correct way for the users. RFID is to navigate the blind people but it needs RFID flags along the user's path (Mohajeri et al., 2011).

In the world, the totally blind persons and the partially blind people use or apply the guide dogs, human guides and the traditional mobility aids such as white canes, aluminum sticks, wood sticks and so on to avoid the obstacles. Although a few of blind people use the guide dogs, those guide dogs need the extensive training course and the fully trained dogs are so expensive and then those dogs cannot be fully believed to overcome the obstacles. Those well trained guide dogs can only be used for about five years. The human guide is also expensive for longer durations, and only the rich men can lend those people for their family member or blind person. Most of the blind people use the white cane or aluminum or wood stick in our country .The visually impaired people swing the aluminum or wood stick and the white cane around their feet to detect the existence of the obstacles near their step. Those sticks and cane have limitation because it can only detect the obstacles when those sticks and canes have contact with the disturbance. The white cane and the traditional sticks cannot sense the obstacle more than 1 meter. Those aids cannot warn to the user with voice message where the obstacle is in the user's path until the blind user has touched and hit with the obstacle. Some incidents might lead to serious injury on the blind users. So, the blind people are requiring the advanced stick or device to walk or move safely in their surroundings (Ashiq et al., 2022; Burta et al., 2020; Chaudary et al., 2023; Fanfeng, 2010; Kalpana et al., 2020; Khan et al., 2021; Messaoudi et al., 2022; Mohapatra et al., 2018; Real & Araujo, 2019).

The main aim of this paper is to create and construct the blind stick with human voice messages depending on the ultrasonic sensors. This paper consists of four main sections: (1) obstacle detection system, (2) obstacle avoidance system, (3) voice-guided system, and (4) lighting system. This blind stick has six ultrasonic sensors to sense the obstacle not only left, right and in front of the user on the ground but also over-hanging obstacle at the level of the user's head. Only one servo motor fixed with front wheel is used to avoid the obstacles easily to secure the user's path.

ISD2590 Chipcorder and vibration motor are applied to produce the voice messages and vibrating alert depending on the ultrasonic sensors. LDR and LDE are carried out for the lighting system to know and see where the blind stick is for the sighted people. This blind stick is useful for the blind people safety. This voice-guided system using mobile robot can be used to allow the blind people to move independently, safely and quickly among the obstacles and hazardous places (Akhil et al., 2022; Al-Fahoum, 2013; Elmannai & Elleithy, 2017; Islam et al., 2018; Rahman et al., 2019; Sumathy et al., 2020; Wei et al., 2013).

OVERALL BLOCK DIAGRAM

This system is composed of ultrasonic sensors, light- sensing circuit, micro-controller, light signal, vibrating motor circuit, servo driver circuit, voice-alarm system and rechargeable battery to be an assistive stick or device for the totally and partially blind persons. In this system, ultrasonic sensors are used to sense and detect the obstacles in front, left and right on the ground of the user's path and even upper side near the user's head. LDR senses the light intensity, for instance; it is day or night or low-light room. LED is used to light when the user goes to the low- light or dark room and walks at night. The first PIC 16F887 is the main controller to control the overall system and the second one is to control the ISD 2590 for six recorded voice messages to be on the voice messages via speaker. The vibrating motor is used to alarm to the user with vibration before hitting with the obstacles. Servo motor is used to turn the front wheel easily to lift or right side based on the sensors. The rechargeable battery is used to supply power to all components, and then only this system will fully

OBSTACLE DETECTION SYSTEM

The obstacle detection system is implemented to sense and detect the obstacle to provide more safely to the blind users. Ultrasonic sensors are used to measure the distance between the obstacle and the stick in front, left and right of the stick for the blind people, and even the upper side for the user's head level. Five ultrasonic sensors are fit at the base house of the stick to detect the obstacles in front, left and right of the blind stick or mobile robot. Another one ultrasonic sensor is placed at the handle of the stick to detect the over-hanging object for the blind user's head. The distance measurement between the stick and the obstacle are displayed on LCD, such as "Up Range =", "Front Range =", "Right Range =", and "Left Range =". In reality, LCD does not need for the blindness. But, why the LCD is used in this system, the distance between the sensors and the obstacles is known to check this mobile robot or blind stick that is correct or incorrect. Those distance measurements for the obstacle detection system are showed in centimeter (cm) on LCD display. The blind user will be alerted by the voice messages from the ISD2590 via the speaker and vibration, and then, the blind stick will be turned the front wheel by using servo motor if the obstacle is near with the stick. Figure.2 shows the obstacle detection system for blindness.

Figure 1. System block diagram

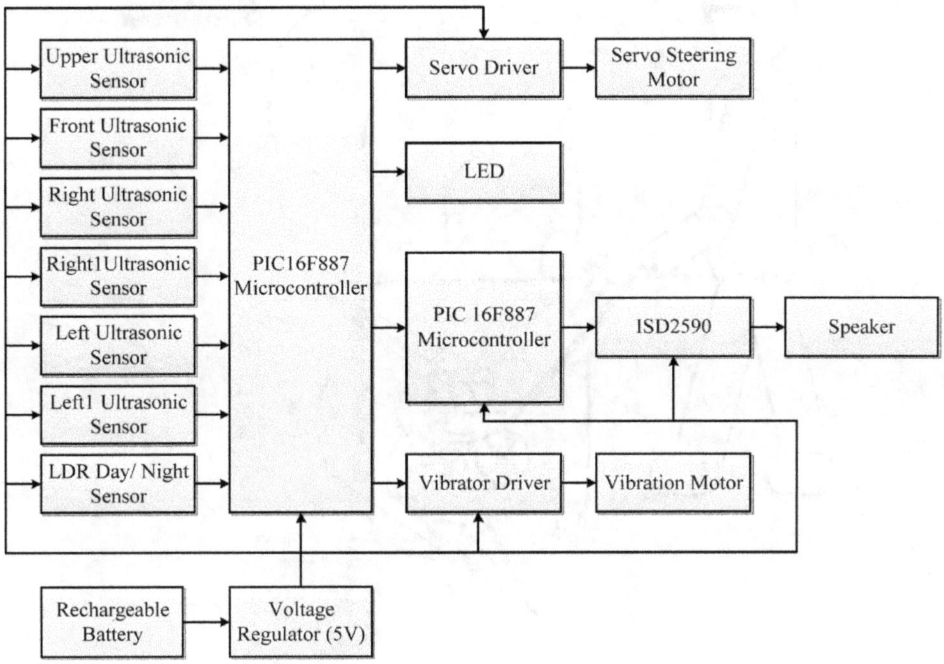

DISTANCE CALCULATION BETWEEN ULTRASONIC SENSOR AND THE OBSTACLE

Distance = Time * Speed
Speed of sound in air = 340 m/s = 34000 cm/s
34000 cm ------------> 1 sec
Then 1cm ------------> 29 micro sec

The ultrasonic burst travels forward and backward between sensor and obstacle. So, it must be taken half of the total distance to calculate the distance of the obstacle.

Therefore Distance = 1/2 * (Time /29) = Time/58 cm

Figure 2. Obstacle detection system for blindness

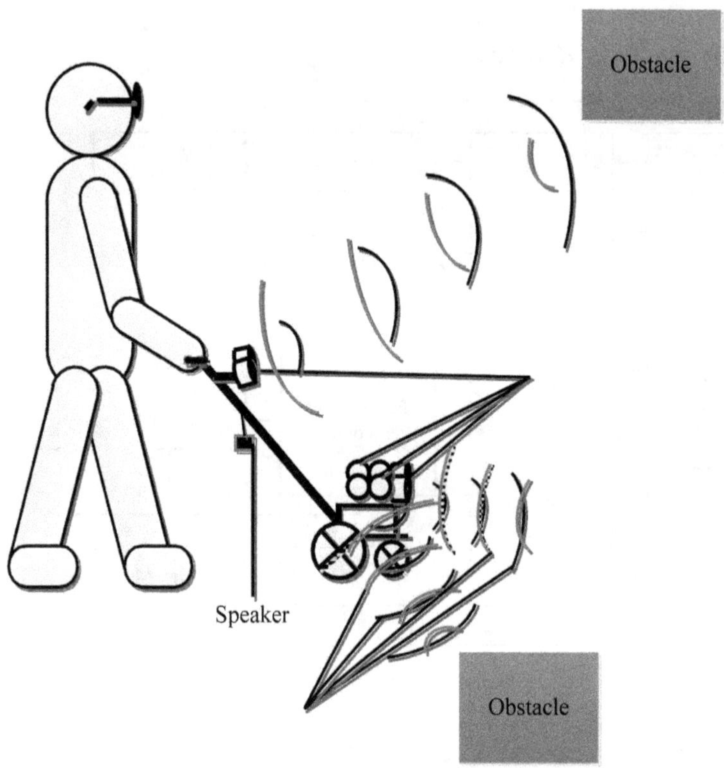

DESIGN OF DISTANCE MEASUREMENT CIRCUIT FOR OBSTACLE DETECTION

For the distance measurement circuit, the ultrasonic sensor is the main element. It is interfaced with the PIC 16F887 controller. For this system, the six ultrasonic sensors are used to know the conditions of the obstacles around the front of, left, right and upper side of the stick or mobile robot for the blindness. The trigger pins of the six ultrasonic sensors are connected to the pin 1 of PORTB for the left, pins 3 of PORTB for right, pin 5 of PORTB for the front, pin 7 of PORTB for upper, pin

5 and 7 of PORTD for right1 and left1 respectively. The echo pins of the sensors are connected to the pin 0 of PORTB for the left, pins 2, 4 and 6 of the PORTB for the right, front and upper, and, pin 4 and 6 of PORTD for the right1 and left1. The controller provides a TRIGGER signal, at least 10us High Level (5V) pulse, to the ultrasonic sensor. The gadget will automatically send eight bursts of ultrasonic sound at 40 KHz. A blockage in front of the module will cause it to reflect the ultrasonic burst. The sensor's ECHO output will be in the HIGH state (5V) for the duration of the time required for sending and receiving an ultrasonic burst if the signal is received. This amount of time is converted to a distance and shown on an LCD. Figure.3 shows the distance measurement circuit for obstacle detection.

Figure 3. Distance measurement circuit for obstacle detection

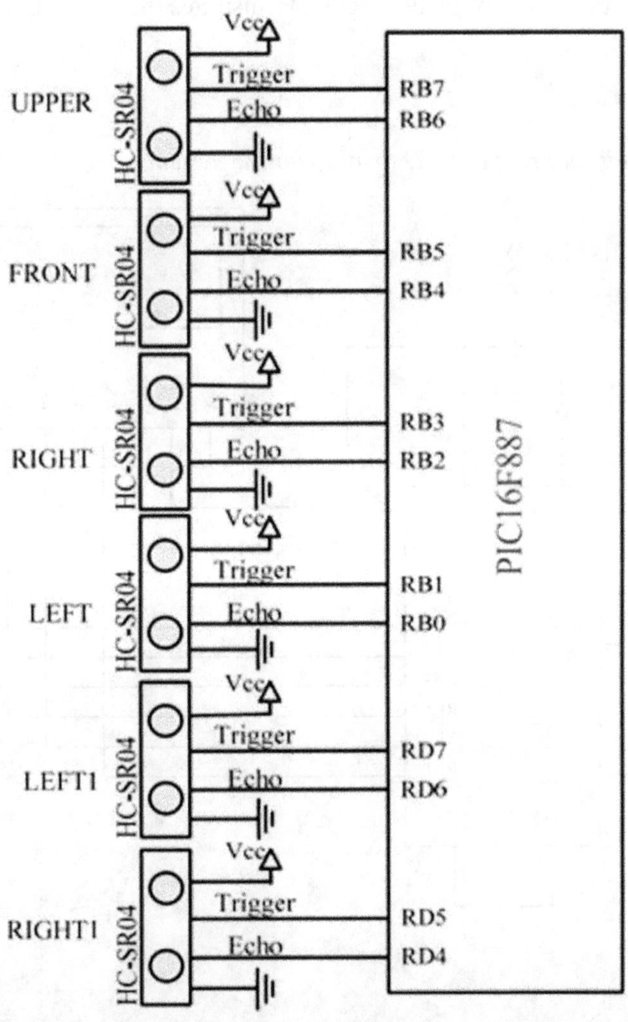

INTERFACING LCD WITH PIC 16F887

LCD is used to display and know the distance between the mobile robot or blind stick and the obstacles at left, right, upper and in front of the vehicle. And, it is also used to display the messages such as "UP RANGE =", "FRONT RANGE =", "RIGHT RANGE =" and "LEFT RANGE =" to check this system that is correct or not. All the data lines of LCD are connected with PORTA. 'EN' pin is connected with RE0, 'DI' (RS) is connected with RE1 and R/W pin is connected with ground. The operation of LCD depends upon these three pin only.Figure4.9shows the circuit diagram of LCD display for the system. For the pins 13 and 14 a crystal oscillator circuit is connected to generate clock signals to the microprocessor to enable its pins. And VSS is grounded and VDD is given to the supply lines. 10kΩ resistor is connected with the MCLR pin and positive power supply. PIC16F887 microcontroller will operate in XT oscillator mode. Figure.4 illustrates the circuit diagram of LCD display for the system.

Figure 4. Circuit diagram of LCD display for the system

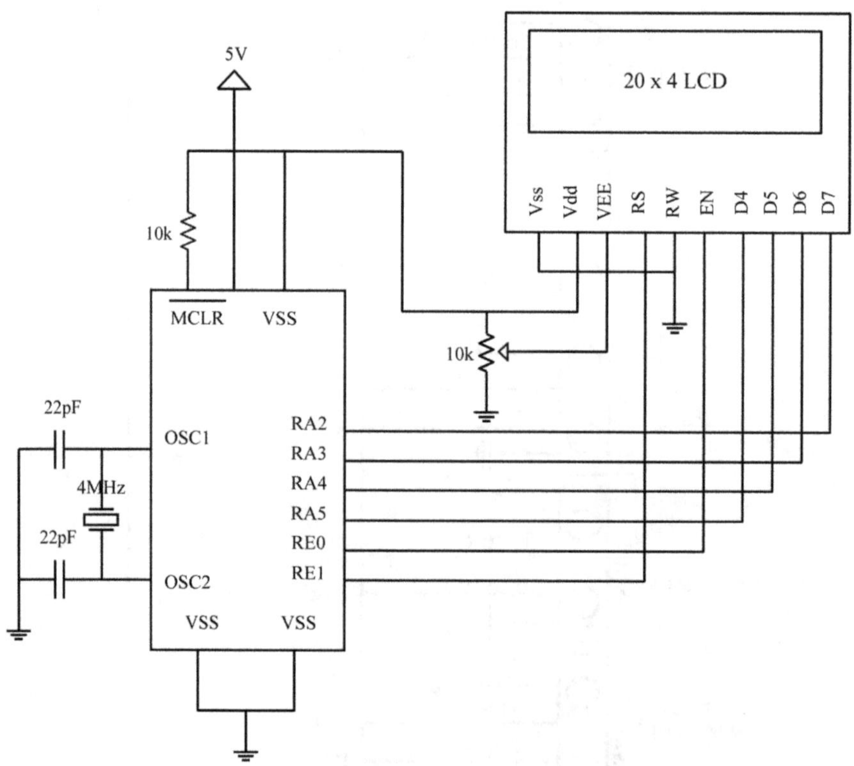

OBSTACLE AVOIDANCE SYSTEM

In this system, ultrasonic sensors are used to detect the obstacles and to know where the obstacle is. And, servo motor is used to turn the front wheel easily based on the ultrasonic sensors. The main PIC16F887 is to decide where the obstacle is from the ultrasonic sensors on LCD and to send the signal to PIC16F676 microcontroller to turn the front wheel equipped with the servo steering motor. Then, PIC16F676 is used to control the servo motor to turn the wheel relatively depending on the sensors. In the obstacle avoidance system, there are two cases. CASE1 is that the right distance from the obstacle is equal to or greater than the left distance from the disturbance and CASE2 is that the left distance from the obstacle is greater than the right distance from the object in the blind user's path. These two cases depend on near or far of the distance between the stick and the obstacles.

DESIGN CIRCUIT DIAGRAM FOR SERVO STEERING MOTOR

For the servo steering motor, PIC16F676 microcontroller is used to control the front wheel of the stick equipped with servo steering motor. The pin2 of the PORTD of the PIC16F887 (main controller) is to control the pin0 of PORTC of the PIC16F676 to turn the servo steering motor to left side (45°). The PORTD3 of the microcontroller PIC16F887 is to control the PORTC2 of the PIC16F676 to turn the servo motor with the wheel to the right side (45°). Figure.5 demonstrates the circuit diagram for servo steering motor. Figure.6 shows the flowchart of the servo steering motor.

CASE 1: RIGHT DISTANCE ^{3} LEFT DISTANCE

In this case 1, there is an obstacle in front of the user and there is also disturbance at both left and right. If the sight distance is greater than or equal to 120cm or right distance is greater than the left distance, the servo motor will turn to right side (45 degree), and, front and left sensors will sense the obstacle. Figure.7 shows the consideration for CASE1 (Right Distance ^{3} Left Distance).

If there is no obstacle in front of the mobile robot or blind stick, the servo motor will turn back to the center position and then the left sensor will sense the object. Left1 sensor will be sensing the obstacle at the left side. If there is no obstacle at left side, the servo motor will turn to the left side. Figure.8 mentions the flowchart for CASE1.

Figure 5. Circuit diagram for servo steering motor

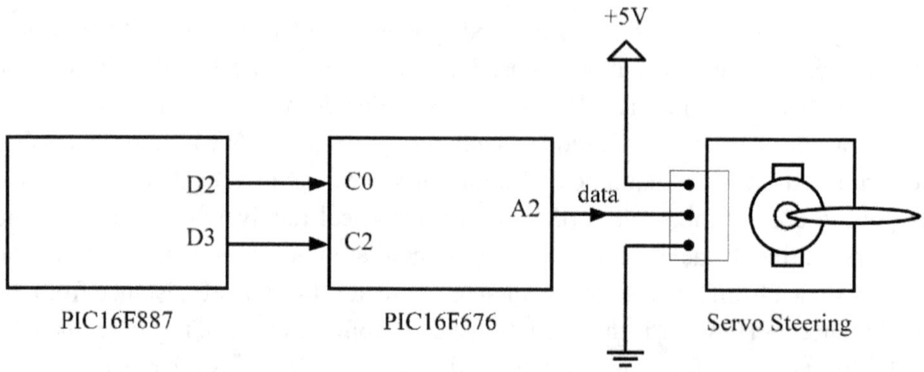

Figure 6. Flowchart of the servo steering motor

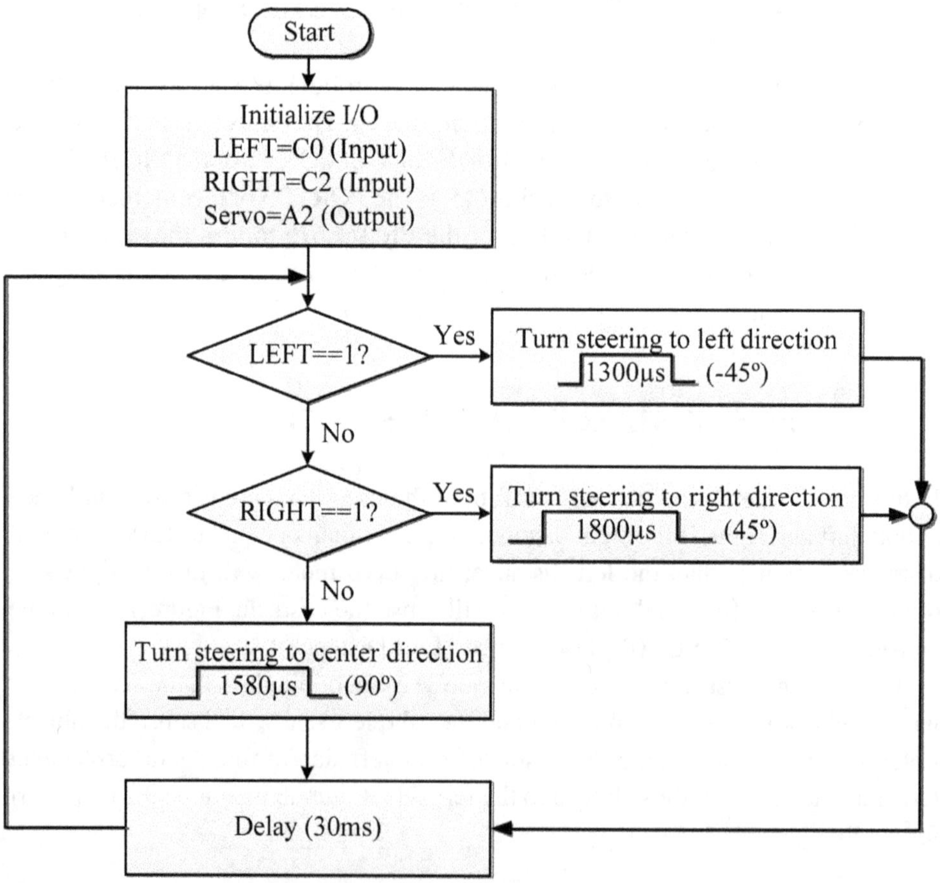

Figure 7. Consideration for CASE1 (Right Distance ³ Left Distance)

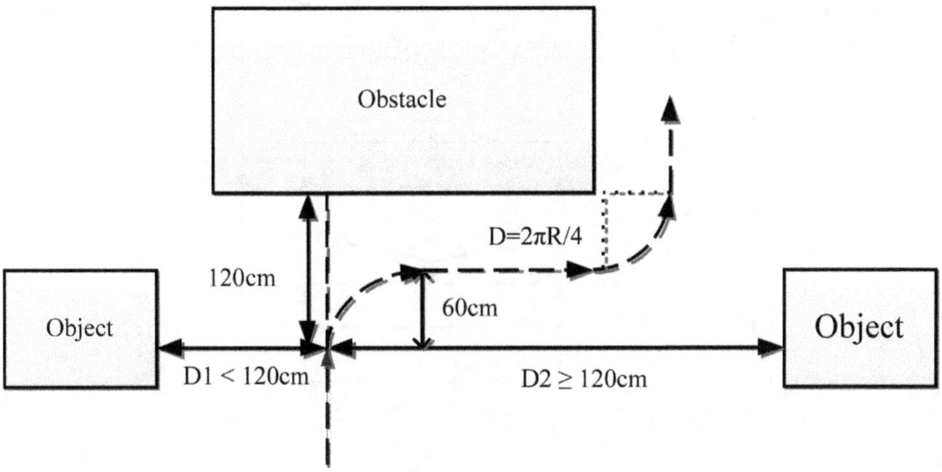

CASE 2: LEFT DISTANCE > RIGHT DISTANCE

In this case 2, there is an obstacle in front of the user and there is also disturbance at both left side and right side. If left distance from the obstacle is greater than the right distance from the disturbance, the servo motor will turn to left (45 degree), and then the right sensor will sense the object. Right1 sensor will be sensing the obstacle at the right side. If there is no obstacle at right side, the servo motor will turn to the right side. Figure.9 shows the consideration for CASE 2 (Left Distance ˃ Right Distance). Figure.10 illustrates the flowchart for CASE 2.

VOICE-GUIDED SYSTEM

In this voice-guided system, there are many voice messages or alerts, giving information where the obstacle is for the blindness in from of human voice. ISD 2590 Chipcorder is used to provide the recorded voice messages for the blind user to guide the secure way and to avoid the obstacle in front of the user's path-way with voice messages. The totally and partially blind people need the sounds from their surrounding in order to provide information for avoiding the obstacle or object and finding the safety way. The ISD2590 device has storage duration of 90 seconds and it is used an ideal for record and playback application with the natural voice form, providing high quality voice reproduction.

Figure 8. Flowchart for CASE1

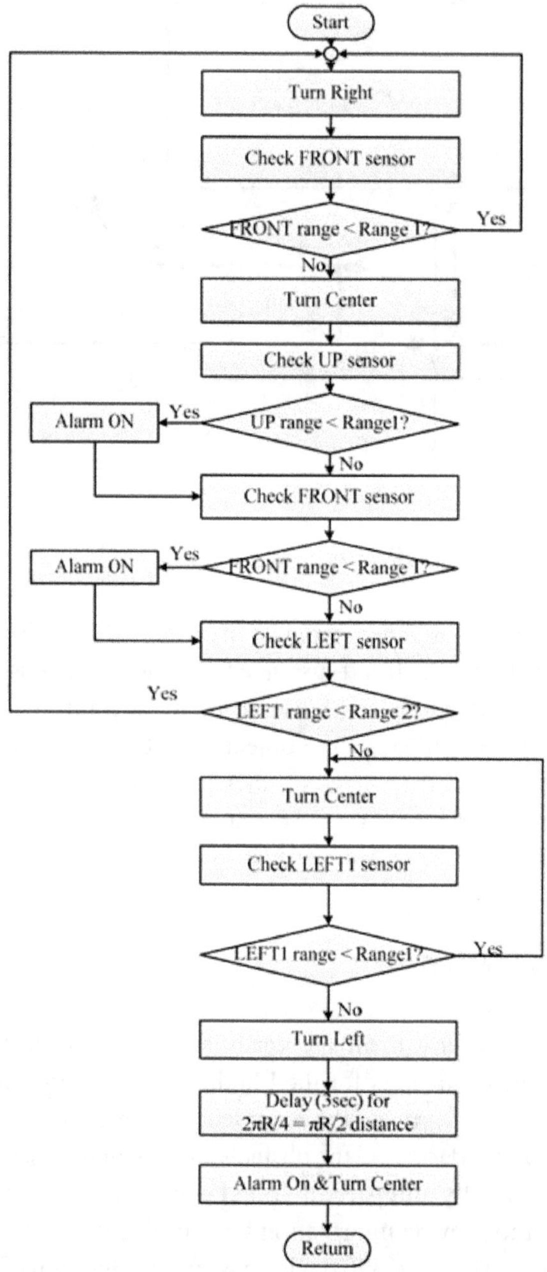

Figure 9. Consideration for CASE 2 (Left Distance > Right Distance)

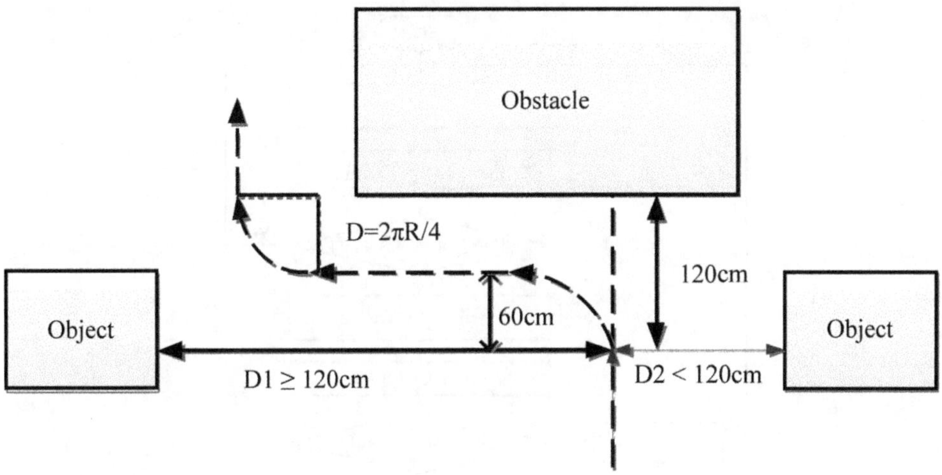

DISTINGUISH OPERATIONAL MODE AND ADDRESSING MODE FOR ISD2590

The two input functions (A8 and A9) can distinguish Operational Mode and Addressable Mode.

Table 1. Distinguishable Modes for Recording and Playing Back for ISD2590 Device to be multi-message

A8	A9	Distinguishable Modes
0	0	Addressable Mode
0	1	Addressable Mode
1	0	Addressable Mode
1	1	Operational Mode

ADDRESSABLE MODE FOR ISD2590

There are some facts for addressable mode for ISD2590.

- Actual number of messages and their duration depends on address selection within the memory address space.

Figure 10. Flowchart for CASE 2

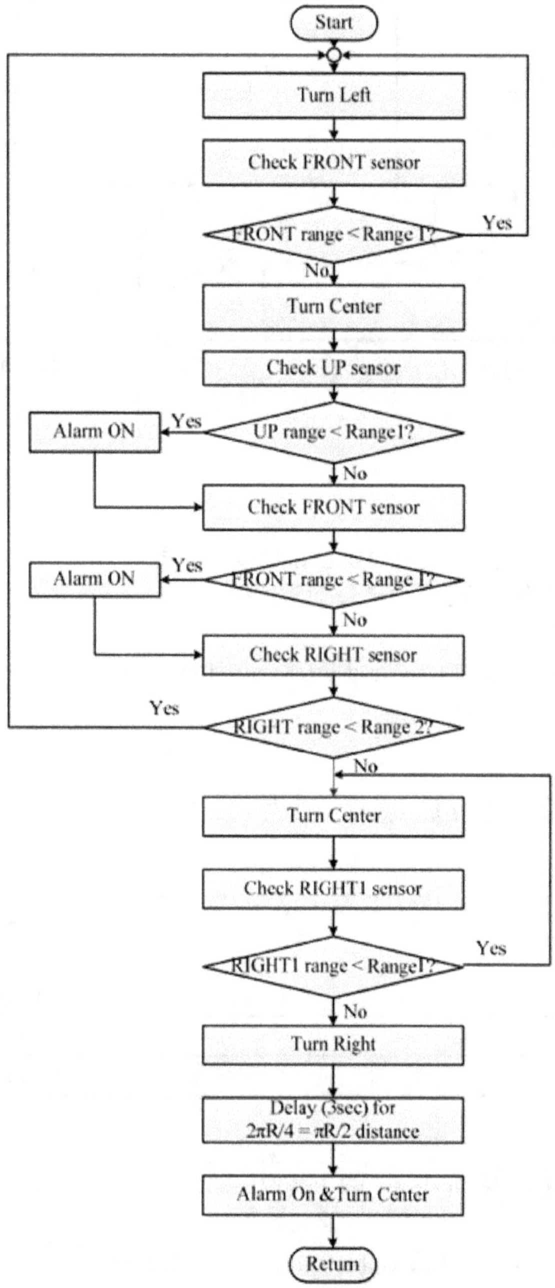

- Addressing is made with a binary coding at A0-A9 to the rule specified in Table 4.2.
- Longer message will be automatically loaded into successive memory rows, as necessary, until the process is completed or memory end is reached.

Table 2. Memory Row for Addressable Mode

Digital No. of Memory Row	Start Time Operation
0	0 sec
1	0.15 sec
2	0.30 sec
.....	.....
599	89.85 sec

DESIGN AND CONCENTRATION OF ISD2590 DEVICE FOR ADDRESSABLE MODE

The 480K storage cells in the ISD 2590 device are intended to store data for 90 seconds at an 8.0 KHz sampling rate. The storage capacity of this is almost four times that of the ISD1000A family. Two extra address pins have been added to allow for the same addressing resolution. ISD2590 devices have an address space that can be divided into 600 increments. The Operational Modes are mapped to certain higher addresses. Any more addresses are incorrect. Figure.11 shows the design of circuit diagram for ISD2590 to record and playback.

DEVICE CONTROL STEPS FOR ISD2590

Table 3 gives the device control steps for ISD2590. Figure 12 shows the flowchart of record and playback function for ISD2590. Figure 13 illustrates the flowchart for record function. Figure 14 demonstrates the flowchart for playback function.

Table 3. Device Control Steps for ISD2590

Figure 11. Design of Circuit Diagram for ISD2590 to Record and Playback

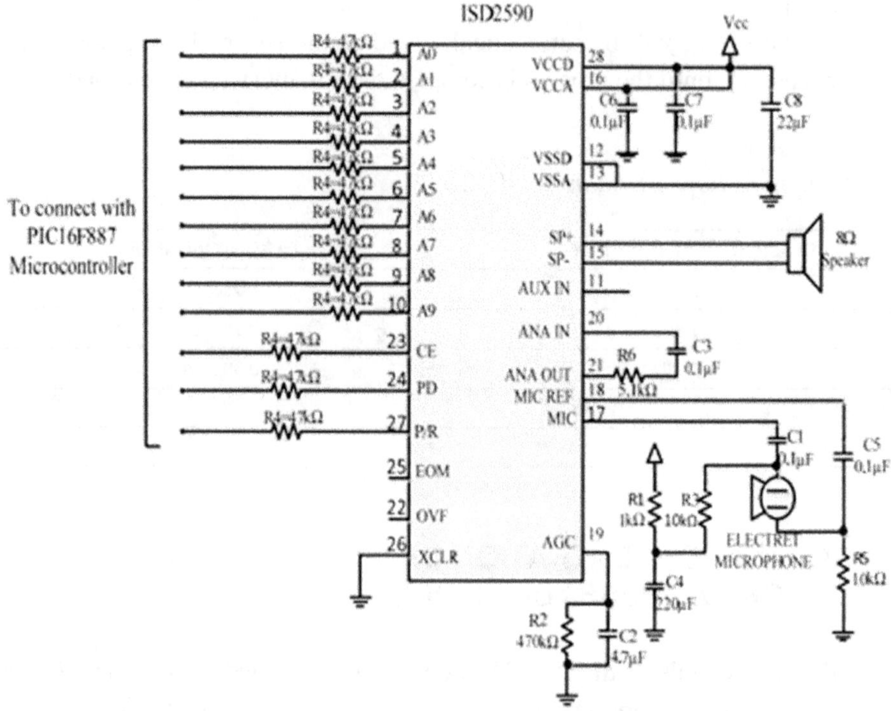

Table 3.

Function	Control Step	Action
RECORD	1. PD=High, CE = High	Chip disables in power down mode. Chip activates/enables in power down mode. Chip starts record mode at configured address. Chip stops recording.
	2. Address set up A0-A9 (0-599) 3. P/R= Low 4. PD = Low	
	5. CE = Low	
	6. CE = High	
PLAYBACK	1. PD =High, CE =High	Chip Disables in power down mode.
	2. Address set up A0-A9 (0-599) 3. P/R= High 4. PD= Low 5. CE=Low trigger pulse	Chip activates/ enables in power down mode. Chip starts in play mode at configured address.

Figure 12. Flowchart of Record and Playback Function for ISD2590

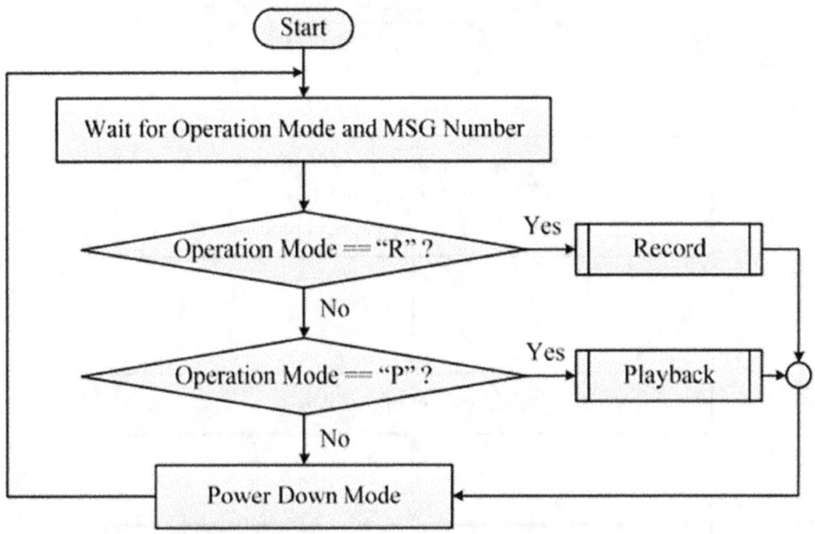

LIGHTING SYSTEM

The lighting system is not effective to the blind users, but it is very useful and good for the sighted people because this lighting system can help to avoid the walking stick before the ordinary people do not hit suddenly. That's why, this blind stick can safe even in a dark room or at night. For this lighting system, light dependent resistor (LDR) and light emitting diode (LED) are used. LDR is the resistor and its resistance varies with the intensity of light incident on it. LDR has high resistance in low light condition or darkness and low resistance in well light or daylight situations. Therefore, the resistance of LDR relies on the light intensity falling or reaching on its surface. LDR is used for this lighting system to sense the light on or off and to light the LED when there is no light for the sighted people to know and avoid or overcome the blind people's stick before the walking stick is not hit or knocked down unfortunately.

Figure 13. Flowchart for Record Function

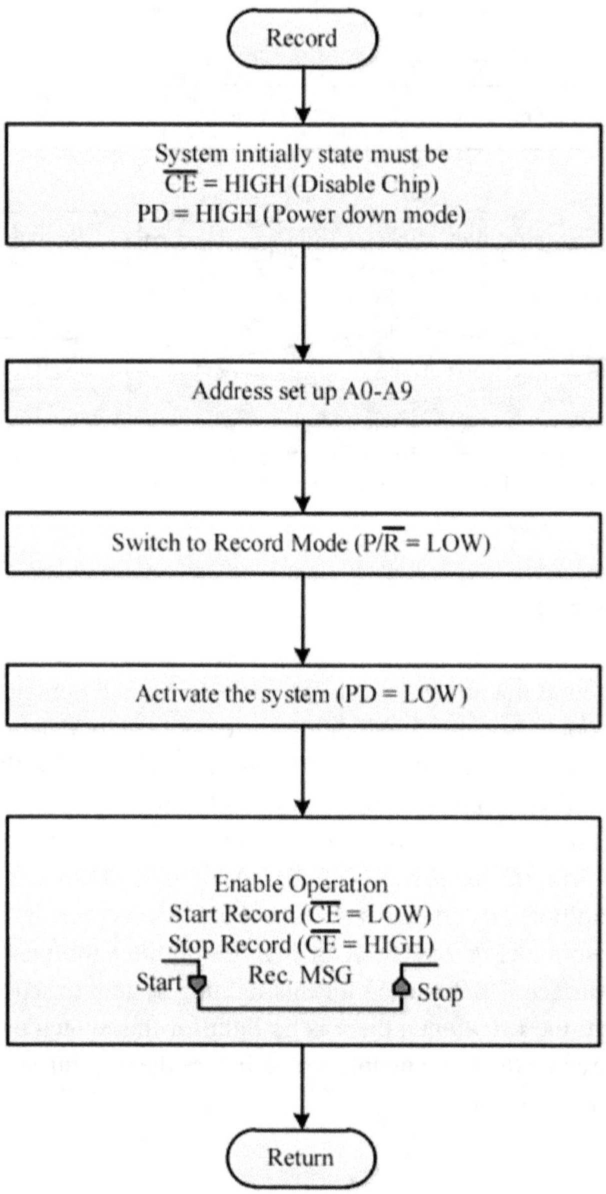

Figure 14. Flowchart for Playback Function

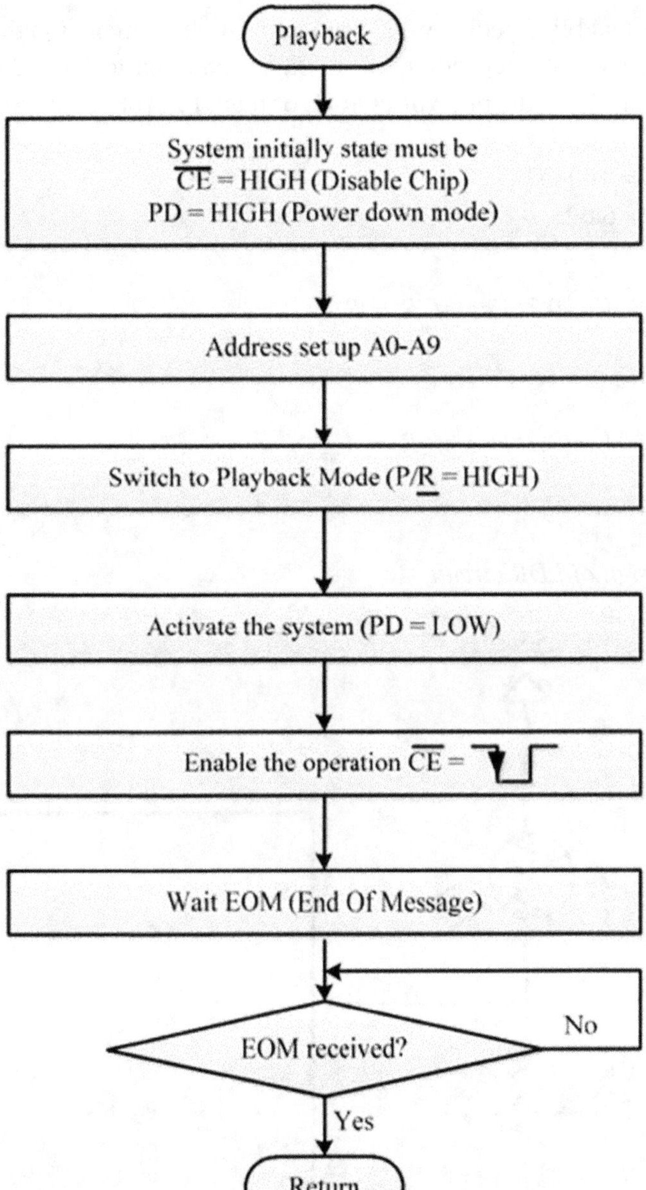

DESIGN OF LDR CIRCUIT

When the stick and blind people walk or go into a dark room or at night. LDR will sense the light intensity depending on the light condition to light ON the LED. This LDR circuit fits at the pin 2 of PORTE of main PIC16F887 microcontroller.

Rmax of LDR = 140 kΩ
Rmin of LDR = 6kΩ

$$V_{R2} = \frac{R_2}{R_1 + R_2} * 5V = 0.25Vcc \quad if \; R_1 = R_{max}$$

$$V_{R2} = \frac{R_2}{R_1 + R_2} * 5V = 0.25Vcc \quad if \; R_1 = R_{max}$$

Figure 15. Design of LDR circuit

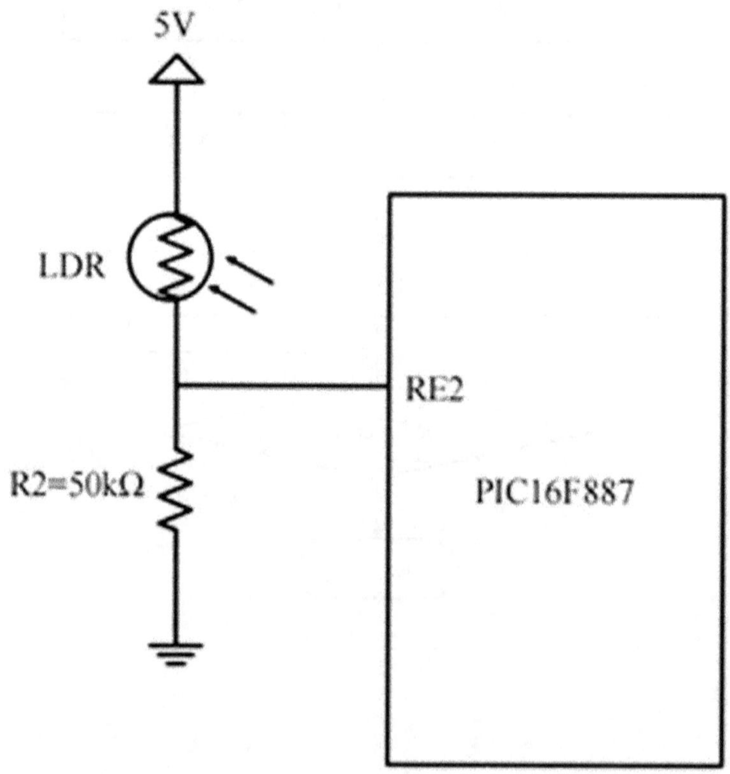

HARDWARE CONSTRUCTION OF THE STICK

The blind stick is about 41-inch (3.6 feet) high, the wheel of it is about 5.6-inch high and the width of the sheet of the stick's base is about 12-inch (1 foot). The blind stick has three wheels to push it easily. The two wheels are free wheels and the only front wheel is equipped with the servo motor to turn the wheel easily without effort to the left or right side depending on the sensors' sensing or detecting. A push rod or stick is fixed to the sheet which is the place to put the main circuit board, three ultrasonic sensors and rechargeable battery. And, the small sheet is fixed near the handle of the stick to place the upper ultrasonic sensor to sense the over-hanging objects for the blind user. The height of the blind stick relies on the people's height for normal person. Figure.16 shows the design for the hardware construction for the stick. Figure.17 illustrates the sensors layout for hardware construction.

Figure 16. Design for the hardware construction for the stick

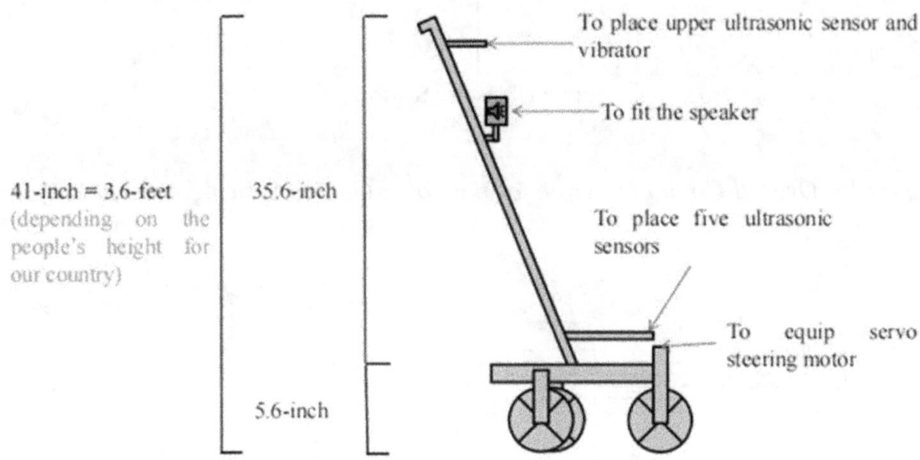

OPERATION OF THE VOICE-GUIDED SYSTEM FOR BLIND PEOPLE USING MOBILE ROBOT

The main operation of the system is to detect the obstacle based on the six ultrasonic sensors, to avoid the obstacle by turning the front wheel equipped with servo steering motor when any obstacle is detected, to alarm and notify with the recorded voice messages from the ISD2590 device via speaker and to give vibration alert from the vibration motor and to light the LED in the low-light room and at night. These systems will operate automatically.

Figure 17. Sensors layout for hardware construction

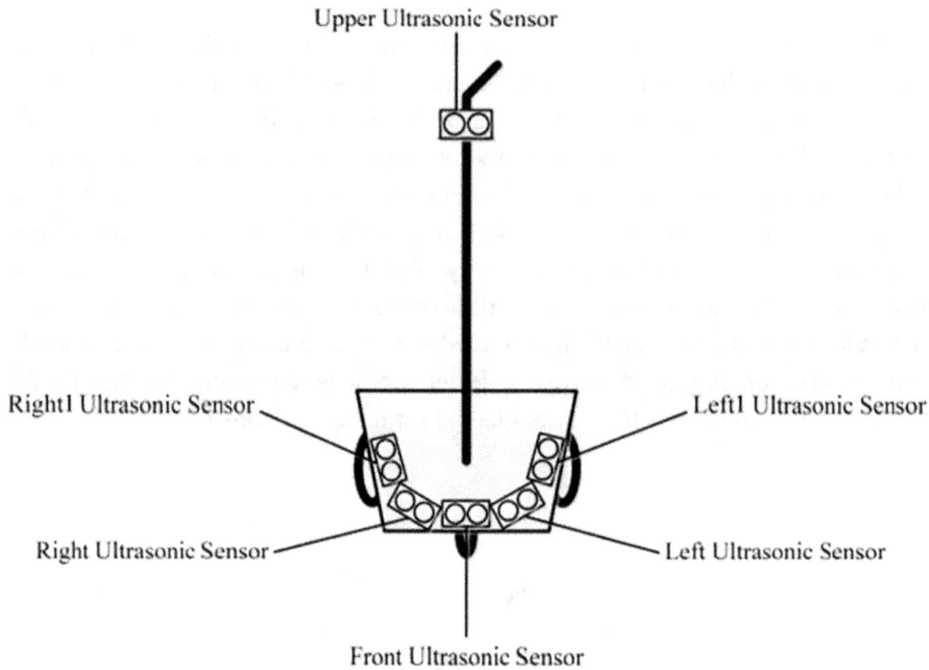

Figure 18. Overall Circuit Diagram for the Voice-Guided System for Blind People

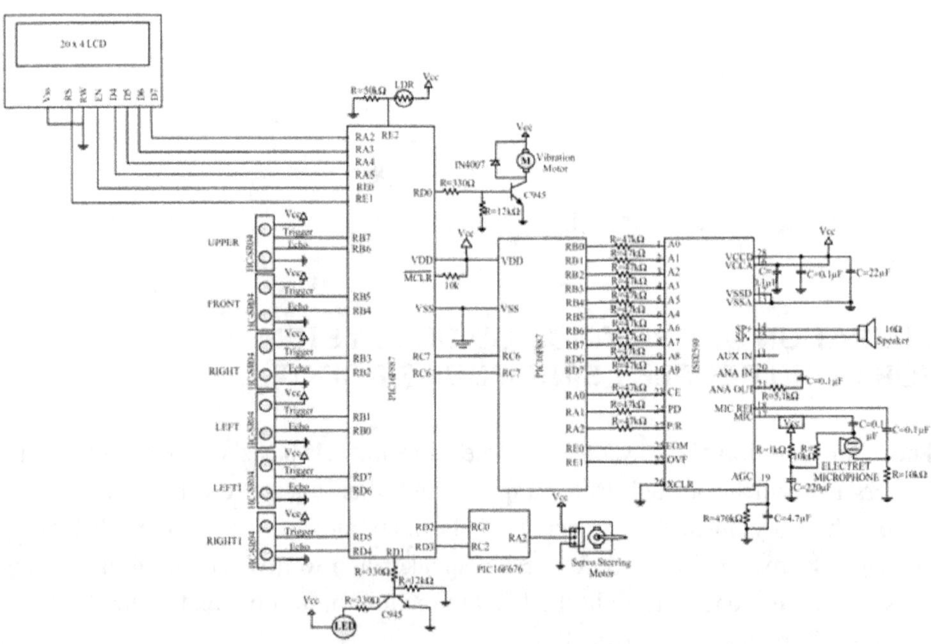

As the input sensors, LDR and ultrasonic sensors are used. Two PIC 16F887s and a PIC 16F676 are used to control the whole system, voice-guided system and to control the servo motor respectively. At output section, LCD, LED, speaker, vibrator and servo motor are used. In this voice-guided system for blind people using mobile robot, all ultrasonic sensors sense or detect the obstacles and disturbance between 120cm from the blind stick or mobile robot for the blind user. That distance (120cm) is enough to avoid the obstacle before accident. If any obstacle is detected in front of the user, the front wheel with servo steering motor will turn to the right side, the suitable voice message will play and the vibration motor will also operate for the blind user. If there is an over-hanging object at the user's head level, the servo motor for the front wheel will not turn to any side but the appropriate voice message and vibration alert will provide to the user to avoid that obstacle before hitting with an object. In this system, there are six voice messages to be security the user. Those messages will be produced according to the situation of the six sensors and the condition of the front wheel. Although the lighting system is not useful for the blind users, the sighted people need the light to know where the blind people's stick exists in the dark room or at night. So, the lighting system is effective both for the blind people and for this mobile robot or blind stick in the dark condition.

The proposed system also includes an embedded system which detects the obstacles at left, right, upper and in front of the mobile robot for blind people. This system uses six ultrasonic sensors to detect the obstacles around the front of the stick, at left, right and upper side. The distance measurements between the mobile robot and obstacles from the ultrasonic sensors are displayed on LCD. When an obstacle is detected at the front, the voice message is "Obstacle is at the front side!", if there is an obstacle at the upper side, the voice message is "Obstacle is at the upper side!", when an obstacle is left or right side of the mobile robot or blind stick, the voice message will be "Obstacle is at the left side!" or "Obstacle is at the right side!", respectively and the front wheel equipped with servo motor will be turned to the left or right direction depending on the ultrasonic sensors and according to the CASE1 and CASE2. Figure.18 shows the overall circuit diagram for the voice-guided system for blind people.

TEST RESULT FOR OBSTACLE DETECTION AND AVOIDANCE SYSTEM

In this research, there are two test results with my blind stick. One is the test result without blind or visually impaired people and another one is the test result with the visually impaired people. Figure.19 shows the test result for no obstacle condition.

Figure 19. Test Result for no obstacle condition

In this photo, there is no obstacle in front of, left, right and upper of the stick. So, the obstacle avoidance and voice-guided system do not operate in this case. Figure.20 shows the test result for straight forward condition.

Figure 20. Test result for straight forward condition

In this one, although the obstacles exist at both left and right side, there is no obstacle in front of the user and the stick can pass through the path-way easily. So, the voice-guided system and avoidance system will not be operated in this situation. Figure.21 illustrates the test result for turning left condition.

Figure 21. Test result for turning left condition

Figure 22. Test result for turning right condition

The obstacles exist in front of the stick and right side of the stick. So, the servo steering motor turns the front wheel to the left side. Figure.22 demonstrates the test result for turning right condition.

The obstacle exists in front of the stick. So, the servo steering motor turns the front wheel to the right side.

TEST RESULTS WITH THE BLIND PEOPLE

The above photo shows that there is no obstacle in front of the blind stick and visually impaired person. So, the front wheel fitted with servo motor will not turn to any side and the voice message will not provide to the user. Then, the vibrator will not operate too. Figure.23 shows the test with a visually impaired person (no obstacle condition in front of the blind stick).

Figure 23. Test with a visually impaired person (no obstacle condition in front of the blind stick)

The above photo shows that the obstacles have both front and left side of the blind stick or cane. Therefore, the front wheel equipped with servo steering motor turns to the side to avoid the obstacles for the blind user. And then, if there is no obstacle at left side of the stick or cane, the front wheel turns to the left side to reach the original direction for the user. Figure.24 mentions the test with a visually impaired person (Both front and left side have the obstacles).

Figure 24. Test with a visually impaired person (Both front and left side have the obstacles)

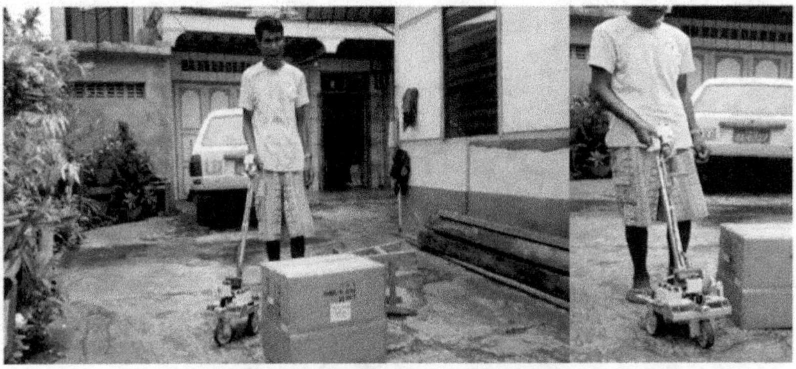

Figure 25. Test for the lighting system in the dark room

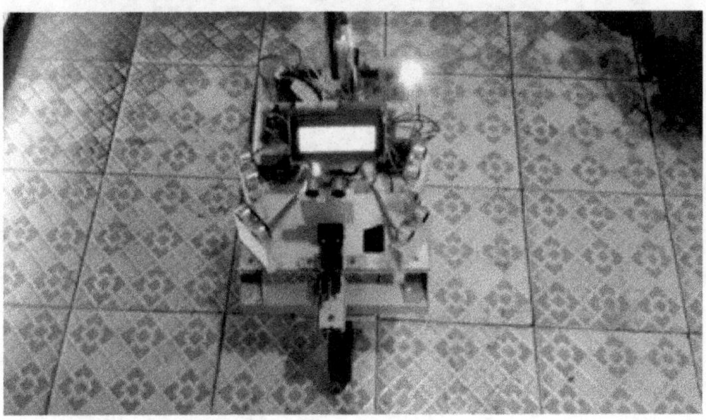

Figure.25 is the photo which is captured in a low light room to test the light signal. In that low light condition, LED is ON, then only, the sighted or ordinary people will notice where the blind people and their stick locate. And then, the sighted people can avoid the stick and blindness before accidence.

In Figure.26, the obstacles exist at both front and right side. So, the front wheel equipped with servo motor turns to the left side to be safe the user without any conscious effort. And then, the suitable voice message is produced for the blind user to know where the obstacle is. The following Table.4 shows the result of actual distance measurement and real distance measurement by using ultrasonic sensor. Data is taken from minimum value of 20 cm to the maximum of 400cm.

Figure 26. Test with a visually impaired person (Both front and right side of the stick have the obstacles)

Table 4. Results of Distance Measurement

Actual Distance (cm)	Real Measurement (cm)
20	20
40	40
60	59
80	78
100	99
120	118
140	139
160	158
180	177
200	198
220	218
240	237
260	259
280	278
300	297

DISCUSSIONS AND CONCLUSION

The important destination of this research is to design and construct a voice-guided system for the totally and partially blind people using mobile robot. This research is to avoid the obstacles and harmful situations for the blind people and, to provide the required voice messages and vibration to the users when there is an obstacle in the user's path-way. So, this stick was designed with 41-inch (3.6 feet) height that is depending on the people's height for our country, Myanmar. It was constructed with aluminum stick which is to handle comfortably, to lift the stick quickly without effort and to join with the based house of the stick that is fit with sensors, circuit board and rechargeable battery and then it was fit with three plastic wheels to push easily and the blind users can feel the steering command from the front wheel with servo steering motor through the handle and follow the stick without any conscious effort.

This system was implemented with six ultrasonic sensors and light sensor (Light-Dependent-Resistor (LDR)) as detecting and sensing units, two PIC16F887 and PIC16F676 microcontrollers as processing units, and LCD, LED, servo steering motor, ISD2590 and speaker as output units. Although laser, radar and ultrasonic sensors are used for detecting the obstacles, ultrasonic sensor is the most suitable one for this research and it is cheaper than the other active sensors. For the obstacle detection system, ultrasonic sensor was chosen to sense and detect the obstacles not only in front of, left and right side of the stick on the ground for the user but also over-hanging objects near the user's head level. There are some limitations in this system. Some kinds of road situations can affect the system's ability to detect the objects, such as dips and down stair on the user's path. Despite of the ultrasonic sensors can detect the obstacles within 400cm (4meter), the obstacle avoidance system and voice-guided system will operate to the reaching on the pre-defined distance under 120cm (4 feet) for the blind users. For the obstacle avoidance system, servo motor was used for turning the wheel easily. It was fit at the front wheel under the stick's based house to turn the wheel to left or right side based on the sensors. For the voice-guided system, ISD2590 Chipcorder was applied to produce the recorded voice messages with Myanmar language to the blind users for our country to avoid the obstacles before accident. Despite there are nine segments for nine messages in the ISD2590, only six voice messages are recorded in it and if necessary, the left three messages can also be recorded. This stick or mobile robot with voice-guided system is more suitable and useful for the blind users. And then, the vibrator will operate if the stick is near or less than 60cm (2 feet) from the obstacle. For the lighting system, LDR was used as a light sensor and LED was used as a light signal. It is not effective for the blind people but the light signal is noticeable and useful for the sighted people in the low-light room or dark room and at night to avoid the stick of the blind person before the ordinary people so not hit the blind stick unexpectedly, so, the blind users and their sticks can be safe in the dark condition from someone's hit.

There are two PIC16F887 and PIC16F676 microcontrollers to control the blind stick successfully. Why three microcontrollers were used in this research work, main PIC16F887 gives the pulse to six ultrasonic sensors, so it cannot give or support pulse for the servo steering motor, and, in reality, although there is no need LCD for the blind user, this research work needs LCD because the distance measurements between the obstacle and sensors are known, then only, this blind stick or mobile robot can be checked with actual distance from the sensors and obstacles. One PIC16F887 was used as a main controller and another one was used for the voice-guided system to control the ISD2590 to provide the voice messages according to the sensors. PIC16F676 was used as servo steering motor controller. So, each microcontroller is useful to complete the whole system successfully and is the main key to create and control the voice-guided system for blind people using mobile robot. To implement the whole system, PIC BASIC PRO programming language was chosen to use compared with MikroC programming language because the program code is short to write for this research work.

This mobile robot or blind stick has many ultrasonic sensors because it does not suppose as prototype and it supposes as the actual blind stick for the blind users to apply or use in their real daily routine. If only three sensors equip at left, right and in front of the stick, there is a problem in avoiding system because the front wheel turns to the left or right side to avoid the obstacle, then, if there is no obstacle, the front wheel will turn back to the center position automatically. In this condition, the wheel or back side of the stick hit to the wall or other obstacles although the left or right sensor escapes from the obstacle. Unfortunately, the blind user can also hit to the wall or obstacles. That's why, five sensors are equipped at the based house for the stick to reduce the above error condition to safe the blind people and the stick.

REFERENCES

Akhil, P., Akshara, R., Athira, R., Kamalesh Kumar, S. P., Thamotharan, M., & Shobha Christila, S. (2022). Smart Blind Walking Stick with Integrated Sensor. *The 2nd International Conference on Innovative Research in Renewable Energy Technologies (IRRET 2022)*. https://doi.org/10.3390/materproc2022010012

Al-Fahoum. (2013). A Smart Infrared Microcontroller-Based Blind Guidance System. *Active and Passive Electronic Components*, *2013*, 1–7. doi:10.1155/2013/726480

Ashiq, FAsif, MAhmad, M. BZafar, SMasood, KMahmood, TMahmood, M.TLee, I. H. (2022). CNN-Based Object Recognition and Tracking System to Assist Visually Impaired People. *IEEE Access: Practical Innovations, Open Solutions, 10*, 14819–14834. https://ieeexplore.ieee.org/document/9698080. doi:10.1109/ACCESS.2022.3148036

Burta, A., Szabo, R., & Gontean, A. (2020 July 27-28). *Object Recognition Development for Android Mobile Devices with Text-to-Speech Function Created for Visually Impaired People*. 2020 Fourth World Conference on Smart Trends in Systems, Security and Sustainability (WorldS4), London, UK. https://ieeexplore.ieee.org/document/9210381

Chaudary, B., Pohjolainen, S., & Aziz, S. (2023). Teleguidance-based remote navigation assistance for visually impaired and blind people—Usability and user experience. *Virtual Reality (Waltham Cross)*, *27*, 141–158. doi:10.100710055-021-00536-z

Elmannai, W., & Elleithy, K. (2017). Sensor-Based Assistive Devices for Visually-Impaired People: Current Status, Challenges, and Future Directions. *Sensors (Basel)*, *17*(3), 565. doi:10.339017030565

Fanfeng, Z. (2010 November 4-6). *Application Research of Voice Control in Reading Assistive Device for Visually Impaired Persons*. 2010 International Conference on Multimedia Information Networking and Security, Nanjing, China. https://ieeexplore.ieee.org/document/5670908

Islam, M. I., Raj, M. M. H., Nath, S., Rahman, M. F., Hossen, S., & Imam, M. H. (2018 April 6-8). *An Indoor Navigation System for Visually Impaired People Using a Path Finding Algorithm and a Wearable Cap*. 2018 3rd International Conference for Convergence in Technology (I2CT), Pune, India. https://ieeexplore.ieee.org/document/8529757

Kalpana, S., Rajagopalan, S., Ranjith, R., & Gomathi, R. (2020 July 3-4). *Voice Recognition Based Multi Robot for Blind People Using Lidar Sensor*. 2020 International Conference on System, Computation, Automation and Networking (ICSCAN), Pondicherry, India. https://ieeexplore.ieee.org/abstract/document/9262365

Khan, S., Nazir, S., & Khan, H. U. (2021). Analysis of Navigation Assistants for Blind and Visually Impaired People: A Systematic Review. *IEEE Access: Practical Innovations, Open Solutions*, *9*, 26712–26734. doi:10.1109/ACCESS.2021.3052415

Messaoudi, M. D., Menelas, B.-A. J., & Mcheick, H. (2022). Review of Navigation Assistive Tools and Technologies for the Visually Impaired. *Sensors (Basel)*, *22*(20), 7888. doi:10.339022207888

Mohajeri, N., Raste, R., & Daneshvar, S. (2011, July 6 - 8). An Obstacle Detection System for Blind People. *Proceedings of the World Congress on Engineering 2011*. https://www.iaeng.org/publication/WCE2011/WCE2011_pp1602-1605.pdf

Mohapatra, S., Rout, S., Tripathi, V., Saxena, T., & Karuna, Y. (2018, May 11-12). *Smart Walking Stick for Blind Integrated with SOS Navigation System.* 2018 2nd International Conference on Trends in Electronics and Informatics (ICOEI), Tirunelveli, India. https://ieeexplore.ieee.org/document/8553935

Rahman, M. A., Sadi, M. S., Islam, M. M., & Saha, P. (2019, November 28-30). *Design and Development of Navigation Guide for Visually Impaired People.* 2019 IEEE International Conference on Biomedical Engineering, Computer and Information Technology for Health (BECITHCON), Dhaka, Bangladesh. https://ieeexplore.ieee.org/document/9063201

Real, S., & Araujo, A. (2019). Navigation Systems for the Blind and Visually Impaired: Past Work, Challenges, and Open Problems. *Sensors (Basel)*, *19*(15), 3404. doi:10.339019153404

Sumathy, B., Pavithran, K. M., Nizam, N., & Surya, V. A. (2020 October 2-3). *Smart Guidance System for Blind with Wireless Voice Playback*. IOP Conference Series: Materials Science and Engineering, Volume 1012, International Conference on Robotics, Intelligent Automation and Control Technologies (RIACT 2020), Chennai, India. https://iopscience.iop.org/article/10.1088/1757-899X/1012/1/012045/pdf

Wei, Y., Kou, X., & Lee, M. (2013, October 20-23). *Development of a guide-dog robot system for the visually impaired by using fuzzy logic based human-robot interaction approach.* 2013 13th International Conference on Control, Automation and Systems (ICCAS 2013), Gwangju, South Korea. https://ieeexplore.ieee.org/document/6703877

Chapter 8
Mobile-Based Weather Forecasting and Visualization Using Augmented Reality

Anil Kumar Biswal
https://orcid.org/0000-0001-7341-216X
Udayanath College of Science and Technology (Autonomous), India

Pratibha Bharti
NIST Institute of Science and Technology (Autonomous), India

Ankit Kumar
NIST Institute of Science and Technology (Autonomous), India

Swati Ranjan
NIST Institute of Science and Technology (Autonomous), India

Milan Samantaray
Trident Academy of Technology, India

ABSTRACT

The aim is to design a 3D augmented reality (AR) weather visualization to investigate future tragedy and daily forecasting. Imagine a world where all physical objects could be made, destroyed, customized, and used in endless ways. Augmented reality (AR) and virtual reality (VR) can create such a place. The background for the study is thus: to analyze, learn implementing and building of an AR Weather application using the OpenWeatherMap Weather API. OpenWeatherMap provides access to local and global weather data via application programming interface (API). OpenWeatherMap serves millions of customers around the world by providing a wide selection of weather-related goods with varying degrees of detail and precision. The research question is if smartphone- and tablet-based 3D AR weather visualizations can be effective tools to enhance current weather forecasting techniques. The methods used in this chapter were making use of AR weather App using unity, AR Foundation, and OpenWeatherMap to stimulate the visualization.

DOI: 10.4018/978-1-6684-8582-8.ch008

INTRODUCTION

As a result of using observation data, science, technology, and scientific knowledge, weather forecasting is the prediction of the weather. In early day's weather forecasting was done by the help of observation. Weather forecasting was first done using a computer in 1955 (Sermet et al., 2020) (Ntagiou et al., 2022). It predicts cloud cover, precipitation, wind speed, temperature, hours, days, and month before they happen for a given current location (Bogomolova et al., 2020).

Since the user may physically move real items with 6 degrees of freedom and virtually interact with objects with the same level of control, this definition allows for the inclusion of 2D overlays while maintaining the aim of having interactions in 3D (Aditama et al., 2020). A visual representation of a virtual object is required, but other senses (hearing, touching, tasting, and smelling) may be added as well. In doing so, we reduce the need for expensive sound equipment that can provide spatialized sound. In addition, it places no limitations on the technology employed to produce the illusions, so that undiscovered technologies can be included into the framework in the future.

Both the physical world and the items within it are simulated in virtual reality. That's why it offers a virtual experience that can be much like the real world or entirely alien to it. Despite its seeming insignificance, this modification modifies the feasibility of building such a system(Sassi, M. S. H. et al., 2021). No longer do we have to worry about the finer points of object environment alignment and supporting a viewfinder to make the real environment work in tandem with the objects. The user's stance is now the ground truth rather than being relative to the actual environment's orientation, which greatly simplifies the system (Bhanipati et al., 2021). Because of these factors, it is technically possible for an augmented reality system to provide a virtual environment by masking or otherwise replacing the user's actual surroundings.

Anemometers (wind speed and direction) and hygrometers (humidity) are used to determine the weather conditions for weather forecasting reports (Meister et al., 2022). Weather forecast includes changes in the earth surface due to atmospheric changes such as ice, snow, tides, floods etc. from the mid-20th century, the digital computer had made it possible to calculate the atmospheric change by the initial condition (Seo. et al., 2018). Generally, radar is used to detect precipitation, but it has also been used to detect larger clusters of insects.

MOTIVATION

In this conspicuous changes, most existing approaches only make use of tabular data, simple 2D plots, or conventional desktop 2D Geographic Information System(GIS) tools that are not optimized for mobile use. The purpose of weather Forecasting using AR is to reduce weather related losses and enhance social benefits by providing advance reports based on current location. By using mobile app one can get up to date with the useful information about weather condition and accordingly act on it. As this generation each and every person has access to internet therefore anyone can get the information at any point of time. Realistic simulations of environmental conditions for in-depth research have become possible because to the integration of new technologies, such as web-based platforms with real-time knowledge creation capabilities, into environmental problem solving. By visualizing the past, incident we have planned to develop a mobile application which can detect the current situation by visualizing the climate condition we can reduce this type of incident.

LITERATURE SURVEY

In the following sections, we provide an overview of the possibility of employing 3D AR weather visualizations to improve representations based on weather theory through a review of AR work. Currently we are using (i) **Doppler radar** (ii) **Satellite data** (iii) **Radiosondes** (iv) **Automated surface-observing systems(v) Supercomputers** (vi) **AWIPS etc. for the purpose of weather forecasting.** Identifying the current gaps that may be addressed with 3D AR visualizations are discussed here in our article. Previously people used to visualize 3D information and conclude to the users. Several immersive AR applications are discussed in the literature about AR in this chapter.

This work employs learner-centered instructional design to improve weather teaching regarding thunderstorms through the use of augmented reality capabilities. The primary goal of learner-centered design is to coordinate instruction so that students make a smooth progression from preexisting knowledge through learning experiences to formative assessments to summative assessments and ultimately to the achievement of the learning goal. Discovering the context in which the target audience will be consuming the design is the first step. The second part of the design process involves choosing a final learning objective and a test to measure that objective. Finding a sequence of learning experiences that leads to mastery of the learning goal's sub-tasks is the third step in the design process. The fourth step in the design process is to provide formative tests that students can use to evaluate their own performance on the smaller tasks and receive constructive criticism.

The printed learning materials are an interactive adaptation of the instructional design. The idea is to strategically insert digital resources into the text itself. The student reads the document and interacts with the embedded digital elements. The digital materials are helpful in expanding the student's knowledge of the subject. Digital content is taken in by the learner in the same way that other visual information is (e.g., charts and graphics), but with the added benefits of depth, immersion, and interaction.

A chapter on thunderstorms from an aviation meteorological textbook served as inspiration for the printed content. We start off this chapter by discussing the prerequisites for thunderstorm formation. The numerous varieties of thunderstorms, from single cells to multicellular clusters to multicellular lines to supercells, are then discussed. Then, it goes on to detail what causes a thunderstorm to move laterally across the earth. Lightning, gusty winds, downbursts, turbulence, icing, hail, and tornadoes are only some of the threats to aviation that are detailed afterward. And finally, it lays out the rules for what to do and what not to do during a thunderstorm. The article concludes with a knowledge quiz based on the commercial pilot exam, which includes questions about weather.

Embedded augmented reality (AR) meteorological content in the form of 3D markers elevates the print text. By registering 3D markers, we can use the WeatherXplore augmented reality (AR) programme to superimpose digital information on top of printed materials. Unlike the 2D page, the spatial and temporal dynamics of meteorological occurrences are visualized in the 3D AR digital overlay. This augmented reality (AR) 3D data is provided inside the context of the text, allowing the student to read the standard weather education materials and also experience the AR upgrades. The AR material covers topics including the lifespan of a thunderstorm cell, what a microburst is like, how it can affect a plane's flight, and how to avoid thunderstorms. In the following paragraphs, we'll go over the augmented reality material.

AUGMENTED REALITY AND VIRTUAL REALITY

The ability to blend computer-generated content with the actual environment is what makes augmented reality (AR) one of the most talked-about technologies today. At the consumer level, there are a wide range of APIs that enable creation of AR experiences, most of which utilize mobile devices like smartphones or tablets. In addition to visual, auditory, haptic, somatosensory, and olfactory sensory modalities, the content can address multiple senses (Joshi, N. et al., 2021) (Samantaray et al., 2021). AR uses digital elements such as visuals, sound, and other sensory stimuli to enhance real-world environments. In particular, mobile computing companies and companies engaged in business applications are taking advantage of this trend (Prophet, J. et al., 2018). 3-D visualization can be used as a substitute to make weather forecasting using AR possible. Semi-immersive virtual worlds, such as those used in augmented reality (AR), augment the real world by superimposing computer-generated perceptual information (Soni et al., 2022) (Singh et al., 2021). AR is used in aviation training for flight procedure training, cabin crew training, and maintenance training (Laera, 2020) (Singh et al., 2020). Military pilots utilise augmented reality to aid in navigation and combat operations (Colley et al., 2022) (Fukuda et al., 2019). These AR systems help user's complete tasks and develop mental models which is depicted in figure 1.

Figure 1. Using Augmented and Virtual Reality (AR/VR) to Support Safe Navigation on Inland and Coastal Water Zones

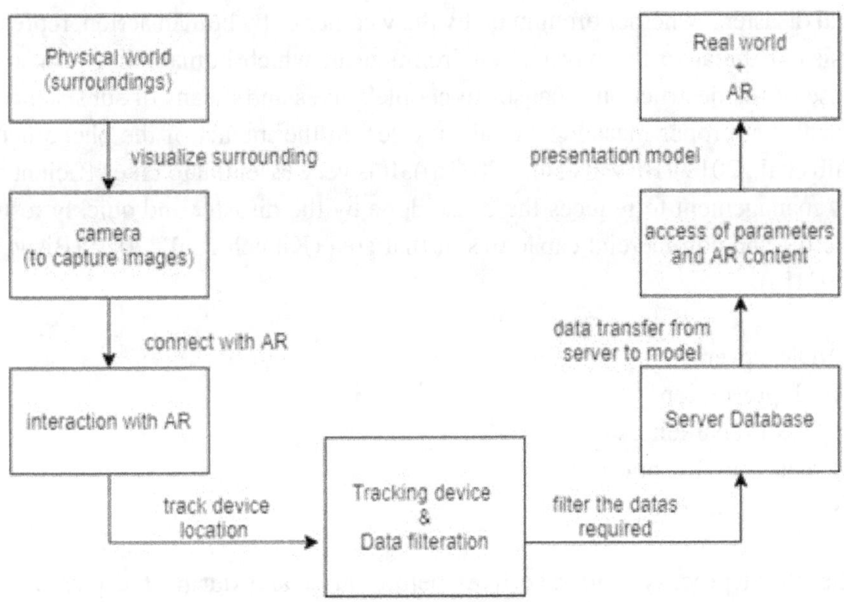

AR/VR APPLICATION FOR DISASTER PREPAREDNESS AND MANAGEMENT

A disaster is an event, whether sudden or gradual, that results in widespread human, material, economic, or environmental loss, beyond the ability of the affected community or society to cope using its own resources (Carter et al., 2022) (Singh et al., 2022). Disaster management is "the body of policy and administrative decision, the operating activities, the actors and technologies that pertain to various stages of disaster eta level" (Bagassi et al., 2020).

As opposed to AR's real-world setting with an integrated but non-dominant virtual object that can react to the user, VR's totally virtual environment with virtual elements obscures the physical environment and stimulates the physical object. By simulating real-world sensory input with computer-generated signals for sight, hearing, and touch, AR and VR equip researchers and rescue teams with useful tools for studying and responding to emergency situations (Gonzalez Vargas, J. C. et al., 2020) (Matviienko et al., 2022).

WHY NATURAL DISASTER MANAGEMENT MUST BE SUPPORTED BY AR/VR?

Natural disasters, whether brought on by the weather or by human action, represent a persistent threat to the physical environment in which humans live and work. Because of the destruction it causes to people's lives and means of subsistence, it is crucial that proper measures be taken to lessen the impact of the phenomenon (Saquib et al., 2019)(Biswal et al., (2021a)). It is very essential to take efficient and timely management to reduces the cause done by the disaster and quickly restore productive and commercial capacities of that area (Kikuch et al., 2022)(Biswal et al., (2021b)).

- NDM covers
- Risk prevention
- Cross Preparedness
- Emergency Response
- Catastrophe recovery

The NDM process is to save lives before, after and during the phenomena. During disaster, the communication and visualization is necessary, therefore AR and VR cannot be ignored (Mathews et al., 2021)(Biswal et al., (2021c)). VR uses computer-generated imagery, sound effects, and movable stimuli to simulate a variety of real-world and fictitious environments. Virtual reality allows the user to imagine, manipulate, and relate unexpectedly complex facts. Instead of building a virtual world, AR enhances the one we now inhabit (Davari 2022) (Kaźmierczak et al., 2021).

AR/VR APPLICATIONS

i. Environmental Visualization and Data Analysis
   - Data growth increases with increase in IOT devices
   - Environmental data increases with increase in development and investment in sensors
   - Information are gathered with awareness in surroundings
   - Good use of dataset helps in large scale like analyze, access, communication, and for stakeholders
   - Presentations are given with the help of either web-based applications or app based application
   - AR VR helps in future data retrials, depth visualization, on site analysis, given by case study.

ii. Environment data retrieval and sensing
   - guide user for complex measurements, data collection.
   - provide accurate survey.
   - example find water stage using geometry-based approaches find nearby obstacles using image processing

iii. Hydrological Simulations and Disaster Education
   - mainly used in disaster sciences
   - education of hydrological processes

iv. Mixed reality (XR)
   - With XR, we can test out classroom setups before committing to them in the real world.
   - Example of web-based hydrological simulation system- here with this we can get the details of different flood scenarios and implement some strategies to overcome the destructions and heavy damages

FEATURES OF AR/VR

- It provides unique and immersive experience.
- It shows great potential for applications in environmental.
- It's useful for making choices, learning more, and spreading information.
- Improvements in the precision, size, and cost of mobile sensor technologies are largely responsible for these changes.
- network technology makes faster data reporting.
- It provides opportunities in future research.

USING AUGMENTED AND VIRTUAL REALITY (AR/VR) TO SUPPORT SAFE NAVIGATION ON INLAND AND COASTAL WATER ZONES

For safe navigation in water can observe water even during night using AR and VR

Mobile GPS-enabled cameras provide a real-time view of users' route coupled with GNSS (Global Navigation Satellite System)

Approach:

Unity

Using OPENWEATHERMAP

AR Foundation setup

Programming and understanding(dart)

For smaller applications where we require fewer details of weather, we can make use of Open.WeatherMap.org. Open.WeatherMap.org provides widely used free weather API's.

After creating account, an API key is provided for authentication purpose of the API request.

METHODOLOGY

First step to start is to get API key for authentication purpose for API request.

After signing up and creation of an account in OpenWeatherMap.org.

After email verification you will get the key.

API takes parameters - (latitude, longitude, and API key).

We'll need a JSON parser to convert it into our usable data types.

Now create an empty game object with the name WeatherManager.

Within Unity, the Input. Location gives complete cross-platform API needed for fetching the device location.

Next, we pass in the latitude and longitude to this routine which uses UnityWebRequest to perform a GET call on the API which will then return the JSON response.

Finally, JSON response and display it onto our Canvas view text objects.

Create a new canvas and mark it as World Space.

Create a new Image as it's child and make it color black with transparency which will act as a background for the UI View.

RESULT ANALYSIS AND DISCUSSION

There have been vast improvements in the accuracy and reliability of weather forecasts over the past twenty years, but still weather conditions are only able to be forecasted or predicted up to a certain point. AR presentations is a more interactive and descriptive way for the users to visualize the before coming catastrophe. A consensus format is needed to allow easy integration of structural and terrain-related data to support global, generalized simulation platforms' vision of global and realistic representation of real-world features. The collected meteorological data can be used to configure factors including the kind and amount of precipitation, cloud cover, and visibility which is depicted in figure 2.

Figure 2. Smartphone-based stream stage measurement

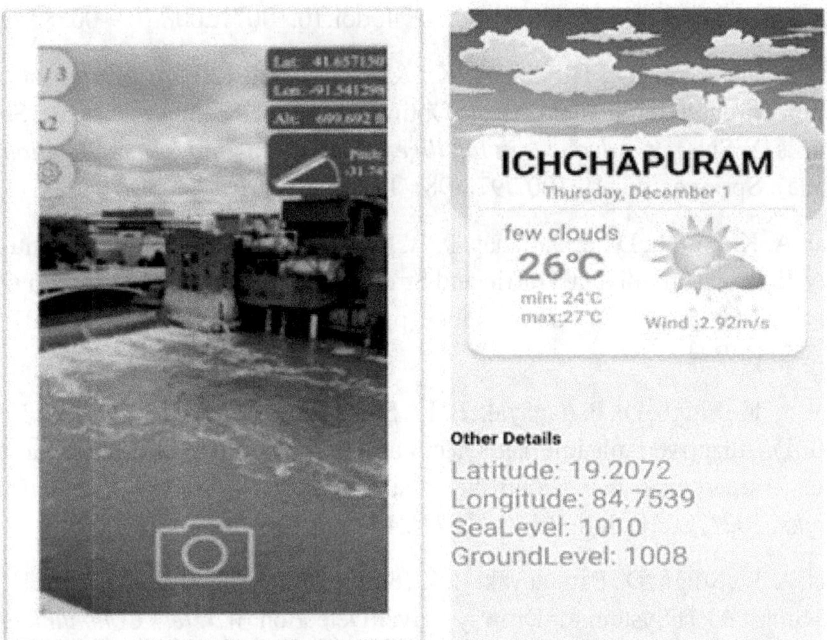

CONCLUSION

AR has a great application in knowing environmental disaster management, it also helps in disaster management educations. Conventional data sensing approaches require a lot of maintenance charges. When it comes to the cost of computational resources, organisations can save money by using data collection methods based on augmented reality. Additionally, they highlight potential advancements and research directions for the future. In the new science of sociohydrology, people are viewed as integral components of the water cycle as they consume, pollute, and use freshwater. AR and VR applications can support this science by supporting water consumption, pollution of freshwater resources, and policies.

REFERENCES

Aditama, P. W., & Setiawan, I. N. A. F. (2020, February). Indigenous bali on augmented reality as a creative solution in industrial revolution 4.0. *Journal of Physics: Conference Series*, *1471*(1), 012008. doi:10.1088/1742-6596/1471/1/012008

Bagassi, S., De Crescenzio, F., & Piastra, S. (2020). Augmented reality technology selection based on integrated QFD-AHP model. *International Journal on Interactive Design and Manufacturing*, *14*(1), 285–294. doi:10.100712008-019-00583-6

Bhanipati, J., Singh, D., Biswal, A. K., & Rout, S. K. (2021). Minimization of Collision Through Retransmission and Optimal Power Allocation in Wireless Sensor Networks (WSNs). In *Advances in Intelligent Computing and Communication* (pp. 653–665). Springer. doi:10.1007/978-981-16-0695-3_61

Biswal, A. K., Singh, D., & Pattanayak, B. K. (2021). IoT-Based Voice-Controlled Energy-Efficient Intelligent Traffic and Street Light Monitoring System. In *Green Technology for Smart City and Society* (pp. 43–54). Springer. doi:10.1007/978-981-15-8218-9_4

Biswal, A. K., Singh, D., Pattanayak, B. K., Samanta, D., Chaudhry, S. A., & Irshad, A. (2021). Adaptive fault-tolerant system and optimal power allocation for smart vehicles in smart cities using controller area network. *Security and Communication Networks*, *2021*, 2021. doi:10.1155/2021/2147958

Biswal, A. K., Singh, D., Pattanayak, B. K., Samanta, D., & Yang, M. H. (2021). IoT-Based Smart Alert System for Drowsy Driver Detection. *Wireless Communications and Mobile Computing*, *2021*, 2021. doi:10.1155/2021/6627217

Bogomolova, K., van der Ham, I. J., Dankbaar, M. E., van den Broek, W. W., Hovius, S. E., van der Hage, J. A., & Hierck, B. P. (2020). The effect of stereoscopic augmented reality visualization on learning anatomy and the modifying effect of visual-spatial abilities: A double-center randomized controlled trial. *Anatomical Sciences Education*, *13*(5), 558–567. doi:10.1002/ase.1941 PMID:31887792

Bogomolova, K., van der Ham, I. J., Dankbaar, M. E., van den Broek, W. W., Hovius, S. E., van der Hage, J. A., & Hierck, B. P. (2020). The effect of stereoscopic augmented reality visualization on learning anatomy and the modifying effect of visual-spatial abilities: A double-center randomized controlled trial. *Anatomical Sciences Education*, *13*(5), 558–567. doi:10.1002/ase.1941 PMID:31887792

Carter, B. A., Iles, G. N., Raju, R., Afful, A. M., Maj, R., Dao, T., Terkildsen, M., Lobzin, V., Bouya, Z., Parkinson, M., Le May, S., Choy, S., Hordyniec, P., Hordyniec, B., Currie, J., Skov, T., & Peake, I. D. (2022). RMIT University's practical space weather prediction laboratory. *Journal of Space Weather and Space Climate*, *12*, 28. doi:10.1051wsc/2022025

Colley, M., Rädler, M., Glimmann, J., & Rukzio, E. (2022). Effects of scene detection, scene prediction, and maneuver planning visualizations on trust, situation awareness, and cognitive load in highly automated vehicles. *Proceedings of the ACM on Interactive, Mobile, Wearable and Ubiquitous Technologies*, *6*(2), 1–21. doi:10.1145/3534609

Davari, S. (2022, March). [DC] Context-Aware Inference and Adaptation in Augmented Reality. In *2022 IEEE Conference on Virtual Reality and 3D User Interfaces Abstracts and Workshops (VRW)* (pp. 938-939). IEEE.

Fukuda, T., Yokoi, K., Yabuki, N., & Motamedi, A. (2019). An indoor thermal environment design system for renovation using augmented reality. *Journal of Computational Design and Engineering*, *6*(2), 179–188. doi:10.1016/j.jcde.2018.05.007

Gonzalez Vargas, J. C., Fabregat, R., Carrillo-Ramos, A., & Jové, T. (2020). Survey: Using augmented reality to improve learning motivation in cultural heritage studies. *Applied Sciences (Basel, Switzerland)*, *10*(3), 897. doi:10.3390/app10030897

Joshi, N., Lakier, M., Vogel, D., & Zhao, J. (2021, December). A Design Framework for Contextual and Embedded Information Visualizations in Spatial Augmented Reality. In Graphics Interface 2022.

Kaźmierczak, R., Szczepańska, A., Kowalczyk, C., Grunwald, G., & Janowski, A. (2021). Using AR technology in tourism based on the example of maritime educational trips—A conceptual model. *Sustainability*, *13*(13), 7172. doi:10.3390u13137172

Kikuchi, N., Fukuda, T., & Yabuki, N. (2022). Future landscape visualization using a city digital twin: Integration of augmented reality and drones with implementation of 3D model-based occlusion handling. *Journal of Computational Design and Engineering*, *9*(2), 837–856. doi:10.1093/jcde/qwac032

Laera, F. (2020, November). Augmented Reality for Easy Sailing. In *2020 IEEE International Symposium on Mixed and Augmented Reality Adjunct (ISMAR-Adjunct)* (pp. 312-313). IEEE. 10.1109/ISMAR-Adjunct51615.2020.00088

Mathews, N. S., Chimalakonda, S., & Jain, S. (2021, October). AiR: An Augmented Reality Application for Visualizing Air Pollution. In *2021 IEEE Visualization Conference (VIS)* (pp. 146-150). IEEE. 10.1109/VIS49827.2021.9623287

Matviienko, A., Müller, F., Schön, D., Seesemann, P., Günther, S., & Mühlhäuser, M. (2022, April). BikeAR: Understanding Cyclists' Crossing Decision-Making at Uncontrolled Intersections using Augmented Reality. In *CHI Conference on Human Factors in Computing Systems* (pp. 1-15). 10.1145/3491102.3517560

Meister, P., Miller, J., Wang, K., Dorneich, M. C., Winer, E., Brown, L. J., & Whitehurst, G. (2022). Designing Three-Dimensional Augmented Reality Weather Visualizations to Enhance General Aviation Weather Education. *IEEE Transactions on Professional Communication*, *65*(2), 321–336. doi:10.1109/TPC.2022.3155920

Meister, P., Miller, J., Wang, K., Dorneich, M. C., Winer, E., Brown, L. J., & Whitehurst, G. (2022). Designing Three-Dimensional Augmented Reality Weather Visualizations to Enhance General Aviation Weather Education. *IEEE Transactions on Professional Communication*, *65*(2), 321–336. doi:10.1109/TPC.2022.3155920

Mishra, A., Singh, C., Dwivedi, A., Singh, D., & Biswal, A. K. (2021, October). Network Forensics: An approach towards detecting Cyber Crime. In *2021 International Conference in Advances in Power, Signal, and Information Technology (APSIT)* (pp. 1-6). IEEE. 10.1109/APSIT52773.2021.9641399

Ntagiou, E. V., Klug, J., Luntama, J. P., & Sarkarati, M. (2022). Virtual Reality in Support of Space Weather Forecasting. In *Space Operations* (pp. 325–343). Springer. doi:10.1007/978-3-030-94628-9_15

Ntagiou, E. V., Klug, J., Luntama, J. P., & Sarkarati, M. (2022). Virtual Reality in Support of Space Weather Forecasting. In *Space Operations* (pp. 325–343). Springer. doi:10.1007/978-3-030-94628-9_15

Prophet, J., Kow, Y. M., & Hurry, M. (2018, July). Cultivating environmental awareness: Modeling air quality data via augmented reality miniature trees. In *International Conference on Augmented Cognition* (pp. 406-424). Springer, Cham. 10.1007/978-3-319-91470-1_33

Samantaray, M., Biswal, A. K., Singh, D., Samanta, D., Karuppiah, M., & Joseph, N. P. (2021, December). Optical Character Recognition (OCR) based Vehicle's License Plate Recognition System Using Python and OpenCV. In *2021 5th International Conference on Electronics, Communication and Aerospace Technology (ICECA)* (pp. 849-853). IEEE.

Saquib, N., Kazi, R. H., Wei, L. Y., & Li, W. (2019, May). Interactive body-driven graphics for augmented video performance. In *Proceedings of the 2019 CHI Conference on Human Factors in Computing Systems* (pp. 1-12). 10.1145/3290605.3300852

Sassi, M. S. H., & Fourati, L. C. (2021, October). Deep learning and augmented reality for IoT-based air quality monitoring and prediction system. In *2021 International Symposium on Networks, Computers and Communications (ISNCC)* (pp. 1-6). IEEE. 10.1109/ISNCC52172.2021.9615639

Sassi, M. S. H., & Fourati, L. C. (2021, October). Deep learning and augmented reality for IoT-based air quality monitoring and prediction system. In *2021 International Symposium on Networks, Computers and Communications (ISNCC)* (pp. 1-6). IEEE. 10.1109/ISNCC52172.2021.9615639

Seo, S., Kang, D., & Park, S. (2018). Real-time adaptable and coherent rendering for outdoor augmented reality. *EURASIP Journal on Image and Video Processing*, *2018*(1), 1–8. doi:10.118613640-018-0357-8

Sermet, Y., & Demir, I. (2020). Virtual and augmented reality applications for environmental science education and training. In *New Perspectives on Virtual and Augmented Reality* (pp. 261–275). Routledge. doi:10.4324/9781003001874-17

Singh, D., Bhanipati, J., Biswal, A. K., Samanta, D., Joshi, S., Shukla, P. K., & Nuagah, S. J. (2021). Approach for Collision Minimization and Enhancement of Power Allocation in WSNs. *Journal of Sensors*. doi:10.1155/2021/7059881

Singh, D., Biswal, A. K., Samanta, D., Singh, D., & Lee, H. N. (2022). Juice Jacking: Security Issues and Improvements in USB Technology. *Sustainability*, *14*(2), 939. doi:10.3390u14020939

Singh, D., Prusty, S. K., Sarangi, S. S., Sahoo, S., & Biswal, A. K. (2020). Anxiety, Psychological Effects and Prevention during COVID-19 in India.

Soni, A., Khare, A., Balaji, P. D., Verma, S., Rani, K. A., & Gowrishankar, S. (2022, October). Pest Identification and Control using Deep Learning and Augmented Reality. In *2022 International Conference on Edge Computing and Applications (ICECAA)* (pp. 1-7). IEEE. 10.1109/ICECAA55415.2022.9936053

Chapter 9
A Scoping Review of Current Developments in the Field of Machine Learning and Artificial Intelligence

Santhosh Kumar Rajamani
https://orcid.org/0000-0001-6552-5578
Maharashtra Institute of Medical Education and Research, India & Dr. Bhausaheb Sardesai Talegoan Rural Hospital, India

Radha Srinivasan Iyer
https://orcid.org/0000-0001-7387-4401
SEC Centre for Independent Living, India

ABSTRACT

This chapter gives a broad outline of machine learning and artificial intelligence, and introduces the reader to many novel and latest developments in the field of machine learning. The first half of this compilation provides a comprehensive view of the classical concepts of machine learning. Subsequently, examples of machine learning frameworks are discussed. Deep learning, concepts, models, types, and algorithms in machine learning are elaborated in the subsequent section, followed by a detailed introduction to neural networks, concepts of weights, propagation, and initialization. The final section of this chapter introduces the reader to the fascinating and latest world of cutting-edge applications of machine learning like convolutional neural networks (CNNs), bidirectional long short-term memory (BLSTM), artistic, image-generating AI engines like DALL-E and stable diffusion, music and drama writing AI engines, human-like chatbot ChatGPT, art generation with AI, generative neural network concepts, regenerative neural network, and natural language processing (NLP).

DOI: 10.4018/978-1-6684-8582-8.ch009

INTRODUCTION

Machine learning (ML) is the process of learning meaningful patterns from a given complex data and produce meaningful predictions about the data found in the real-life situation with minimal human involvement. The term was coined by Arthur Samuel in 1959. Like all human experiences, like playing football, the ability of computer to learn improves with experience or training. The data that is used to train or teach the computer is called the "training set". The most common example of ML is the spam filter that filters junk mail from genuine ones that arrive daily at our inbox. Another common example of ML is heuristics, used in the antivirus program that differentiates a virus program from a genuine one. Yet, another example is the personalized suggestions given on sites like Google or Amazon. These ML prediction on our liking, are based on data harvested from our browsing or shopping habits. A set or collection of data is called a dataset (Watters, 2021).

ENSEMBLE LEARNING

Ensemble learning is an advanced machine learning technique utilized to improve the predictive accuracy of a model by combining the predictions from multiple individual models. By training multiple models with different supervised (regression and classification) algorithms, ensemble learning leverages aggregated insights from different classifiers or regressors for enhanced predictive power. This highly versatile strategy functions well for complex real-world problems as it narrows down uncertainties in prediction tasks due to its ability to capture cross-model correlations across data sets. Furthermore, through intuitively deciding which models are good candidates for data extraction based on their overall performance, ensemble learning can help reduce overfitting and the need for substantial amounts of training data (Sullivan, 2017).

BIG DATA

"Big data" means that large datasets that are usually, automatically generated and that cannot be processed using traditional data management software like Microsoft Excel or Access. This term was popularized by John Mashey in 1900s. Essentially, parallel computation and large storage media are needed to process big data. Special software like Apache Hadoop, Java based Hive, Cloudera, MemSQL, Apache Spark, Amazon S3 can handle the volume of big data. This is due to the sheer size of the data produced, that storage and manipulation is cumbersome. For example, it is estimated that Google processes about 15 Exabyte of data per day. It is staggering to note that 15 Exabyte is equal to 15,000,000,000 (109) Gigabytes, for comparison consider a standard pen drive has 64 Gigabyte of data storage capacity, and a large hard disk has an average capacity of 1 terabyte, 15 Exabyte of data would need around 15,000,000 hard disks. Interestingly, it is estimated that only 0.5% of generated data is used for some form of data analysis(Ankur Saxena et al., 2021).

It is amazing to note that within the span of a minute, 575,000 tweets are posted, 240,000 photos are shared on Facebook, 694,000 hours of video are streamed on the YouTube, 5.7 million searches are conducted on Google search, 283,000$ worth of shopping is done by customers on Amazon, and 856 minutes of webinar are hosted on Zoom meeting platform. Such vast quantity of data is processed online per minute (Fontichiaro, 2018).

Distributed File Systems (DFS) are used to store big data where large data is divided into chunks, which are the stored in pieces across multiple devices or nodes connected by a network. Examples are MongoDB, Apache Cassandra, and ElasticSearch, Apache Spark, Apache Hadoop, and Amazon S3 (based on Amazon Web Service).

Big data can be curated and vetted structured data, or it can be raw, dirty unstructured data. This data can in form of various data types like numbers, images, text, audio-files, and videos. Data can be structured or unstructured. Structured data, curated and annotated Number, Dates and String, constitute only about 20% of big data. These can be processed by Relational database like SQL. Unstructured data image, audio, video, excel sheets, emails, word documents, constitutes a major fraction of about 80% of big data. These cannot be directly processed by Relational database like SQL and need cleansing, conversion, and pre-processing.

This differentiation is important as different ML strategies are needed to evaluate different data types like for example, numerical data is modelled in ML using mathematical linear regression and image classification is achieved using mathematical logistic regression (which produces binary True or False outputs). Typical examples of big data sources are from diverse fields like Physics -Computational ñuid dynamics (CFD) data or Particle physics - Large Hadron Collider data, Bioinformatics - human genome project, Climatology – NASA's Climate Simulation, Astronomy- Event horizon telescope or Sloan Digital Sky Survey (SDSS) and Business – Walmart or Amazon (Dinov & Springerlink, 2018)(Bari et al., 2017).

Such large volume of data can be used to analyze using ML and solve problems which could not be possible with smaller datasets. There are five "V" associated with big datasets, namely: volume (bulky datasets), variety (structured, unstructured, numeric, videos and all kinds of data), velocity (fast speed of build-up of data, using high speed internet connection), value (usefulness of data) and veracity (truthfulness of data). ML aims to classify or derive meaning of this big data (Bari et al., 2017).

PREDICTIVE DATA ANALYTICS

Predictive data analytics is a highly effective tool to enable organizations to gain deeper insights into the behaviour of their customers, operations, and systems. By leveraging predictive algorithms and machine learning techniques, companies can uncover patterns in their data that can be used to inform future decisions, improve operational efficiency, and reduce cost. Predictive analytics also helps companies identify potential risks and opportunities, enabling them to make well-informed strategic decisions based on comprehensive analysis of historical data. Through predictive analytics, companies can improve customer experience, segment valuable customer segments for targeted marketing campaigns and monitor business performance over time with greater accuracy (Antoniou et al., 2019).

BLOCKCHAIN TECHNOLOGY

Blockchain technology is a revolutionary network revolutionizing how businesses process, store, and protect sensitive data. The distributed ledger technology allows computers on a decentralized network to digitally store, update, and share ledgers of transactions in real-time, increasing security and transparency. Additionally, because of its design blockchain also makes it nearly impossible for hackers to intercept or alter data stored on the network since each new block is cryptographically secured to the previous block as well as being processed by a consensus mechanism that requires different nodes to agree before any changes take place. This means that users can trust that any transaction they authorize are accurate and secure from tampering or malicious attacks without the need for an intermediate third-party.

Blockchain technology is a powerful network that stores and records digital transactions in a decentralized database that is secure by design. The blockchain, also known as distributed ledger technology, allows all parties involved in the network to securely access records of past activities on the network. A key advantage of blockchain networks is that they are almost impossible to tamper with and the data stored on them can't be altered or deleted without leaving a traceable record of all past transactions and activity. As such, this provides unparalleled levels of security and data transparency for users and businesses alike. Ultimately, blockchain networks provide increased trust between contributors as well as improved privacy, scalability, cost efficiency, and authentication for commercial services (Isaacs, 2017).

ARTIFICIAL INTELLIGENCE

Artificial intelligence is a term coined by John McCarthy in 1956 at the first ever AI conference held at Dartmouth College, Hanover, USA. Depending on autonomy of the AI, it is classified into 2 types: weak AI and strong AI.

Weak AI—also called narrow AI is training of computer to perform a single, specific task like driving a car (CV), predicting a mileage of a car (Linear regression), chatbots/ virtual agents (NLP), Google translate (NLP), Amazon's Alexa (ASR), classifying an X-Ray (Logistic regression), identifying a tumour on CT scan (CV). This is the current level of development of AI(Pearce, 2011).

Strong AI - is making the machine think and behave like a human being, including self-awareness, learning, planning, and improvising. This level of autonomy of AI will be reached in the coming decades(Pearce, 2011).

MACHINE LEARNING

Machine learning is a sophisticated form of artificial intelligence consisting of algorithms that can learn from data and improve their performance over time. It focuses on the development of algorithms used to find patterns in large datasets, then use those patterns to make predictions about future events or outcomes. Through this technology, businesses can automate decision-making processes and increase operational efficiency by utilizing predictive analytics such as customer segmentation, fraud detection, risk management, and recommendation systems. Machine learning can also be utilized for natural language processing applications like voice recognition and text translation. This technology has helped organizations save both time and money by leveraging ever-growing amounts of data so that organizations can benefit from it quickly and efficiently (Sullivan, 2017).

DEEP LEARNING

Deep learning is a type of machine learning that has been used in AI that is a subset of ML, which can input unstructured data and automatically determine the feature set based on the provided dataset(Goodfellow et al., 2016). Features are generally determined by human intervention in conventional ML by observation and interpretation of the dataset. It is an algorithm that learns to do things by processing data and information. Deep learning algorithms are trained with large amounts of data, like pictures, voice clips and text documents (Sejnowski, 2018).

The applications of deep learning are vast, and it can be used in many different fields. In the future it will be able to help with medical diagnosis, self-driving cars, natural language processing, speech recognition and more. Deep learning achieves this feat by using what is called a Neural Network as shown in the figure below (Dargan et al., 2019).

Types of Learning in ML

Supervised learning or Discriminative ML is ML from curated datasets, in which features determined by human speculation of the datasets. Some paradigms in supervised ML are as follows: neural networks, k Nearest Neighbours, linear regression, decision trees, logistic regression, random forest, Naïve Bayesian ML, support vector machine (SVM), Convolutional Neural Network, Boosting algorithms (XGBoost, Gradient Boosting Machine, and Light GBM) and Neural networks (Watters, 2021).

Naïve Bayesian ML, which uses a Markov chain to compute Posterior distribution in a manner of classical Naïve Bayesian reasoning. This is a specific useful algorithm for smaller datasets where the conventional ML fail due to lack of sufficient learning data. Naïve Bayesian ML is thus especially suited for implementing ML on memory limited Android mobile phones. In addition, Naïve Bayesian ML provides insight into the inner working of the underlying model which is essential in fields like healthcare research (effectiveness of a lifesaving drug or survival after a cancer) (Watters, 2021).

Unsupervised learning or Generative ML is learning from unstructured datasets, where patterns are captured as probability densities. The various algorithms used for Unsupervised ML are as follows: K-means clustering, Principal Component Analysis (PCA), Singular Value Decompose (SVD), Boltzmann machine, Auto encoder, Recurrent Neural Network and Sum Product Network. Hierarchical Clustering, and Neural networks are some examples of Unsupervised ML methodologies. A common example of Unsupervised learning ML are product recommendations offered on Amazon site, which are generated using Clustering methodology, predictions based on data of user's shopping habit. A Hybrid ML also exist for example, Deep Neural Network (Watters, 2021).

CONCEPT OF A MODEL IN MACHINE LEARNING (M.L)

A machine learning model is a program that has been trained using training dataset, has achieved a defined accuracy, having learnt to recognize patterns in data and predict or classify data in the test dataset or unseen dataset. As an example, consider the game of chess. In this game, there are more possible moves than there are atoms in the universe. A computer cannot play Chess well because it cannot look ahead and see all the possible moves. However, if we feed a computer with enough data about past games of chess and allow it to learn from them, then it will be able to play Chess at a reasonable level. Machine learning algorithm is a mathematical methodology used for detecting patterns in a training dataset. Some common algorithms are Linear regression, logistic regression, random forest, decision tree and XGBoost (Rajamani & Iyer, 2022).

Models are the mathematical objects that are built after learning from datasets, using statistical and probabilistic methods. Models are used for prediction and are the workhorse of ML. Artificial Intelligence AI is intelligent behaviour of machines, which mimic the human actions and interactions. Machine Learning is one of the paradigms of Artificial Intelligence. Natural language processing, Deep Learning, Computer vision are other paradigms of artificial intelligence (Elomaa et al., 2002).

TYPES OF MACHINE LEARNING MODELS

There are 2 types of models in ML Ecosystem namely: pre-trained models and self or custom-trained models. Pre-trained model files are ready made models which can be downloaded from sites like Kaggle.com and can immediately classify or predict using test or unseen data (Rajamani & Iyer, 2022). Self-trained model requires us to train the model using dataset which we generate ourselves or acquire from the internet. These models must be fine-tuned using various algorithms until they achieve a desired level of accuracy (Pazzani et al., 1991).Machine learning ML algorithm is a mathematical methodology used for detecting patterns in a training dataset. Some common algorithms are Linear regression, logistic regression, random forest, decision tree and XGBoost (Jobin et al., 2019).

DECISION TREES (DT) ALGORITHMS

Decision trees (DT) algorithms are a sequence of binary trees which branch into two branches or edges, based on some rule. The branch is called a leaf node. Functions are used to minimize node purity. Gini index is a function used on leaf node to determine the accuracy of the split. This index ranges from 0 to 1, zero where all observations belong to a class and 1 where every observation is random noise. For optimization or best decision of a leaf node we need limit of Gini as close to zero as possible(Rokach & Maimon, 2015).

Step by step process of categorizing an object by going through the process of Splitting the decision-making process into 2 pathways which are the 2 branches of the Decision node. The terminal outcomes are called Leaf nodes. This is one of the most common algorithms of ML.

The first decision node of the entire network is called the Root node. Pruning is the process akin to Pruning of the branches, where some of the redundant branches of a Decision node are removed. There are 2 types of DT, namely: 1. categorical variable decision tree -used to classify categorical variables 2. continuous variable decision tree- decision for continuous variables (Sullivan, 2017).

CROSS-VALIDATION PROCESS OF THE MODELS

Cross-validation is process of testing the model's performance on a test set after training on the training dataset. Problem of "overfitting" refers to the process by which the model becomes very fine tuned for the training data but performs poorly on test data set. This is a case where model fails to generalize based on training dataset. "Underfitting" is where model does poorly even on training dataset(Dargan et al., 2019).

Bagging/ Bootstrapping technique

Bagging is a technique that is used in machine learning to combine the predictions of multiple models. Bagging or Bootstrap aggregation is a machine learning algorithm where a subset of data is generated using Bootstrapping procedure to improve the performance and accuracy of ML model. The idea is to take the predictions of all the models and then use some voting algorithm to come up with a final prediction. There is a paradigm called "*Out-of-Bag error*" (OOBE) feature. This algorithm is only available in the R language, but not currently available in Python (Lior Rokach & Maimon, 2015).

Bootstrapping is a resampling technique where we take samples from our original dataset and then create new samples by sampling with replacement from those samples.

In other words, bootstrapping helps us to make inferences about our data by using the same data repeatedly which makes it an excellent tool for estimating confidence intervals.

There are many ways in which bootstrapping can be done, but one of them is called "bagging." This technique works by taking random subsets of your original dataset and training on those subsets. These subsets are called "bootstrap samples" or "bagging samples" or "training sets" (Shrestha & Mahmood, 2019).

VALIDATING A LINEAR REGRESSION MODEL

In Linear regression problems mean squared error (MSE), the mean absolute error (MAE), and the R2 are used in model validation. In regression classification problems, true positive rate (TPR), true negative rate (TNR), and accuracy (ACC) are used in validation. Tabulation of TPR, TNR, positive predictive value PPV, negative predictive value NPV and accuracy is called Confusion matrix. Area Under the receiver operating Curve (AUC) is trade-off between TPR and TNR, is another common parameter used in model validation. Error Function thus effectively predicts the errors in our model's predictions or classifications. This is the accuracy of the ML model. There are many general ML frameworks like Scikit-learn and NLTK natural language processing tool kit which are used for general purpose ML. Examples of machine learning (M.L.) frameworks (Shrestha & Mahmood, 2019).

COMPUTER VISION (C.V)

Computer vision (C.V) is a subfield of artificial intelligence focused on enabling computers to interpret and process visual data to understand the environment and various objects present. Using algorithms, computer vision seeks to automate tasks that require complex image analysis and would be too tedious or difficult for humans. It has become increasingly important with the emergence of digital technologies, as it can be applied virtually anywhere from medical diagnosis to self-driving cars (Liu, 2020). The development of computer vision has significantly impacted many areas such as robotics, facial recognition, autonomous vehicles, and machinery automation. Its potential for automating processes, completing sophisticated tasks quickly, and providing valuable insights continues to attract interest from researchers and businesses alike (Keller et al., 2016).

SWARM ALGORITHM

A swarm algorithm is a system of many agents that work together to solve a problem. The agents can be both artificial and organic. They can be used in many different fields such as data mining, robotics, and optimization. The idea behind the swarm algorithm is to find the best solution from many different solutions by having each agent specialize in one area and then combine their results to create a more accurate and complete solution (Pearce, 2011).

Network science/ Graph theory and applications in Machine Learning

Graph theory is a branch of mathematics that studies the properties of graphs and networks. Graph theory is a branch of mathematics that studies the properties of graphs and networks. It has many applications in computer science, chemistry, physics, biology, engineering, and other sciences (Merris, 2001).Graph theory can be used to find the shortest path between two nodes in a network or to study how information spreads through a social network. Graphs can also be used to represent physical systems such as electrical circuits or mechanical devices such as levers. Machine Learning Graph theory is a branch of machine learning. It consists of algorithms that are used to create graphs from data. Graphs are visual representations of the data, which can be used for various purposes. Graphs can be used for clustering, finding patterns in data, and predicting future events. They can also be used for classification and regression problems. Some examples of these problems include predicting the next word in a sentence or determining if two words share the same meaning (Bornholdt & Heinz Georg Schuster, 2006).

BIOINFORMATICS USING MACHINE LEARNING

Machine learning techniques have become increasingly important in bioinformatics, as they present a powerful tool to analyze and interpret the huge amount of data generated from genomic analysis. These approaches can be used to recognize and evaluate patterns that are difficult to detect manually, regardless of the size or complexity of the data sets. Machine learning methods can also reduce the time for decision making by automating processes in relation to gene expression and protein databases. The models developed based on this technology provide robust analyses with high specificity in areas such as sequence alignment, functional annotation, and disease classification. Furthermore, it has become practical to employ machine learning algorithms when refining existing techniques and computational workflows, thus allowing new insights into computationally intensive problems in biology. Ultimately, leveraging machine learning studies explores opportunities across different domains of bioinformatics research such as drug discovery, population genetics, protein structure prediction, phylogenetics inference and genome assembly refinement (Zhang & Rajapakse, 2009).

INTERNET OF THINGS (IOT)

Internet of Things (IoT) are common devices that are connected to the internet and gather data on customer usage patterns and product performance. The internet of things (IoT) is a network of physical objects — including devices, vehicles, buildings, and other items embedded with electronics, software, sensors, and connectivity that enable these objects to collect and exchange data. This creates an environment where objects can be sensed and controlled remotely across existing networks — resulting in improved efficiency, accuracy, and economic benefit (Schwartz, 2016).

IoT has unlocked new opportunities for businesses as well as individuals to capitalize on the increased access to data across various industries. It brings with it the potential to transform entire business models by providing predictive analytics, preventive maintenance alerts and automated operations that can result in a superior customer experience. Additionally, IoT provides the opportunity for people to stay connected with their surroundings through real-time notifications about their community or environment. Examples of this technology include connected appliances like refrigerators which not only help keep food fresh better but also can order groceries when necessary; smart homes equipped with security systems that automate tasks such as locking doors at night; and connected factories that monitor performance levels for rapid intervention if something goes wrong (Clickard, 2019).

R LANGUAGE FOR MACHINE LEARNING

R is a programming language, is the second popular choice for Machine learning. R language (R Studio/Posit) is software environment for statistical computing and graphics. It is widely used among statisticians and data miners for developing statistical software and data analysis. R is also a popular choice for implementing machine learning algorithms and models, due to its rich ecosystem of libraries and frameworks for working with data like: forcats, tidyverse, tinyR. Some of the most used machine learning libraries in R include mlr3, xgboost, caret, randomForest, and gbm. These libraries provide efficient implementations of a wide range of machine learning algorithms, making it easy for R users to build and evaluate models. In addition, R's strong support for visualization and data exploration makes it a useful tool like ggplot2 and plotly, for understanding and interpreting the results of machine learning models. Despite availability of many tools (many more than Python as R has a longer history), free of cost, the language has considerable complex syntax, lack of compact installation and dependency on numerous distributed packages even for accomplishing a minor task, poses a practical hurdle in the actual usefulness of R language (Grolemund, 2015).

INTRODUCTION TO NEURAL NETWORKS

A Neural network is an artificial network of stimulated artificial neuron called Perceptron; these nodes are arranged in layers. This schematic is inspired by real life neurons of animal brains. Neural networks are basis of Natural Neural Network modelling ML framework like Keras, TensorFlow, Tensor board, Pytorch and Caffe2.

Weight of connection or link refers to the strength or weakness of link or influence between the individual neurons. These weights are adjustable and change in the process of learning. A burst of signal in the network is called Propagation.Data via input nodes produces a burst of activity in the network called Forward propagation. The adjustment involves both forward propagation of signal inputs through the network, as well as backward propagation of error terms generated by comparing actual output with expected output. In this way, the algorithm fine-tunes the weights and biases so that outputs more closely match desired results(Gurney, 2018).

Figure 1. Input nodes (yellow) feed data into the Neural network and Output nodes (white) harvest information from the network. There are few layers of processing nodes (red, blue, and green) which process the data based on predetermined algorithms. These nodes that are coloured red, blue, and green, are processing layers of nodes are called Hidden layers. The arrow represents the flow of information much like the synaptic connection of a real neuron.
Source: Gurney (2018)

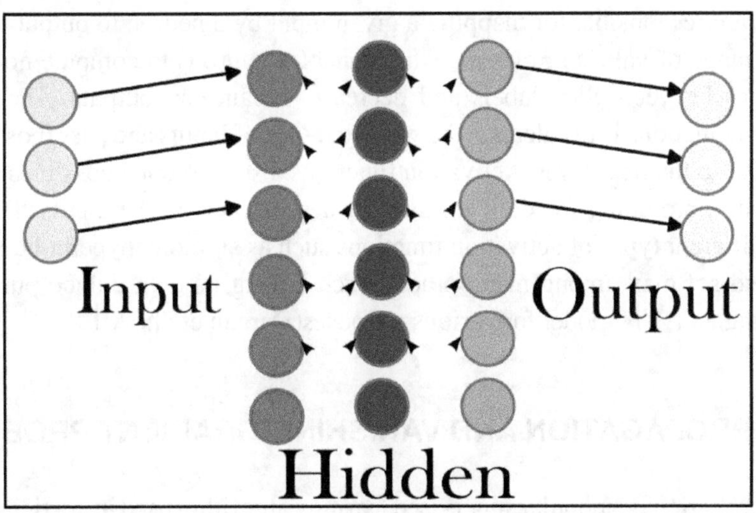

FORWARD PROPAGATION

Forward Propagation is an algorithm that involves taking the data forward through a network of nodes. In this process, each node multiplies the input with a weight and adds bias to generate a single output value. This value is then passed on to the next layer and the process is repeated until an output layer value is calculated which can be used for prediction or classification. Forward Propagation is one of the key components in training neural networks as it eliminates the need for manual adjustments of weights and bias for every node-link combination. It also provides insights about how changes in parameters can affect overall performance (Yang & Shami, 2020).

ACTIVATION FUNCTIONS IN NEURAL NETWORKS

Activation functions are an important component in artificial neural network architecture. To determine the output of a layer we take the weighted sum of all the weight of the links and add a value called the Bias. This weighted sum is referred to as Activation. This Activation plus Bias is fed into a function called Activation function. This function acts like a gate and determines whether the signal will propagate to next layer or get discarded. Back propagation is the adjustment of weights of the links, to minimize the error in learning. Here the signal propagates in reverse direction.

They are responsible for mapping a given input by a neuron to output within a specific range of values or classes, which enables neurons to compute non-linear boundaries between class labels and decrease variance in outputs. The role of activation functions is to calculate the weighted sum of inputs and pass those values as an input to the next layer. Activation functions also introduce non-linearity into the network, providing freedom from linear dependence on the inputs. There are several different types of activation functions such as sigmoid, hyperbolic tangent, ReLU and softmax among many others; each having its own unique purpose in developing an ANN model for various purposes(Dargan et al., 2019).

BACK PROPAGATION AND VANISHING GRADIENT PROBLEM

Back propagation is a popular supervised learning algorithm used in artificial neural networks. It is based on a cyclic process of optimization that adjusts the weights and biases to minimize a given cost function, thereby making the network better able to accurately respond to input signals. Back propagation has been widely used in various supervised applications such as image recognition, speech translation, and medical diagnosis(Sims, 2006).

During the process of Back propagation, the weights are updated in proportion to a partial derivative of the error function. Partial differentiation is done with respect to the current weight. An issue encountered in early stages of development of ML was that this derivative may become smaller and smaller tending to zero, effectively preventing any modification of weight, a classic ML problem known as 'vanishing gradient problem'. The exact opposite of this problem is 'exploding gradient problem', where derivative becomes so large that it starts tending to infinity (Stamile et al., 2021).

Figure 2. Scheme of forward and backward Propagation of signal in a neural network. Process of Back propagation, the weights are updated in proportion to a partial derivative of the error function which is L and derivative of L is L'.

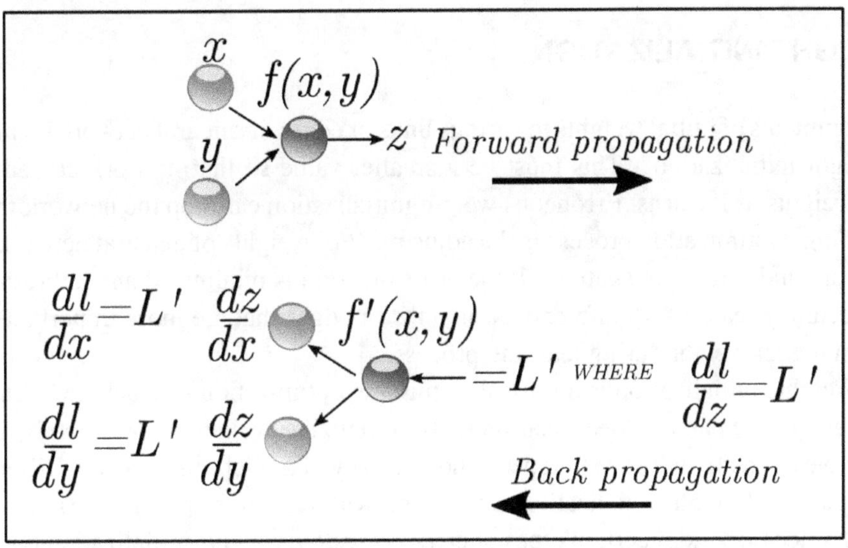

Since, there is an exponential growth computational power of CPUs, GPUs and Cloud computing, the processor can compute smaller and smaller gradients to effectively jumping over the small gradient limitation and continue learning despite the Vanishing gradient, making Back propagation feasible (Xiaojin Xhu & Godberg, 2009).

Currently Back propagation is being investigated for efficiency, in correlation with biological system like our memories, which are Neural network can learn continuously without a pause, to update weights by Back propagation. Geoffrey Hinton in his preprint paper released in Dec 2022 has proposed that twice - Forward-Forward Algorithm maybe be more efficient and much simpler for learning. This comprises of one forward pass with positivedata and another forward pass the other with the negative data (Geoffrey Hinton in preprint paper "Forward-Forward Algorithm," under review, Dec 2022).

GPUs *Graphical Processing Units* comprise of multiple cores that can process large amounts of image data in parallel, parallel execution in form of threads, typically in computer games and video editing. GPU are programmed using frameworks such as NIVIDIA CUDA (https://developer.nvidia.com/cuda-zone) and OpenCL. Many ML applications like TensorFlow can utilize GPUs for enhanced performance.

Cloud computing refers to online storage, transmissions, and processing of data over fast and large bandwidth internet connection without storing information on the local computer. A cloud storage provider like the Amazon web services (AWS) supplies the users with required space and processing on-demand.

WEIGHT INITIALIZATION

Assignments of initial weights to various links or edges of a neural network is called "weight initialization". This must be a smaller value so that network can adjust the weights as it learns. Erroneous weight initialization can stop the network from learning. Optimization process is the adjustment of weights of a neural network by forward and Back propagation till the error function is minimized and a threshold of accuracy reached. Reinforcement is links or edges that are more important are given higher weight during learning process.

Weights initialization is a critical step in the optimization of machine learning models, particularly in deep learning. By initializing weights appropriately, it is possible to greatly improve the speed and accuracy of model convergence. Different approaches of weight initialization can also deal with vanishing or exploding gradients and reduce the risk of certain types of errors. As such, weights initialization plays a pivotal role in providing stability that enable training neural networks with greater accuracy and efficiency. Careful consideration must be taken when selecting an appropriate weight initialization strategy for a given problem as the choice can drastically alter optimization performance (Xiaojin Xhu & Godberg, 2009).

LATEST CUTTING-EDGE EXTENSIONS OF NEURAL NETWORKS

The most sophisticated and cutting-edge methods and tools in the area are referred to as the state of the art in machine learning. These strategies can be quite successful at resolving complicated issues since they frequently draw on the most recent research and innovations. Deep learning, which includes training enormous neural networks on data, is one of the state-of-the-art methods used in machine learning, as is reinforcement learning, which involves teaching agents to act in a way that maximizes rewards. In several fields, including image and audio recognition, natural language processing, and gaming, these techniques have produced outstanding results.

The subsequent section elucidates the bleeding-edge, latest extensions of neural networks as of December 2022.

CONVOLUTIONAL NEURAL NETWORKS (CNNS)

Convolutional neural networks (CNNs) are a powerful type of neural network that has become the go-to for many computers vision projects. The key differentiator is their use of convolutional layers, which allow them to work with multi-dimensional data. By converting inputted visual information into numerical representations that can be processed mathematically, CNNs can recognize patterns and making predictions in complex data sets. Beyond image processing, these networks can also be used for natural language processing and self-driven cars due to their ability to quickly identify features. They are uniquely adept at working with unstructured datasets and have found success in areas such as facial recognition and detections of objects in images (Yang & Shami, 2020).

BIDIRECTIONAL LONG SHORT-TERM MEMORY (BLSTM)

Bidirectional Long Short-Term Memory (BLSTM) is a recurrent neural network architecture that can effectively remember patterns in data over long periods of time. Unlike regular LSTMs, the BLSTM model can analyse both past and future information simultaneously. This makes it especially powerful for recognizing complex relationships between objects and events in time series data, such as sentiment analysis tasks or speech recognition applications. By incorporating this form of deep learning into artificial intelligence applications, machine learning models can better generalize long-term patterns and make more accurate predictions. Furthermore, rigorous testing has been conducted to demonstrate its excellent performance on a variety of natural language processing tasks including document classification and text generation. Ultimately, BLSTM proves to be an exemplary technique for amplifying the accuracy of advanced deep learning applications.

AI STARTS TALKING DRAWING PICTURES - STABLE DIFFUSION

A deep learning text-to-image model called Stable Diffusion was released in 2022. Stable Diffusion is a type of deep generative neural network, conceptualized by the LMU Munich-based CompVis research team. The Stable Diffusion model allows you to create new images from scratch by using a text prompt that describes which elements should be included or excluded from the output. Through its diffusion-denoising mechanism, the model can redraw existing images to incorporate new elements described by a text prompt (e.g., guided image synthesis)(Charlton, 2022).

Figure 3. Hyper-realistic outputs of Stable Diffusion, produced by prompting the AI engine using simple text prompts. Incredible that these 2 images were drawn by AI without any human intervention using Machine learning. These are public domain CC0 images according to the programmers of Stable diffusion engine.

DALL-E IMAGE SYNTHESIS PROGRAM

Dall-e, developed by OpenAI is an artificial intelligence system that uses natural language processing and computer vision to generate images based on text descriptions. It can draw complex scenes with a wide variety of objects and colours, such as rooms filled with furniture or modern cityscapes. It also works well in combination with other models, such as GPT-3; when these two models are combined, they can be used to generate realistic 3-dimensional images. Dall-e's impressive capabilities make it possible for us to create extremely detailed artworks that would otherwise take much more time and effort to draw manually. It also has potential practical applications, such as in the areas of visualization and augmented reality.

The Dall-e image synthesis program is a cutting-edge technology based on deep learning and natural language processing. Developed by OpenAI, the program can generate incredibly life-like visuals from simple text descriptions. For example, it can render 3D images of buildings when provided with detailed written directions, or accurately depict animals when given their species name. This impressive system is powered by a 12-billion parameter model called GPT-3 and an AI algorithm that recognizes patterns in text input. Perhaps most importantly, Dall-e creates objects that look real - unlike traditional computer-generated images which often do not match up to reality's complexity. Thus, this tool could revolutionize image creation in various industries like architecture and entertainment media (Bok & Langr, 2019).

Figure 4. Sample, output Dall-e image synthesis program which draws images using Artificial intelligence, Dall-e engine takes much longer time than Stable diffusion to render art, but quality of images is very good and realistic than Stable diffusion.

MIDJOURNEY'S ARTISTIC ENGINE VIA DISCORD BOT

Midjourney is Artificial Intelligence Discord bot that aims to provide users with an engaging and interactive communication interface. Like Open Ai's DALL-E and the open-source Stable Diffusion, Midjourney is an independent research lab that develops a private artificial intelligence application that generates visuals from textual descriptions. On July 12, 2022, the tool entered open beta, which it is presently in. David Holz, a founding member of Leap Motion, oversees the Midjourney team. Holz stated that the business was already profitable to The Register in August 2022. Using Midjourney's Discord bot commands, users produce art.

Currently, the only ways to access Midjourney are through a Discord bot on their main Discord, via direct messages, or by inviting the bot to a different server. Users use the /imagine command, like other AI art generator applications, and enter a prompt to produce images. Next, the bot produce art based on user prompt(Solanki et al., 2021).

AI STARTS TALKING – OPENAI'S CHATGPT

ChatGPT (GPT stands for Generative Pretrained Transformer) is an innovative chatbot technology based on the open source GPT-3 language model developed by OpenAI. By utilizing this natural language processing capabilities, ChatGPT can generate responses to customer inquiries that are highly accurate and personalized to the individual's specific context. ChatGPT can also be used in moderation scenarios to detect inappropriate messages and automatically respond with corrective actions. Additionally, it uses sentiment analysis of customers' messages and reacts accordingly making conversational experiences more natural and humane. Finally, its integration into existing CRM systems allows unified management of customer queries while scaling across multiple devices with ease(Pearce, 2011).

AI starts writing dramas- Deep Brain's Dramatron

Dramatron is a tool that may be used by authors to collaborate on screenplays and theatre scripts since it generates long, coherent texts using pre-trained, existing massive language models. For uniformity across the created text, Dramatron employs hierarchical tale generation. Dramatron dynamically creates character descriptions, story points, location descriptions, and dialogue starting from a log line. These generations offer material for compilation, editing, and rewriting by human authors.

Dramatron is intended to be used by writers as a writing tool, as well as a source of creativity and exploration. We worked with 15 playwrights and screenwriters in two-hour user study sessions to co-write screenplays with Dramatron to assess its usability and capabilities.One playwright staged four scripts that were co-written with Dramatron and significantly edited and reworked using Dramatron.

AI STARTS COMPOSING MUSIC- SOUNDFUL

Ada Lovelace (N 1815 –M 1852) is frequently considered as the first computer programmer and known for her algorithm contributions to Charles Babbage's Analytical Engine, the first analogue computer. She believed that the computer could compose music, today indeed with Artificial Intelligence computers can compose music and her wishful thinking of past century has come true.

AI-powered sites like https://soundful.com, are the latest development in this field where the user can quickly and easily create context-specific audio assets to customize their own unique soundful music. Thanks to advancements in neural networks and machine learning algorithms, AI can make use of data sets to generate realistic sounds that mimic real-world instruments as well as newer unheard possibilities. Furthermore, AI has allowed for accurate music stylization and interaction with users leading to more diverse personalized music experiences that aim to replicate creativity offloading a human touch. As this technology continues to grow, it will become easier for companies and individuals alike to access sophisticated musical tools that would be otherwise unattainable or too costly.

REGENERATIVE NEURAL NETWORK

An artificial neural network that can regenerate or heal itself when some parts break down or stop working is known as a regenerative neural network. A combination of hardware and software is utilized to create this kind of network, which is frequently employed in fields like medical technology, transportation systems, and military applications where a high level of availability and reliability is necessary.

The fundamental idea behind a regenerative neural network is that it can automatically identify and isolate any defective components before using a combination of backup mechanisms and self-healing algorithms to repair or replace the defective components without impairing the network's overall performance. This permits the network to continue working at full capacity, even in the state of partial failures or faults, RNN can keep running (Razavi, 2020).

One of the key benefits of employing a regenerative neural network is that, because the network can self-repair and continue operating without human involvement, it can help to eliminate the need for expensive and time-consuming maintenance.

Because there will be fewer downtime and other disturbances, this can also help the network be more dependable and available. Regenerative neural networks can also frequently adjust and learn from their experiences, which can help them develop over time and enhance their performance (Razavi, 2020).

Regenerative neural networks are frequently employed in systems that must have a high level of dependability and availability, such as in military, medical, and transportation applications. Here are some examples of use cases for regenerative neural networks:

Medical devices -To enable the continued operation of a medical device even if some of its components malfunction or fail, such as in implanted devices or monitoring systems.

Autonomous vehicles or drones - To allow the system to function even if some sensors or other components fail in transportation systems like Tesla, such as self-driving vehicles or drones.

Military applications- To allow the system to continue operating even in the event of damage or other disruptions in military applications, such as surveillance systems or unmanned vehicles.

Industrial Control systems -To allow the system to keep running even if some sensors or other components fail in industrial control systems, such as those used in manufacturing or power generation (Razavi, 2020).

To make it possible for financial systems, such as trading algorithms or fraud detection systems, to keep working even when some of its components break down or stop working properly (Razavi, 2020).

GENERATIVE NEURAL NETWORK

An artificial neural network called a generative neural network is made to produce fresh data that is comparable to training data. In applications like picture production, audio synthesis, or natural language processing, where it is intended to produce new data with the same properties as a given dataset, this sort of network is often utilized.

A generative neural network's main principle is that after being trained on a dataset of examples, it leverages that training to produce new data that is like the dataset's examples. This is accomplished by combining neural network topologies and machine learning algorithms, such as generative adversarial networks (GANs) or variation auto encoders (VAEs).

Machine learning models known as "generative models" are used to create fresh data that is comparable to an input. They can be used to produce new data points with the same properties as the training data after being trained on a dataset. In addition to producing new images, texts, or music that are like an input, generative models can also produce new data points that can be added to existing datasets. Variational auto encoders and generative adversarial networks (GANs) are two of the most well-liked generative models (VAEs). Deep learning approaches, which entail training big neural networks on plenty of data, are frequently used to train these models.

GENERATIVE ADVERSARIAL NETWORKS (GANS)

Generative adversarial networks (GANs) models are used to create new data that is comparable to an input. A generator network generates new data, while a discriminator network assesses the generated data to try to detect whether it is true or fraudulent. Together, the generator and discriminator networks are trained, with the generator attempting to produce data that can deceive the discriminator and the discriminator attempting to reliably discern between actual and fake data. It has been demonstrated that GANs can generate a wide range of data, including images, text, and audio, and that the results they produce are quite realistic and resemble real life (Ahirwar, 2019).

VARIATIONAL AUTOENCODERS (VAES)

To produce fresh data that is like an input, generative models called variational autoencoders (VAEs) are used. They can be used to produce new data points with the same properties as the training data after being trained on a dataset. Latent space, which is a representation of the data in a lower-dimensional space, is the foundation of the generative model class known as VAEs. By sampling from the latent space and then translating the data back into the original data space, this enables VAEs to produce new data points. VAEs have been demonstrated to yield very accurate results when used to generate a range of data, including images, text, and audio. It has been demonstrated that VAEs yield highly realistic results when employed to generate a range of data, including images, text, and audio (Bok & Langr, 2019).

DEEP GENERATIVE NEURAL NETWORKS

Deep generative neural networks are a powerful type of artificial intelligence that is capable of learning and producing high-dimensional data. Deep generative neural networks are composed of multiple layers, each with its own set of weights and connections. This allows them to learn the underlying structure of datasets, by learning how to map inputs and outputs into rich dimensional probability distributions. These networks can be used for a variety of applications, such as image recognition, voice generation, natural language processing, and even drug design. As deeper levels of analysis become more commonplace in AI research, these neural network architectures continue to improve accuracy, precision, and speed while simultaneously lowering hardware costs.

Utilizing generative neural networks has several benefits, one of which is its ability to create new data that is like a given dataset. This new data can then be used in a variety of applications. A generative neural network could be used, for instance, to produce new images that are like a training dataset of images or new audio samples that are comparable to a training dataset of audio samples. Applications like those for creating art, music, or languages can benefit from this. Furthermore, generative neural networks frequently learn and adapt over time, which can enhance their functionality and performance. In applications where it is desirable to produce new data with the same properties as a given dataset, generative neural networks are frequently utilized. Examples of specialized applications for generative neural networks include:

Produce new images in image generation that are comparable to a training dataset of images The creation of new artistic images based on a dataset of existing artworks or the generation of photographs from written descriptions could both be uses for this. Produce new audio samples for audio generation that are comparable to a training dataset of audio samples. This might be applied to making music or new sounds for video games or movies, among other things. Produce fresh text in natural language processing that is like a training text set. This might be applied to creating chatbots or creating new text passages based on a specific subject. Create novel chemical compounds that are like existing medications throughout the drug development process. This could be used to find prospective new medications with the needed properties. Produce new stock market data in finance that is comparable to a training dataset of stock market data This might be employed to evaluate trading algorithms or create fresh investment plans(Shrestha & Mahmood, 2019).

NATURAL LANGUAGE PROCESSING (NLP)

The branch of artificial intelligence known as "Natural Language Processing" (NLP) focuses on teaching computers how to comprehend, analyse, and produce human language. It involves analysing, comprehending, and producing natural language text or voice using computer algorithms and software. Natural language processing (NLP) is a form of artificial intelligence that enables computers to interpret, understand and generate human language. NLP incorporates techniques such as machine learning, deep learning, natural language understanding and natural language generation to analyse unstructured data like text and speech to produce actionable output that can be used for automated decision-making. By employing linguistic analysis methods such as text classification, sentiment analysis or automatic summarization, NLP helps to streamline manual processes, unlock insights from large volumes of data and enable the development of more efficient and precise solutions for problems across industries. Although natural language processing (NLP) is a broad field with many distinct techniques and approaches, some typical jobs that are usually connected with NLP include:

Text classification is the process of automatically classifying text into predetermined groups or classes using algorithms. A text classifier could be used, for instance, to categorize email communications as spam or not. Sentiment analysis is the process of automatically identifying the emotional content of text using algorithms. For instance, a sentiment analysis algorithm could be used to assess the positivity or negativity of a movie review.

Part-of-speech tagging is the process of automatically identifying the components of speech (such as nouns, verbs, and adjectives) in a given sentence using algorithms. Part-of-speech tagging is the process of automatically identifying the components of speech (such as nouns, verbs, and adjectives) in a given sentence using algorithms. Named entity recognition is the process of employing algorithms to recognize and categorize named entities (such as individuals, groups, and locations) in text. Utilizing algorithms to automatically translate text from one language to another is known as machine translation. In general, natural language processing is utilized in a wide range of applications, from voice assistants and chatbots to machine translation and text analysis, and it is essential for enabling computers to comprehend and produce human language (Goldberg, 2017).

LARGE LANGUAGE MODEL (LLM)

An artificial intelligence system known as a LargeLanguage model has been trained using a sizable amount of text data, including books, articles, and other written materials. This enables the model to produce text that is human-like and understand the structures and patterns of natural language. For tasks like language translation, text summarization, and producing conversational responses, large language models are frequently used. Large language models such as GPT-3 (Generative Pertained Transformer 3) and BERT are two well-known examples (Bidirectional Encoder Representations from Transformers). These models, which are referred to as "big," may generate extremely accurate and natural-sounding text since they have been trained on a vast amount of data (Goldberg, 2017).

CONCLUSION

This chapter serves as a primer to young student of data science or computer science, introduces the concept of machine learning, artificial intelligence, neural networks, and their functioning. The second half highlights the latest developments in the field of Machine learning, recent and interesting advances in the application of machine learning, like ChatGPT and Stable diffusion. The chapter focuses on simplicity, and lucid language which should be easy to grasp for anyone who is interested in data sciences.

Finally, machine learning is an effective technique for forecasting and spotting trends in data. It has been effectively used to solve a variety of issues, including natural language processing, recommendation systems, and picture and speech recognition. The future of machine learning appears bright and promising, even though there are still numerous obstacles to be solved, such as improving the interpretability of models and resolving ethical issues. It has the potential to revolutionise a variety of sectors and enhance our daily lives with further study and development.

REFERENCES

Ahirwar, K. (2019). *Generative Adversarial Networks Projects*, *18*, 45. Packt Publishing Ltd.

Antoniou, C., Dimitriou, L., & Pereira, F. B. (2019). *Mobility patterns, big data, and transport analytics: tools and applications for modeling*. Elsevier.

Bari, A., Chaouchi, M., & Jung, T. (2017). *Predictive analytics for dummies*. John Wiley,and Sons.

Bok, V., & Langr, J. (2019). *GANs in Action*, *34*, 40. Simon, and Schuster.

Bornholdt, S., & Schuster, H. G. (2006). *Handbook of Graphs and Networks*. John Wiley & Sons.

Charlton, M. (2022). *All Stable Diffusion Artwork*, *22*, 34. Independently Published.

Clickard, C. (2019). *The internet of things*. Norwood House Press.

Dinov, I. D. (2018). *Springerlink, Online Service. Data Science and Predictive Analytics: Biomedical and Health Applications using R*. Springer International Publishing. doi:10.1007/978-3-319-72347-1

Elomaa, T. & Helsinki, E. (2002). *Machine learning: proceedings*. Springer, C.

Fontichiaro, K. (2018). *Big Data, 10*, 20. Cherry Lake Publishing.

Goldberg, Y. (2017). *Neural network methods in natural language processing*. Morgan & Claypool Publishers. doi:10.1007/978-3-031-02165-7

Goodfellow, I., Bengio, Y., & Courville, A. (2016). Deep learning, (p. 44-90). The MIT Press. https://www.deeplearningbook.org/

Grolemund, G. (2015). *Data Science withR*. O'Reilly& Associates Inc.

Gurney, K. (2018). *An Introduction to Neural Networks., 152*, 180. CRC Press.

H., D, & Pazzani, M. J. (1991). *Concept formation: knowledge and experience in unsupervised learning,* (pp. 4-12). Morgan Kaufmann.

Hinton, G. (n.d.). Forward-Forward Algorithm: Some Preliminary Investigations.

Jobin, A., Ienca, M., & Vayena, E. (2019). The global landscape of AI ethics guidelines. *Nature Machine Intelligence, 1*(9), 389–399. doi:10.103842256-019-0088-2

Keller, J. M., Liu, D., & Fogel, D. B. (2016). *Fundamentals of Computational Intelligence, 34*, 78. John Wiley & Sons.

Liu, S. (2020). [John Wiley & Sons.]. *Engineering Autonomous Vehicles and Robots, 19*, 21.

Merris, R. (2001). *Graph theory*. John Wiley.

Montenegro, A. A., Carvalho, P. C. P., Velho, L., & Gattass, M. *Space carving with a hand-held camera.* Proceedings of the XVII Brazilian Symposium on Computer Graphics and Image Processing, (pp. 396~403). John Wiley & Sons. 10.1109/SIBGRA.2004.1352986

Pearce, Q. L. (2011). *Artificial Intelligence, 22*, 29. Lucent Books.

Rajamani, S. K., & Iyer, R. S. (2022). Development of an Android Mobile Phone Application for Finding ClosedLoop, Analytical Solutions to Dense Linear, Algebraic Equations for The Purpose of Mathematical Modelling in Healthcare and Neuroscience Research. *NeuroQuantology: An Interdisciplinary Journal of Neuroscience and Quantum Physics, 20*, 4959–4973. doi:10.6084/m9.figshare.c.6156024.v1

Razavi, M. (2020). Neural *Regenerative Nanomedicine, 4*, 9. Academic Press.

Lior Rokach, Maimon, Oded Z. (2015). *Data mining with decision trees: theory and applications*,,16~41, World Scientific, Cop.

Saxena, A, Rani, U., Agarwal, I., & Jangade. R. (2021). Introduction to Big Data, Chapter 1 Introduction to Big Data. In Ankur Saxena, Nicolas Brault, & Shazia Rashid (Eds.), Big Data and Artificial Intelligence for Healthcare Applications, (p. 3-21). Taylor and Francis..

Schwartz, M. (2016). *Internet of Things with ESP8266: build amazing Internet of Things projects using the ESP8266 Wi-Fi chip*. Packt Publishing.

Sejnowski, T. J. (2018). *The deep learning revolution*. The MIT Press. doi:10.7551/mitpress/11474.001.0001

Shrestha, A., & Mahmood, A. (2019). *Review of Deep Learning Algorithms and Architectures*. IEEE. doi:10.1109/ACCESS.2019.2912200

Sims, E. (2006). *Deep learning*. SpecialistSchools,and Academies Trust.

Solanki, A., Nayyar, A., & Naved, M. (2021). *Generative Adversarial Networks for Image-to-Image Translation, 10*, 23. Academic Press.

Stamile, C., Marzullo, A., & Deusebio, E. (2021). Graph *Machine Learning, 11*, 26. Packt Publishing Ltd.

Sullivan, W. (2017). *Machine learning for beginner's guide algorithms: decision tree & random forest introduction*. Healthy Pragmatic Solutions.

Watters, A. (2021). *Teaching Machines, 19*, 24 MIT Press..

Xhu, X., & Godberg, A. B. (2009). *Introduction to semi-supervised learning*. Morgan & Claypool.

Yang, L., & Shami, A. (2020). On hyperparameter optimization of machine learning algorithms: Theory and practice. *Neurocomputing, 415*, 295–316. doi:10.1016/j.neucom.2020.07.061

Zhang, Y., & Rajapakse, J. C. (2009). *Machine Learning in Bioinformatics, 74*, 84.

Chapter 10
An Approach for Mobile Application Design Using Figma

Akshat Jain
CHRIST University (Deemed), India

ABSTRACT

With the increased mobile usage throughout the world, there is an enormous demand for mobile applications that provide not only good functionality but also good experience to the user, which is why it is important to have a good user interface for a mobile application. In this chapter, various components of an app are designed and illustrated in Figma software to make the work easier for a developer to create them. It also demonstrates the usage of Fig-ma. Meanwhile it shows the design thinking behind adding the correct colour schemes, right font, using proper white space, proper placement of buttons, and other design principles.

INTRODUCTION

This research consists of the tools and features to create a good user experience as well as a good user interface design for mobile applications and also for other similar design such as website design, graphic design, product design, game design. This research includes the usage of Figma software's features, plugins, community, components, and prototype. Design is the pre-requisite of mobile application development as you need to be sure what interfaces your users are going to see in order to use your app. An App design should be ready before you start off with the development. Colour scheme, pages, features, images, icons, illustrations, buttons,

DOI: 10.4018/978-1-6684-8582-8.ch010

text are some of the things you should design first in Figma or any other alternate software (Adobe XD). You might ask "Why should I waste time on designing when I can directly jump onto the development and finish the work faster?". Well, the answer to that question will take take more time if you directly jump into the coding of the app before thinking how your app should look and how the flow of your app should go about (Yu et al., 2017). Suppose you code a button and its functionality only to realize that it isn't required or is in the wrong section, you would have to re-code in order to fix that problem, but when you are designing an app, it becomes relatively easier to make changes there was no coding is involved and the interface can be fixed in minutes.

It reduces the load off the mind while you are creating logic and functionalities of the app. Once you complete the designing process, you just have to make your app interface workable. All the assets and files would be available to you in a single design file. You also get code snippets from the design software which makes your task smoother as you don't have to worry about coding the special effects in your mobile app (Zhang et al., 2021).

WHAT IS FIGMA?

Figma is a modern interface design tool which is free, online. It is used to create, collaborate, and prototype your creative designs and UI interfaces while working on a mobile application. It is an easy to share software through which your teammates or managers could see the design and comment the changes or critiques required to fix or improve the design of the app. It works on cloud which means you don't have to worry about your files getting lost or deleted due to your system failure as it is saved on cloud. It offers Autosave feature through which you don't have to worry about saving the files manually. Best part is that you don't need to install it on your system, it works on browsers directly. Just open any of the browser (e.g., Chrome, Firefox, Brave) and go to Figma's website and run the software directly. It is lightweight and loads the content files fast and efficiently (Ye et al., 2017).

LET'S GET STARTED WITH FIGMA

When you Open Figma in your browser or app, it will ask you to create an account or log in if you have an existing account. You can Sign up through your Google account and get started quickly. Once you log in, you'll get your dashboard where you need to create a new design for the mobile app screen. Once you log in, you'll get your dashboard where you need to create a new design for the mobile app screen. Dashboard is depicted in figure 1.

Figure 1. Figma Dashboard after logging in

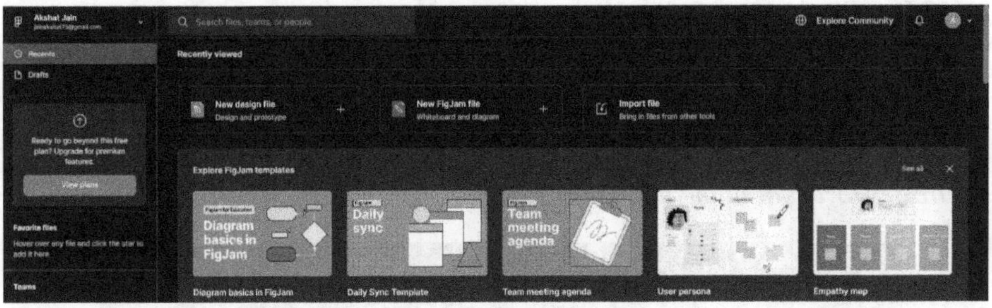

You'll get a blank screen to perform your designing tasks. In the centre, you will see the file name as **Untitled**, double-click on it and change the name of the file according to your requirements. Figure 2 shows new design file created in Figma.

Figure 2. Figma New Design file named as Untitled

Let's discuss the features of Figma before getting into designing a mobile screen.

TOOLSIN FIGMA

Let's discuss the important tools that we are going to require in order to design our mobile interface.

1. Move Tools
   a) Move tool – Move tool is used to select an element and position it anywhere on the screen. Position of Move tool is shown in figure 3.

Figure 3. Move tool in the toolbar

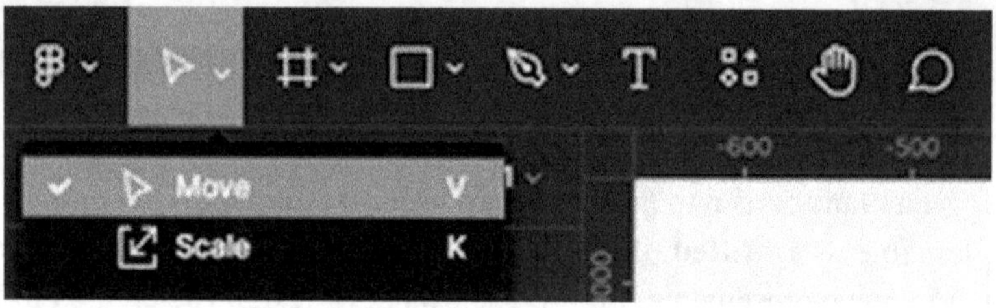

   b) Scale – Scale tool is used to increase or decrease the size of elements in a design. Refer figure 4 for location of scale tool.

Figure 4. Scale tool in toolbar

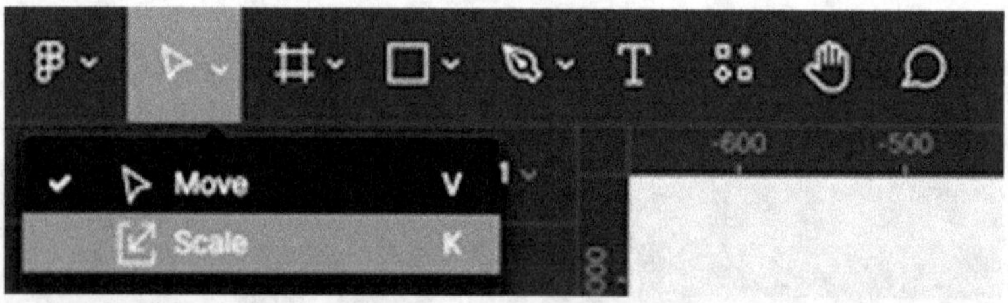

2. Region Tools
   a) Frame - Frame creates a canvas or a screen for you to design into it. Frame is one of the most essential elements that you're going to use in Figma. Every design that you make will be created in a frame (Ma et al., 2017). Although it is not necessary to have every element in a frame, but it is surely recommended. Refer figure 5 for Frame tool's location.

Figure 5. Frame tool in region toolbar

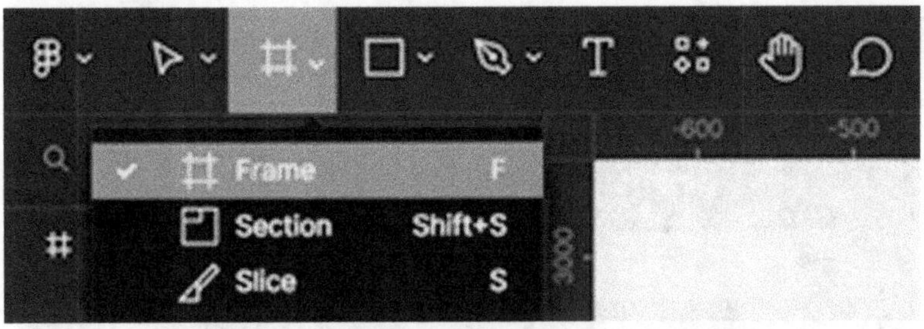

3. Shape Tools
   a) Rectangle – Creates a rectangle on the screen which you can use as a button, section, gradient and much more. Location of rectangle is shown in figure 6.

Figure 6. Rectangle tool in shape toolbar

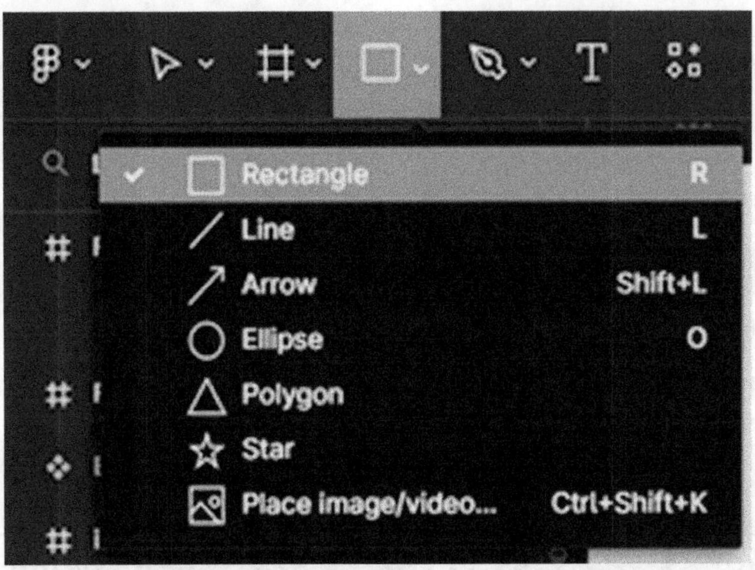

   b) Line – Creates a single line on the screen which you can use for creating icons, divisions. Figure 7 represents the location of line tool.

Figure 7. Line tool in shape toolbar

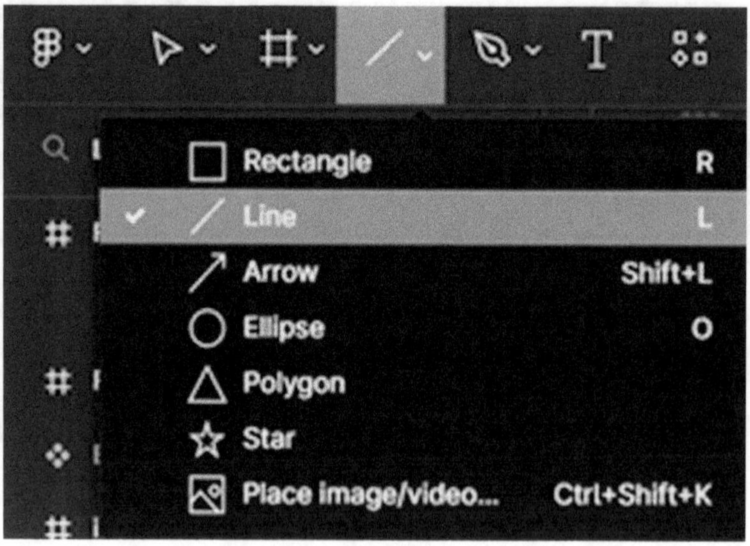

c) Arrow – Arrow is a line with a single sided pointer which can be used as back button, arrow key, or simply indicating the flow of app. Figure 8 shows the location of arrow tool.

Figure 8. Arrow tool in shape toolbar

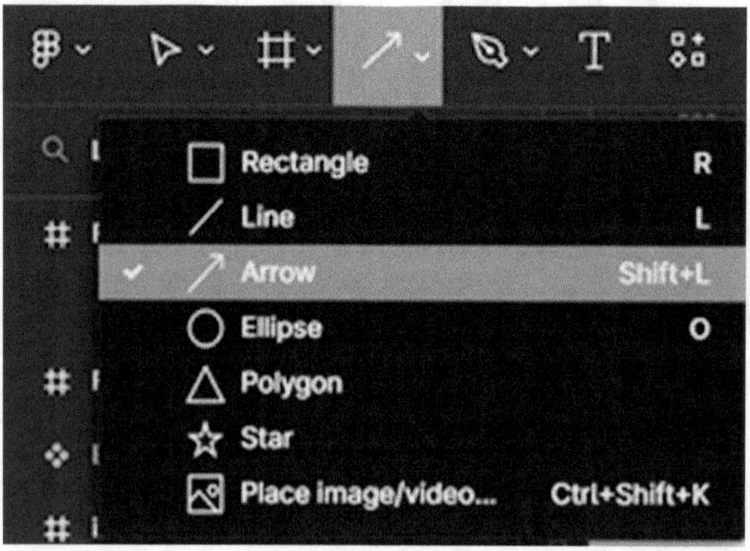

d) Ellipse – Creates a Circle or oval according to your requirements. It can be used for icon backgrounds, circular buttons, graphics. Location of ellipse tool is given in figure 9.

Figure 9. Ellipse tool in shape toolbar

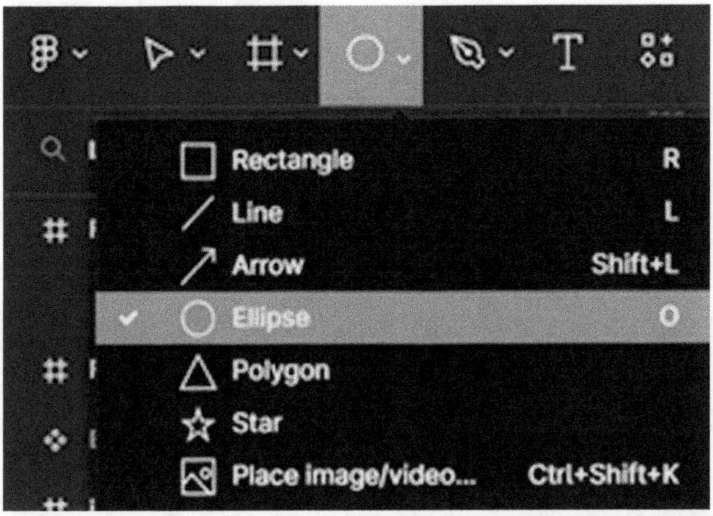

e) Polygon – Create shapes with n number of side like triangle, quadrilateral, pentagon and much more. Figure 10 shows the location of polygon tool.

Figure 10. Polygon tool in shape toolbar

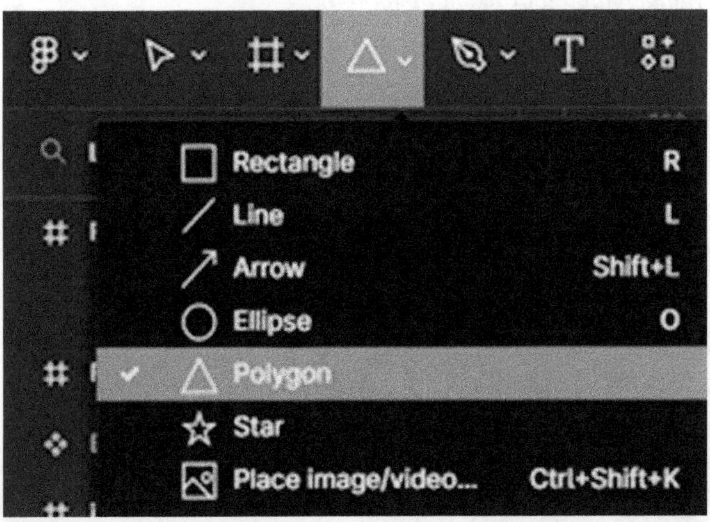

f) Star – Creates a star shaped element which can have n number of sides according to your requirements. It can be used as stars for customer review icons. The location of Star tool is shown in figure 11.

Figure 11. Star tool in shape toolbar

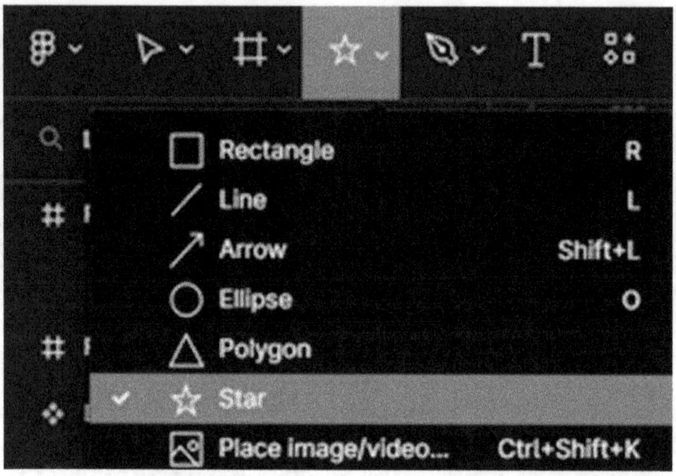

g) Place image/video – Add images, illustrations, GIFs, videos and much more into your screen design. It can embed JPG, PNG, SVG, GIF, WEBP and other image formats. Figure 12 shows the location of Place image/ video toolbar.

Figure 12. Image/video tool in shape toolbar

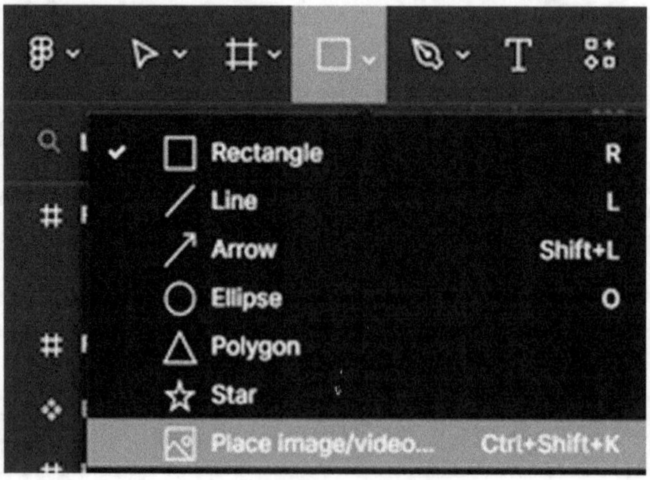

4. Drawing Tools
   a) Pen – It is used to create any complex shape or trace a particular drawing to create it into a vector image. It can be used to create and edit icons and illustrations. Figure 13 shows the location of the pen tool in the toolbar.

Figure 13. Pen tool in drawing toolbar

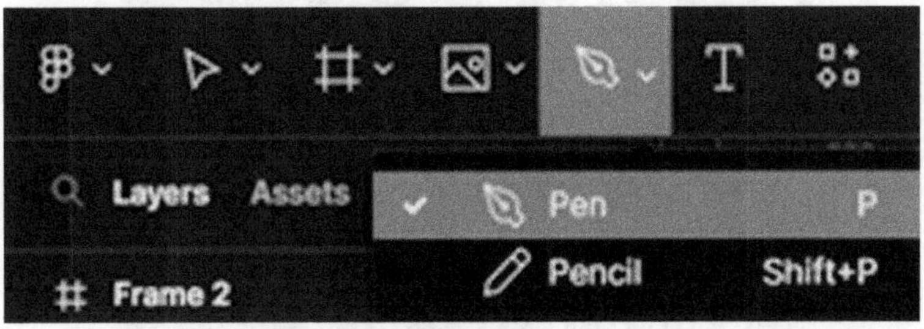

   b) Pencil – It is used to draw around in the screen to create some hand drawn lines and elements. It can be used to create rough sketches of the ideas that you get for the interface design. The location of pencil tool is shown in figure 14.

Figure 14. Pencil tool in drawing toolbar

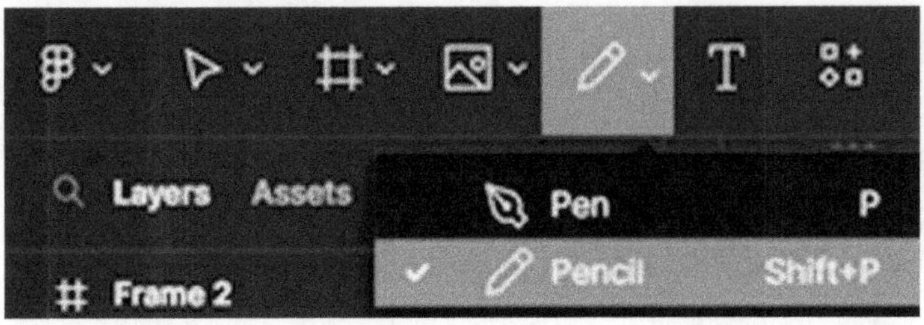

5. **Text –** Text tool is used to create a text box anywhere in the screen. You can write text relating to the app on various screens, buttons and cards. The location of text tool is given in figure 15.

Figure 15. Text tool in the toolbar

6. **Resources –** It provides you with various menus such as Components, Plugins, Widgets so that you can add complexity and more features into your design. Figure 16 displays the location of resources tool.

Figure 16. Resources in the toolbar

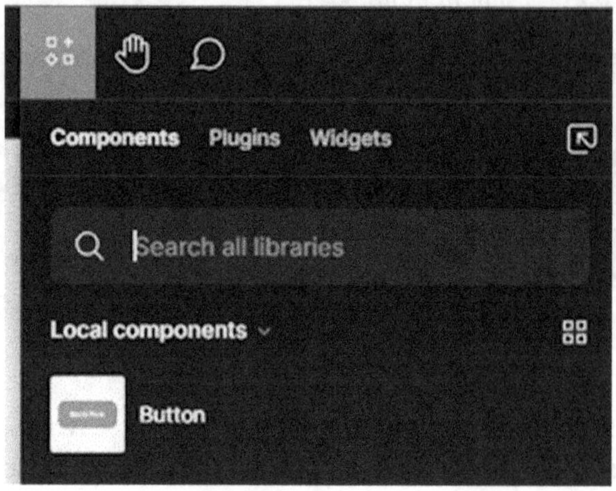

7. **Hand Tool –** It is used move around the screen. If you are using a mouse, hold middle button to move around the screen. The location of hand tool is given in figure 17.

Figure 17. Hand tool in the toolbar

8. **Add Comment –** It is used to add comment anywhere in the design file so that you can work on it later or maybe convey a message or correction in the design through the comment option. Figure 18 depicts the location of comment tool.

Figure 18. Comment tool in the toolbar

DESIGNING THE INTERFACE

Let's discuss the steps required to create a basic mobile interface design for your app. We will use one sample design in order to illustrate how to go through the process of designing an interface. Figure 19 displays the interface design for a hotel finding and booking app.

Figure 19. Mobile Interface of a hotel booking app

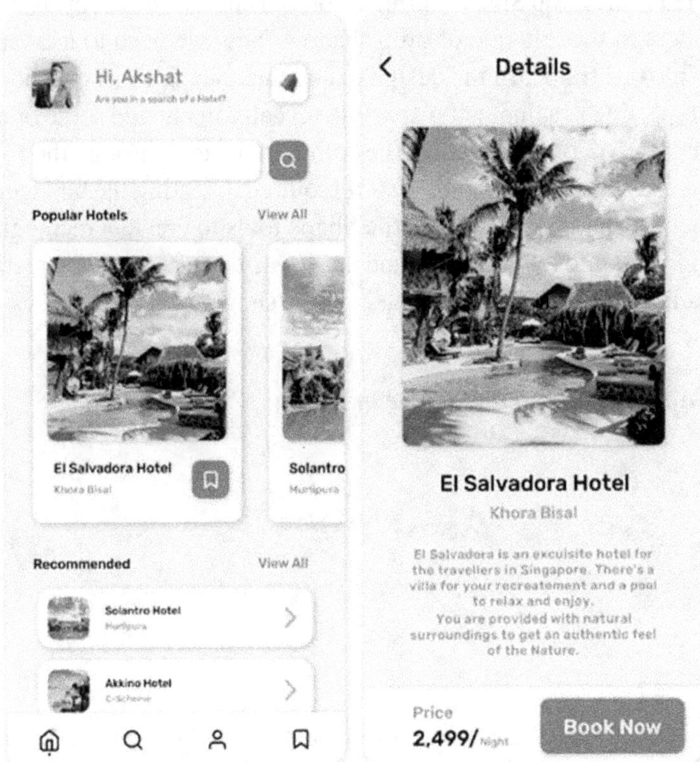

Step 1: Select the frame tool and create a frame. Once you click on the tool, you'll get various options in your right panel known as the **design panel** to select what type of design frame you want to insert. It contains a variety of frames including various smartphone models such as iPhone 14, iPhone 14 Pro, Android Large, Android small and a lot more. Select Android small/large or any frame you want. After placing the frame in the screen, you can change the dimensions of the frame according to your requirements (Lin et al., 2020).

Step 2: Add elements into the frame by selecting the image option from the Shape Tools and add a profile image on the screen. Once added, you can resize the image according to your requirement. You can use the crop tool in the top centre menu to crop your image. After adding the image, we are left with beautifying the image. You can see that in the design panel there are various options to perform on the image. Look for the corner radius icon through which we can make the corners of the picture rounded which gives a premium and aesthetic look. Next thing to add is the Drop shadow (Zhou et al., 2019). To find drop shadow, just scroll down and find Effects Menu wherein you can click the Plus icon to add an effect. The default effect selected is the Drop Shadow effect. You can alter the effect by using the effect settings present in the left side of drop shadow. Now we need to use the Text tool to add text into our frame. In the design panel, there are options to modify the font style, size, type, line spacing, letter spacing, text alignment and some other features (Zhang et al., 2013). You can change the colour of the text through the Fill property and can also add transparency to the text if required. Creating the bell icon is almost similar to adding the image, just use the Shape tools to create a rectangle and then add an icon image into the rectangle and apply various effects to the rectangle such as corner radius and drop shadow. The result is shown in figure 20.

Figure 20. Addition of Text and Image in Figma

Step 3: Create a search bar by Adding a rectangle and applying effects on it. And then add text in the rectangle and be sure to make it a prompt text or a placeholder which specifies what the search bar does. You should keep the text colour light e.g., grey. Then add the Search icon by using Plugins from Figma. To use the plugins, just right click anywhere on the screen and look for plugins, then select Find more Plugins and then search for icon plugins to add into your interface (Tang et al., 2015).

Some of the best plugins for icons are Iconify and Feather icons. Plugins are not only used for icons but are used for a variety of elements such as getting stock photos, removing background, generating dummy text and a lot of never-ending features. After the completion, the section should look similar to figure 21.

Figure 21. Adding search bar, placeholder and icons using plugins

Step 4: If you have got the idea of using these elements then other parts of your design are moreover the same things with various different dimensions such as creating cards, lists. The created of cards are displayed in figure 22.

Figure 22. Card section containing rectangle, text and icons

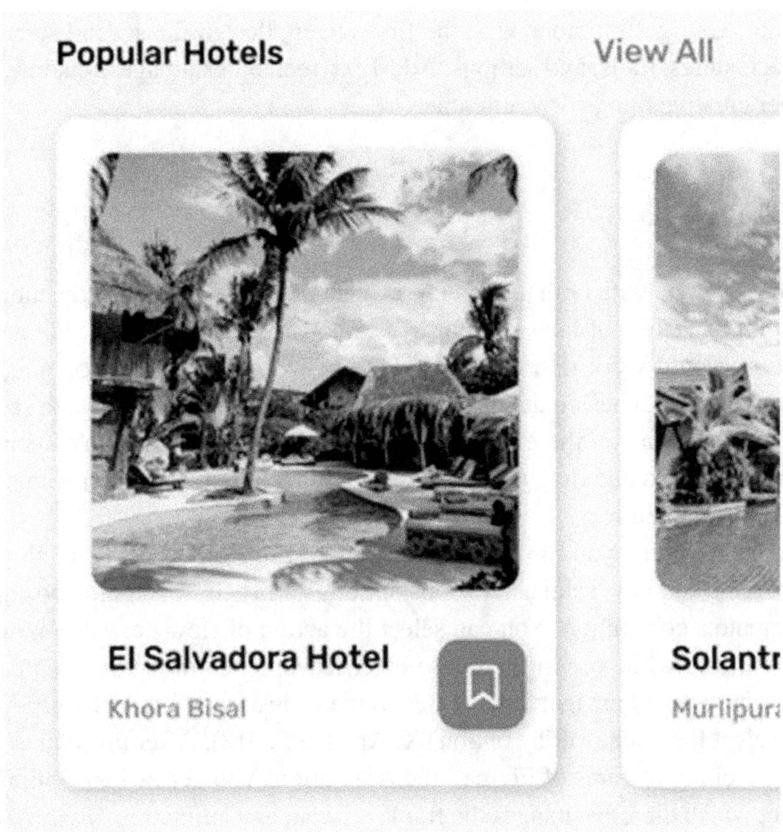

Step 5: Create a bottom navigation bar by using the rectangle tool and add various icons by using the plugins. Be sure to use same style icons and have the same size for all (S. Y. Shi et al., 2018). You can alter the sizes of items from the design panel. You can just place all icons in a line and then use the Tidy up option for arranging them equidistant from each other. The result should look like figure 23 shown below.

Figure 23. Navigation bar containing icons

Step 6: After getting done with the first screen, design the second screen with the same features, tools, and settings. Add Text, rectangles, images, shadow effects and other creative things of your choice.

PROTOTYPING

Once you are done with both the screen, we can move to prototyping the interfaces. Prototyping is a flow of how the app will look if it was built in real life and how the buttons would work (Ho et al., 2017). To get started with prototyping select the prototype option beside the design panel and then you will be able to see some circles or Plus icons on the elements when you hover over them. You can select one of those circles and drag to create a line connecting it to another frame. Refer figure 24 shown below.

Here in the figure, you can see how clicking on the Hotel card is directing me to another frame and how I can use various options for my flow. A dialog box appears as you create a connection. You can select the action of flow i.e., on tap, on drag, while hovering, while pressing, mouse enter/leave, touch down/up. You can also select if you want to navigate to another frame or just open it up as a pop-up. For pop-up select the open overlay option (X. An et al., 2018). After this comes one of the most exciting features of Figma - the Animation. You get various options such as instant which directly changes the frame without any animation, dissolve option shows a slight transition effect in the change of frames. It dissolves the first frame and opens up another one with a slight delay (Yuan et al., 2018). Smart Animate is one of the best effects as it changes the frames in a way that it makes the user feel like the elements are animating and changing position and size just like a video. This feature automatically detects the changes in elements and generates animation. In Smart Animate you get various options such as Ease in, Ease out, Ease in and out. You can also select the time for the animation, by default it is 300ms i.e., 3 seconds.

Figure 24. Prototyping using Figma

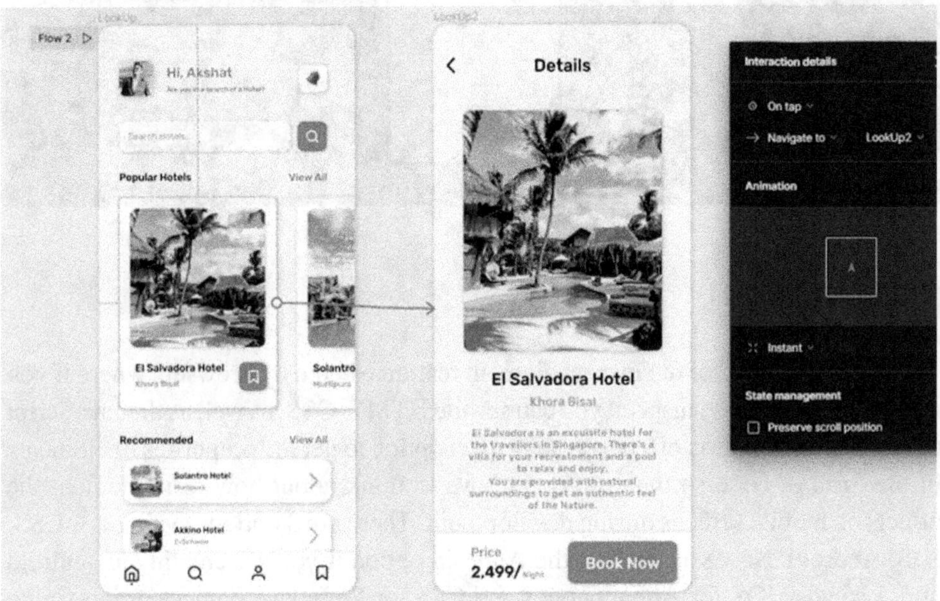

PRESENTING THE PROTOTYPE

Once you are done with prototyping a design, just look for a present buttonon the top right corner to present and see how it would be visible in an actual mobile screen. Once you click on the present button, you will get a new tab with the prototype presentation of your design (Wang et al., 2021). If the size is too big, be sure to go to Options menu in top right and select Fit to screen. Your presentation screen should look like figure 25.

You can click on buttons or cards to see if your prototype connections are working as you want them to be. You can also alter any design or prototype and it will reflect here automatically.

Figure 25. Presentation of Prototype

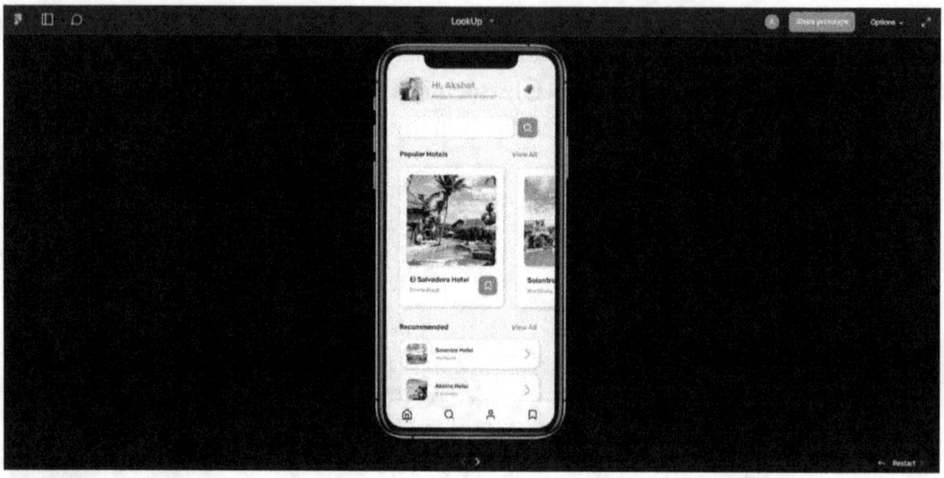

INSPECT MENU

You might be aware of the inspect element feature offered by browsers where if you right click and choose inspect, you can see the HTML, CSS, as well as the JavaScript code. Similarly, Figma offers you an inspect option to get the properties of elements in the form of code so that you don't have to think about how to implement the effects on mobile screens during development. There are 3 options available – CSS, IOS, Android. Now you can use the Android option if you are creating an android app. Likewise IOS for Apple phones and CSS for Web development. Refer figure 26 for various options available (Yu et al., 2022).

ADVANCED FEATURES

Now that we are done with the basic features of Figma, we can move ahead to learn the advanced features such as constraints, auto-layout, layout grids, components.

Let's get started with each one of them

1) Constraints – Constraints can be assumed as a string tied to an element. It can be from left, right, top, bottom. It is used to set the position of an element on the frame. When you click on any element in a frame, you can see a constraints menu in the design panel. As default, constraints are left and top which means if you increase the size of the frame from right and bottom, there would be no change in position of the element. If you change constraint settings to center

and center then it will remain in center no matter how much size of frame you increase. The visual is presented in figure 27.

2) Similarly, you can set constraints to right and bottom if you need any element to be fixed at the right and bottom side of the frame.

Figure 26. Using inspect menu to get CSS, IOS and Android code

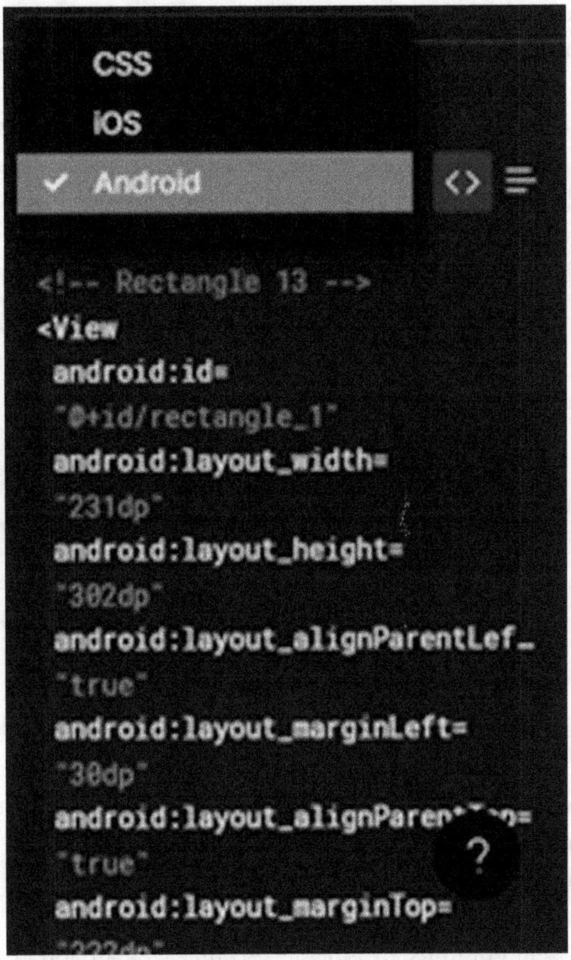

Figure 27. Usage of Constraints on Text

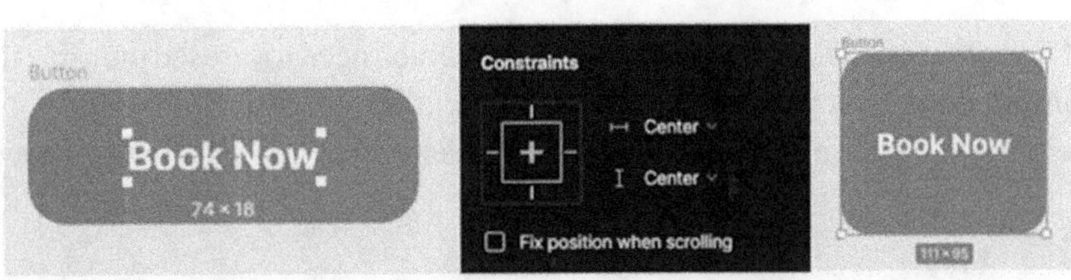

3) Auto-layout – Let's learn about auto-layout which makes our work easy in terms of aligning multiple items in a line, getting a fixed padding for elements. It can be used to create various buttons from a same rectangle and text without worrying about resizing the button or increasing/decreasing the width of button. You need to select both the objects by using your mouse click and holding the shift button to select multiple items at once. Once you select both items, select Auto-layout in the design panel. Click on the plus icon. Now you need to select whether you want vertical or horizontal layout. Then you have the option to increase/decrease the spacing between items. You can also set padding of items both vertically and horizontally (Lee et al., 2019). You can set the item in Center, left top, right top, top, left, right, bottom, bottom left, bottom right. You can also go to advanced options by clicking 3 dots and then you can select the spacing mode to Packed or Spaced between. The Spaced between option can be used to set the icons in navigation bar easily without worrying about getting all the icons in the right position. Figure 28 represents Auto-layout usage.

Figure 28. Center Alignment of Text using Auto-layout

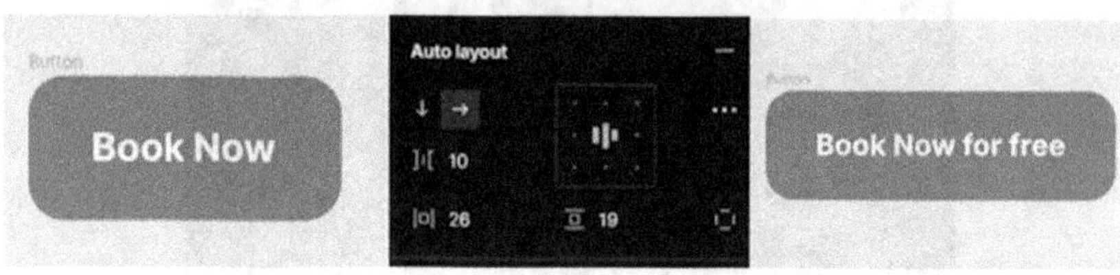

Here the button automatically adjusted its width according to the text.

4) Layout Grids – These are various grid lines which help us build a stable and professional user interface. It is available in 3 types – grids, columns, rows. Figure 29 displays layout grids.

Here you can see, columns are selected so that we can add consistency to the design and add elements according to this layout. You can change the count. Color. Type, width, margin and gutter of the grids. The gutter is the space between the grids.

5) Components – These are various elements which need to be used multiple times in the design. Suppose you want to create a hover effect for a button. Now, either you'll have to create different frames every time in order to show the

hover animation. Instead of that, just create a component variant of that button and use it everywhere you want and you can also edit the button according to your requirement and it would still provide you the same features. Figure 30 shows a normal button with text in center.

Figure 29. Column grid for better alignment of elements

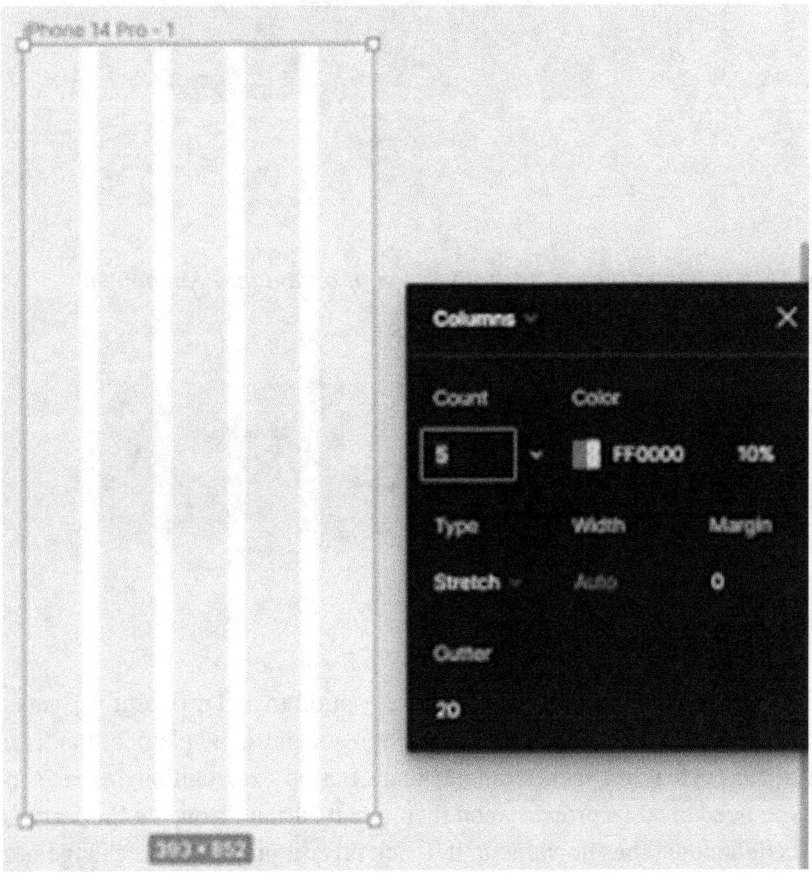

You need to select the element that you want to create a Component set of. And then in the top center section, you will get an icon as "Create Component", Once that is done, you will see your element turn into Purple which indicates that now it is a component. The component icon and indication of component is shown in figure 31.

Figure 30. A button with text

Figure 31. Component icon and Conversion of button into component

After this, we need to look for properties option to add different variants of the component. Variants are just different versions of the same element that can have slightly different features such as color change, button size, rotation. In the Properties menu, you need to add a property and then select variant. Now the Property1 of the button is default and has no value in it. Click on edit property and change property name and values. In this example, let's set the name as hover and value as False. Property option is depicted in figure 32.

Now you get a plus icon on your button to add variants to create various versions such as the hover state of the button and other animation styles. Figure 33 shows the Plus icon for creating a new variant.

Add a variant and set the hover property to True. Now change the background color of the button to dark blue. And then go to Prototype menu and create a connection between both the variants and select the "while hovering" option. Refer 2 variants of a button in figure 34.

Figure 32. Editing variant property of a component

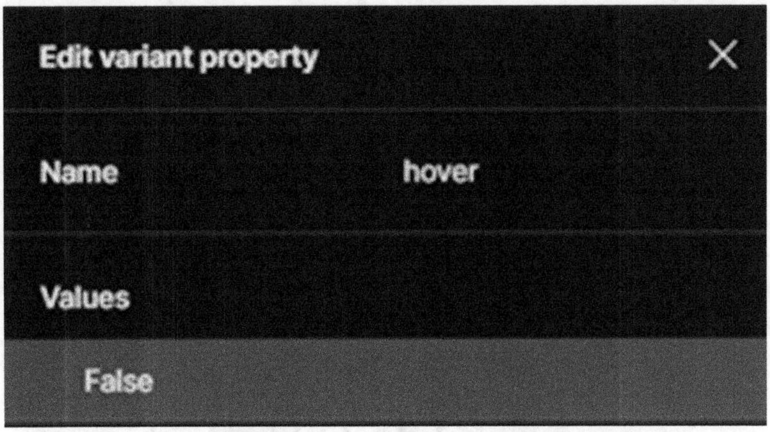

Figure 33. Plus icon to add new variant

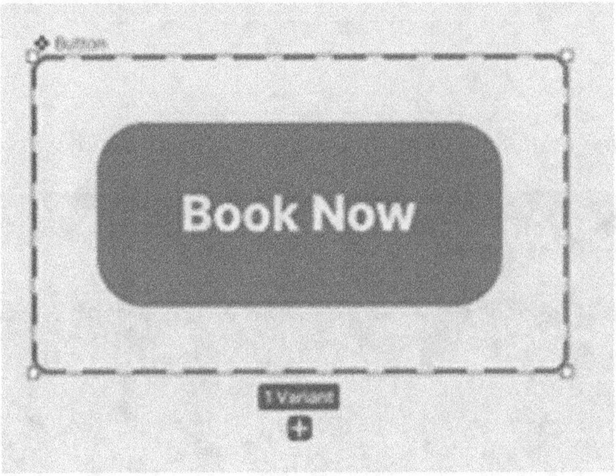

Once the component variant is created, you can use it from the Assets menu on the left along with Pages. The button can be used on any frame by just dragging and dropping it into the frame. It creates an instance of a button and it can be changed according to your requirements. It won't affect the actual component but will imbibe the all the effects that you put in the component set.

If you want to use the button for some other purpose and need different effects on it. You can select the instance and detach it by selecting the Detach Instance button available from the three dots in the Design Panel. Visual of detaching an instance is illustrated in figure 35.

Figure 34. Different variants of a button for normal and hover state

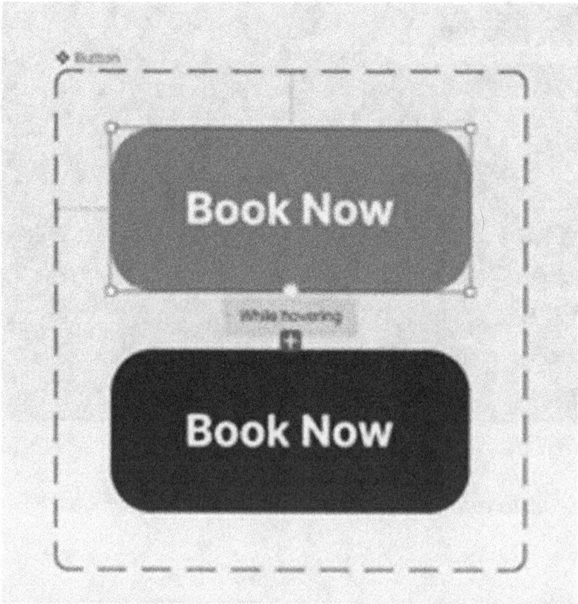

Figure 35. Detach Instance option for removing connection with component

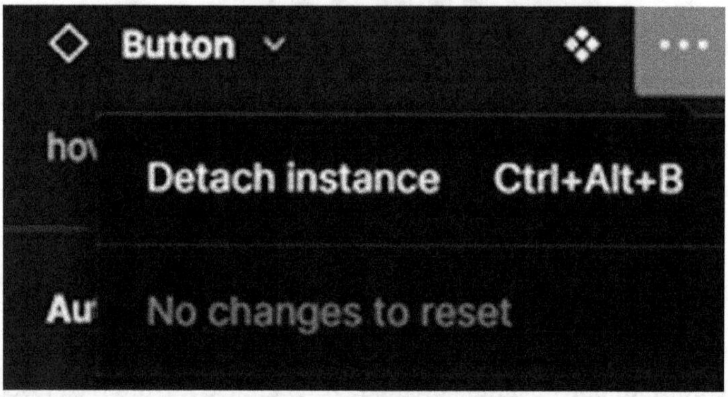

BETTER USER EXPERIENCE ON MOBILE

The usage of your app will increase only if the user experiences a smooth flow of the app and easy to understand illustrations, text, buttons, and other components. So here we will discuss some of the design tips to make your app more interactable.

1. Buttons: Buttons are the key elements of your app that provide the Call to action for user. Their size, placement, border, state, hierarchy are all an important aspect of mobile app design.
    a. Standard Size – The standard height of the button should be minimum 44pt. Width of the button can vary according to the text in it.
    b. Button Placement – Try to put the call-to-action buttons (CTA) in the thumb's reach so that user can access it easily.
    c. Border Radius – Also called corner radius should be appropriately according to the seriousness of the app. Rounded corners seem more friendly while sharp corners feel more professional.
    d. Visual Hierarchy –Primary button should have high emphasis and be your most prominent call to action button, secondary button should have medium emphasis, and tertiary button should be used with least importance. Make primary button filled, secondary button stroked, and tertiary button as only text.
    e. Button Text –You should have single line text in button as it wouldn't look good in multiline. If your text breaks in 2 lines, reduce it. Make sure you are designing your button according to the responsiveness and how your buttons might look on smallest devices.
2. Color Scheme: Always make sure that you use good color schemes to make your app look more attractive. Do not use too much colors in a single screen, try to use different shades of a single color along with a contrasting color for text elements. For e.g., use various shades of blue for the whole design and use black or white for text. You can find a lot of good color palettes at Colorhunt.co and coolors.co. One example of color scheme is shown in figure 35.

Figure 36. A Color scheme for using in app interface

3. Images: Images play a very vital role in your designs as they are the most visible section in most screens and are used to visually represent any function. Images need to be matching with the color scheme of the app. You can get a large variety of images from websites like Unsplash, pexels and pixabay. Be sure to check if the images have a copywrite or not. If yes, then avoid using it in your commercial projects as it might create some legal issues. You can use images for your personal projects though. Figure 37 shows an example of image downloaded from Unsplash.

Figure 37. Image from Unsplash

4. Icons: Icons play the most important part in any app. Always be sure to use icons of similar types. For e.g., Use filled icons for active state, stroked icons for inactive state. The consistency in type of icons is necessary in mobile app design otherwise it may confuse the user. You can get an amazing range of icons from feathericons, icons8, Google icons. You can also use animated icons in your mobile design as it make your app more interactable. One of the places to find interactive animated icons is lordicon. Icons are displayed in figure 38.

Figure 38. Various Icons from Feather Icons

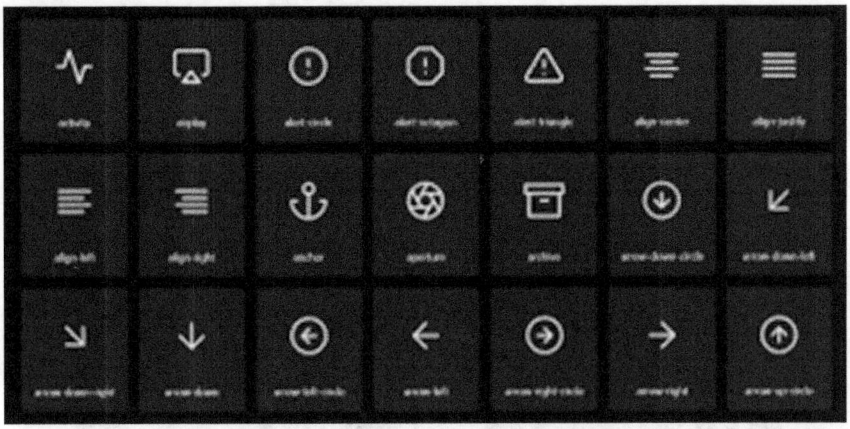

5. Illustrations: Instead of using real life images, many designers prefer to use vector images and illustrations to make the app look more visually appealing. Illustrations are soothing to the eye and offer a better understanding of any particular feature. Illustrations can be 2D as well as 3D. Some of the best illustrations and vector graphics can be found at freepik, undraw, storyset. Storyset even provides you with the feature of animating the illustrations. In Figma, any SVG file can be edited i.e., you can change the color, position of any item in illustration, add/remove items from illustration.
6. Margins and White Space: Margins from all corners must be kept in mind while designing. The white space means the blank portions in a design, the sections containing nothing in them, are an important part of design as it's never a good practice to fill the whole screen with elements, it makes the user feel awkward and it may reduce the usage of your app. So, remember to always have appropriate empty spaces in your design to make the user feel comfortable.

COMING OF AGE DESIGN STYLES

1. Glassmorphism: Glassmorphism is a design method that unifies and combines all of the UI's applications of the "frosted glass" appearance. It basically provides you with a shape that looks like glass, you can add elements on or behind the glass. The process to create a glassmorphic effect on a normal rectangle in Figma is shown below:

a. Create a frame and give it a dark color. Add some element like circles or any other object on the frame. Refer figure 39 for visuals.

Figure 39. Rectangle and Circle creation in a frame

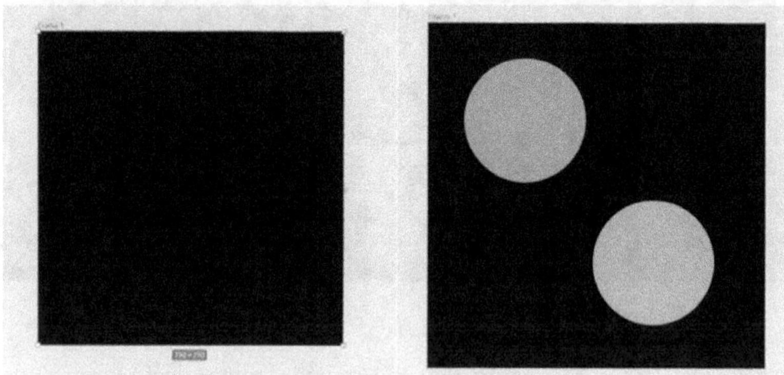

b. Create a rectangle and add it in the center and middle of the frame and add some corner radius. Select the rectangle and go to design panel and in the Fill menu, you will see that the fill is 100%. Select white color and decrease the fill to around 30%. The effect and transparency can be seen in figure 40.

Figure 40. Transparency effect on rectangle

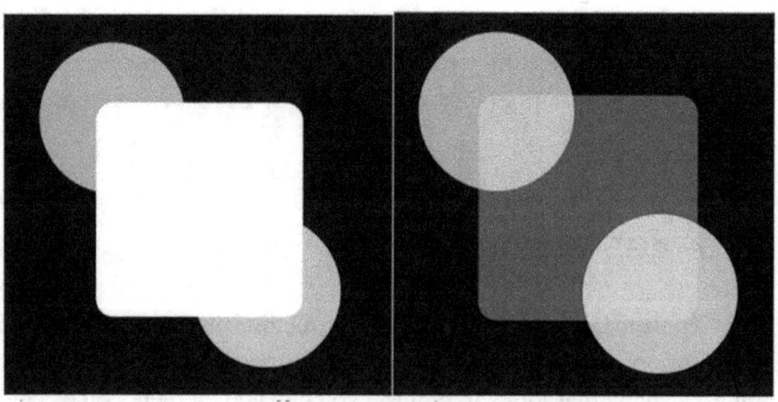

c. In the design panel, add effect. Default effect selected is the drop shadow, click on it and change the effect to background blur. Then modify the effect and increase the blur to almost 60. The effect is visible in figure 41.

Figure 41. The Blurry effect in background using Background Blur

d. Now you need to add a stroke to the rectangle. Change the color style from Solid to Linear. Add the left color as white with 60% opacity and right color as white with 30% opacity. The color tab is shown in figure 42 through which you can change the color style to Linear.

e. Now change the direction of the linear color in a diagonal direction as shown in figure 43 below.

Now your Glassmorphic rectangle is ready to use, you can add content on the rectangle and use it anywhere in your app.

Figure 42. Adding Linear gradient to stroke

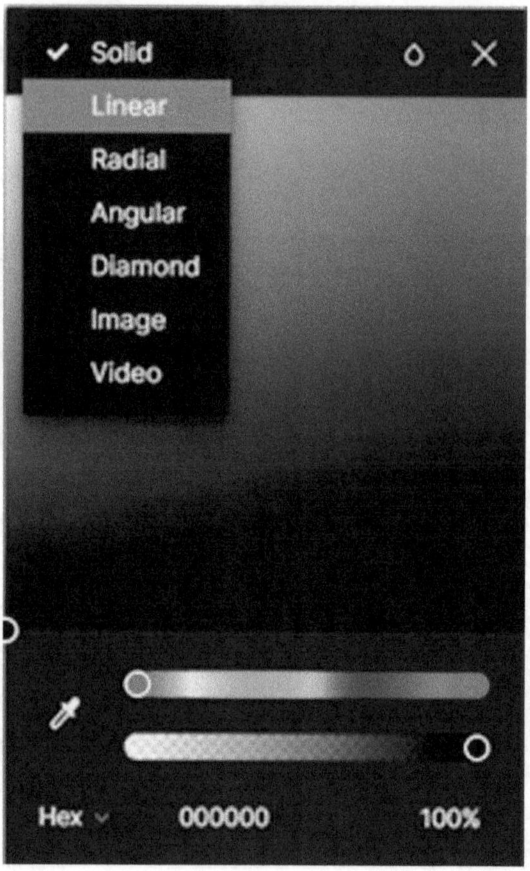

Figure 43. The Final Glassmorphic effect on a rectangle

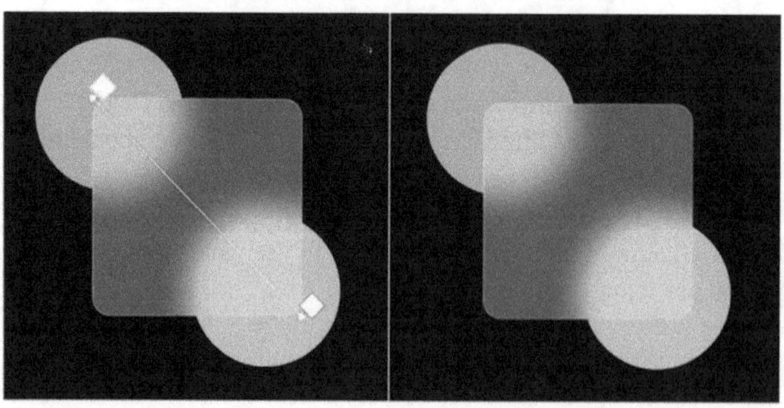

2. Neumorphism: Neumorphism is a minimalist design typewhich makes the user think that some part of the screen has bumped up or pushed inside. Theelements can be identified from the background by the Drop Shadow and Inner Shadow effects. The color of both element and background is typically the same. The process to create a neumorphic effect on a normal rectangle in Figma is shown below:
    a. Create a frame and add a light background to it. Create a rectangle and place it in the center of the frame and add corner radius, also change the color to the same as the background color. Refer figure 44 for visuals.

Figure 44. Frame and rectangle creation

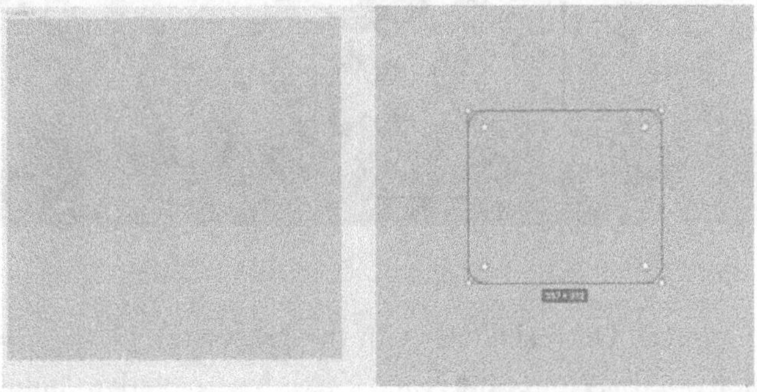

    b. From the design panel, add Effects to the rectangle. Use Drop shadow with the properties X=4, Y=4, Blur=20, Opacity=20%.Refer figure 45 for values.

Figure 45. Drop shadow on rectangle

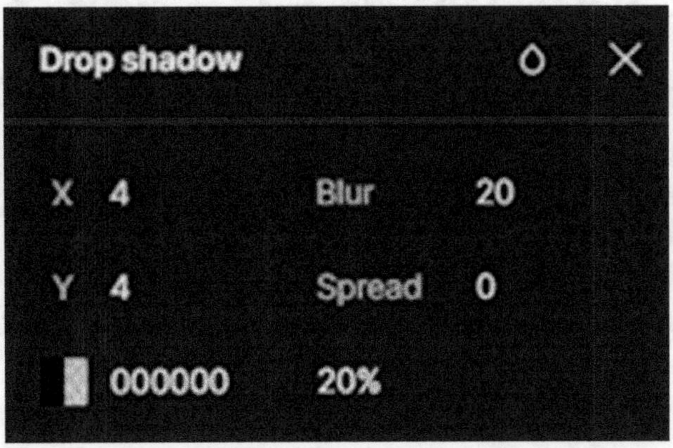

c. Use an Inner shadow with the properties X=4, Y=4, Blur=20, Opacity=20%. Refer figure 46 for values.

Figure 46. Inner shadow on rectangle

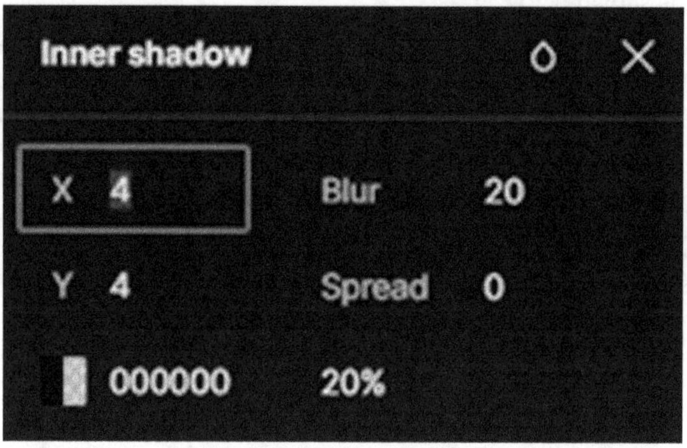

d. Use another Drop shadow with the properties X=-4, Y=-4, Blur=20, Opacity=20%. Notice the X and Y axis have negative values. Refer figure 47 for values.

Figure 47. Drop shadow on rectangle with negative values

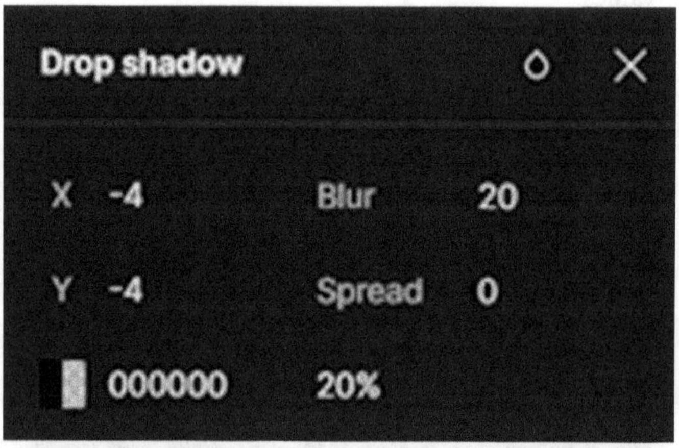

e. Use another Inner shadow with the properties X=-4, Y=-4, Blur=20, Opacity=20%. Notice the X and Y axis have negative values. Refer figure 48 for values.

Figure 48. Inner shadow on rectangle with negative values

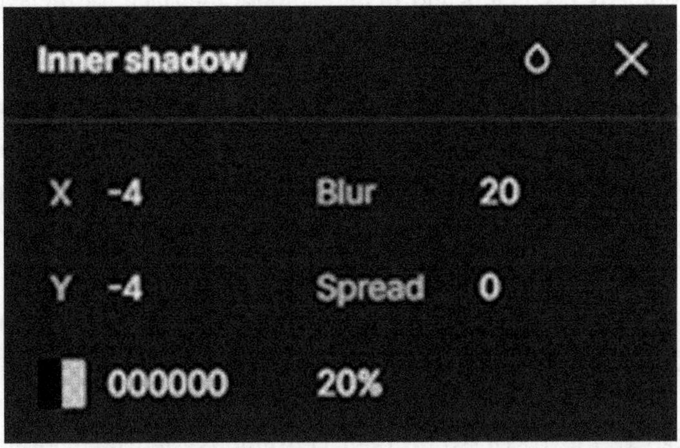

f. Now, you will get some visual as shown in the image. You can hide both drop shadow effects to give it an inner pushed look if required. See the effect in figure 49.

Figure 49. Final Neumorphic effect on the rectangle

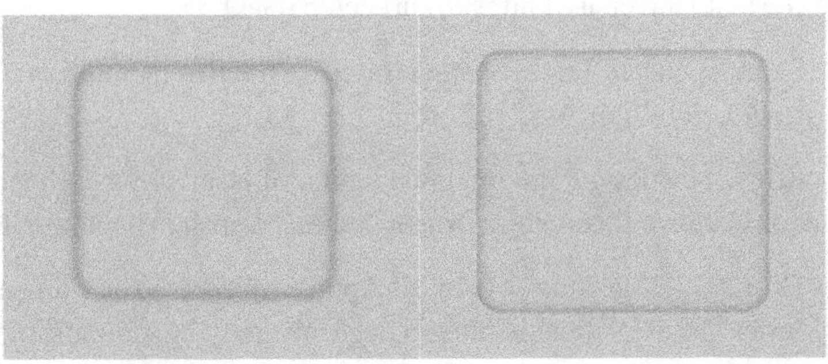

Now your Neumorphic rectangle is ready to use, you can add content on the rectangle and use it anywhere in your app.

CONCLUSION

The Mobile app designing is one of the most important aspects of Mobile app development, a good and well thought design is required to create a wonderful app. Figma is a tool used by 4 million people to create interface designs and prototypes for mobile development, web development, graphic designing, product designing. Everytech company needs a beautiful and smooth interface for their app to grow their customer retention and gain more profits. The demand for User Interface and User Experience designers are increasing and the average package offered by various companies are about $75,000 in United States.By learning the features of Figma, you can create not only designs for your apps but also a career in designing industry.And even if you don't choose design career, it will benefit you in your development journey as you will understand the design created by the designers faster and implement it in the app.

REFERENCES

An, X., & Zhao, J. G. (2018). Design of self-help guide system for intelligent tourism. *Industrial & Science Tribune*, *17*(13), 77–78.

Lee, Y.-C. "Analysis of Service Experience of the Community Sports Center Mobile Apps Based on a Service Design Perspective", Master Thesis, 2019, [online] Available: https://reurl.cc/dV7r06

Lin, P.-W. (2020). Research on the Marketing Strategy of Small Gyms: A Case Study of Small Gyms in Metropolitan District of Tainan City [Master Thesis, National Kaohsiung Normal University.] https://reurl.cc/2b8GaO

Ma, L. (2017). Independent Travel System Based on Intelligent Solution. *Computer Systems & Applications*, *26*(3), 57–62.

Shi, S. Y. (2018). The design and implementation of Bank Online Shopping mall system based on Cordova Technology, [Master's thesis, Lanzhou Jiaotong University].

Shih-Hsun, Ho. (2017). The Study of Fitness Applications based on Competence Set Analysis. [Master Thesis, National Chung Hsing University. https://reurl.cc/bzG8dM

Tang, S. W., Dai, A. L., & Chen, Y. Z. (2015). Research on Cross platfrom APP for Minxi's Pig Disease Base on CORDOVA and Its Implementation. *Journal of Wuyi University*, *34*(12), 60–64.

Wang, X. (2021). Design and Implementation of Cross Platform IntelligentTravel APP Based on Cordova. *Railway Signalling & Communication Engineering*, *18*(7), 40–44.

Ye, J.-H. (2017). *Development of App for Strength Fitness Products* [Master Thesis, National Kaohsiung University of Applied Sciences https://reurl.cc/KxXY2g

Yu, D. S., Lu, Q., & Leng, Y. J. (2017). Design and implementation of self service tourism management system based on B/S mode. *Inner Mongolia Science Technology & Economy*, *21*(19), 61–62.

Yu, X., Jiang, M., & Liu, A. (2022). Design and Application of Self-service Cultural Tourism Information Platform in the Mobile Information Age. *2022 International Conference on Information System, Computing and Educational Technology (ICISCET)*. 10.1109/ICISCET56785.2022.00059

Yuan, X. Y. (2018). Problems and countermeasures of catering management in tourist attractions. *Yangtze River Series, 40*(26).

Zhang, H. C., & Li, D. J. (2021). The theoretical interpretation and test of the integration of leisure sports with tourism phenomenon *Journal of South-Central Minzu University*, *40*(5), 543–550.

Zhang, Q., Zhang, L., & Qiang, Y. S. (2013). Study on intelligent travelling self service system based on Android. *Internet of Things Technologies*, *3*(8), 53–55.

Zhou, Q. D. (2019). Construction of smart tourism management platform based on mobile Internet technology. *Electronic Technology & Software Engineering*, *8*(3), 2–2.

Chapter 11
Android Applications for Lung Nodules Classification Using Convolutional Neural Network

Karthikeyan M. P.
https://orcid.org/0000-0002-2346-0283
School of CS and IT, Jain University (Deemed), Bengaluru, India

Banupriya C. V.
Department of Computer Science, PSG College of Arts and Science, Coimbatore, India

Kowsalya R.
Department of Computer Science(PG), PSGR Krishnammal College for Women, Coimbatore, India

Jayalakshmi A.
Department of Computer Applications, Hindusthan College of Arts and Science, Coimbatore, India

ABSTRACT

Digital image processing is currently used in various fields of research. One of them is in the field of medicine. In fact, experienced radiologists have difficulty distinguishing the cancerous portions of the blood vessels in the lung or detecting fine nodules that suggest lung cancer on X-ray images. Previous studies have shown that doctors and radiologists fail to detect cancerous patches in 30% of positive cases. Implementation of CAD system to classify and detect parts of cancer has been developed, but the results obtained from this implementation are that there are still many errors in the classification results. Therefore, this study will develop android app image technique to perform the classification process of lung cancer. With this research, it is hoped that the developed algorithm can help doctors and radiologists to detect cancer in a short time with more accuracy. Finally, after 20 iterations, a percentage of 90.65% was attained for the test results' performance in classifying 10 X-ray pictures.

DOI: 10.4018/978-1-6684-8582-8.ch011

INTRODUCTION

A type of cancer called lung cancer typically affects smokers and takes place in the lungs. The two flexible organs in your chest known as your lungs medical image that will be processed in this study is an image taken using the Lung Computed Tomography Scan (CT scan) technology. CT scan technology was developed through the process of scanning every layer of body tissue from various directions using X-ray radiation. The result of scanning using this technology is known as an X-ray image. Due to the significant number of deaths brought on by lung cancer, lungs were chosen as the study's test subjects. When compared to other cancer including stomach cancer (740,000 deaths), liver cancer, and other types, lung cancer claimed 1.4 million lives in 2008, according to data from the WHO's official website. As, quick action, early diagnosis can actually save lives. Despite being the most sophisticated imaging technology in the medical industry, CT scan scans can be challenging for doctors to interpret and spot cancer. Clinicians may benefit from computer assisted diagnostics in order to accurately identify cancerous cells. Numerous computer-aided methods using machine learning and image processing have been studied and tested. Early-stage lung cancer typically manifests itself without any symptoms or warning indications. Lung cancer signs and symptoms typically appear when the disease is advanced. A cough, which is frequently bloody, a headache, hoarseness, chest pain, wheezing, weight loss, and shortness of breath are among the symptoms. B. S. Chhikara et al (2022) discussed about the lung cancer was the cause of 1.79 million fatalities. For both men and women, It is the main reason why people die from cancer. There is an urgent need for methods of diagnosing pulmonary exacerbations and improving the course of chronic lung diseases. Patients must typically visit a pulmonary hospital and use standard spirometers. The price and the time lost make this solution difficult, though. In addition to paying for medical bills, patients may have to wait in line for a doctor for several hours. An android-based application that aids in measuring lung function has been developed to address this issue. Today, everyone uses Android phones, which are quite common devices, to measure lung function. It may lead to quicker healing, faster exacerbation treatment, and lower expenses for medical care. The lung region's uncontrollable cell development will have an impact on a person's chance of survival. Using computer-aided diagnosis (CAD) methods, it is possible to identify lung cancer at an earlier stage. A more advanced kind of artificial neural networks, deep learning uses multiple layers to produce high-order characteristics from its input and then displays the predicted value at the network's top. One deep-learning tool that has been extensively used to computer vision issues is the convolutional neural network (CNN). Since more computed tomography (CT) scan images, it was difficult for radiologists to swiftly and effectively diagnose disease. Deep learning (DL) techniques were created as a result of CAD becoming more and more essential to aid. A more current updated method known as 3-Dimensional CNN (3D CNN) mechanism is generally recognised as being superior to employing 2-Dimensional CNN for lung cancer diagnosis (2D CNN).

This paper is arranged in five parts. Session II explains the existing method and its drawback of techniques in lung diseases prediction. Session III explains the proposed techniques withlung module classification using Convolutional Neural Network. Session IV presents the experimental findings from tests on the images and their analysis, and Session V provides a conclusion.

LITERATURE REVIEW

This session will discuss the current or existing system for classifying lung disorders, which has some drawbacks and produces results with a bit less accuracy.

The most intriguing area of research for scientists is early lung cancer diagnosis, according to several researchers by Miah, M. B. A., and Yousuf, M. A. (2015). The proposed method aims to recognise early-stage lung cancer in two stages. The suggested system uses a number of processes, including picture acquisition, pre-processing, thresholding, segmentation, feature extraction, and neural network detection are some of the techniques used. Lung CT scan data was initially input into the system, and the images were subsequently prepared using image processing methods. To identify lung cancer in its earliest stages, a binaryization method is utilised to turn a picture into binary, which is then contrasted with a threshold value. The lung CT image is segmented and a powerful feature extraction algorithm is used in the following stage method has been introduced.

Masood et al., (2008) has been proposed the nodule was separated into benign and malignant groups. Considering the CT imaging and metastatic data gathered from the medical IoT network, the latter was then further divided into four sub-classes. Thakur, S. K., et al., (2020) says medical experts believe using a lung nodule detected by computed tomography (CT) screening as an example, this statement may be made about how early diagnosis of lung cancer can minimise mortality. Considering the enormous number of CT scans could lower the risk. In recent years, a variety of techniques based on learned methods and craftsmanship have emerged to support radiologists. Through the examination of Android applications employing convolutional neural networks.

Kalimuthu (2021) provides a thorough explanation of the various steps of system classification for the photos. The effectiveness of several image segmentation and classification techniques has been examined using MATLAB. The recommended convolutional neural network method achieves good accuracy, sensitivity, precision rate, and specificity when compared to other image processing systems like SVM, logistic regression, and neural networks.

From a lung cancer screening programme, a database of 38 low-dose CT scans with 50 lung nodules was compiled and examined Armato et al., in 2002. 38 of the nodules had lung cancer, which had not been discovered at the time of the initial clinical interpretation, according to a biopsy. Gray-level thresholding methods were used to identify the three-dimensionally continuous structures within the lungs from the CT data. To decide if a structure became a nodule candidate, computer-extracted volume was used. To differentiate genuine nodules from regions of normal anatomy, the set of nodule candidates was put through a rule-based framework and cascaded automatic classifier.

Various tests are discussed in Hatuwal and Thapa's (2020) discussion. To diagnose and identify the many forms and subtypes of lung cancer, trained pathologists must examine the microscopic histopathology slides during the biopsy. It takes a long time for pathologists and other medical experts to diagnose the many types of lung cancer. There is a high likelihood that patients will receive the wrong treatment and suffer the consequences, including possible loss of life. The following sections provide details on the proposed system's data collecting, data formatting, model training, testing, and prediction steps. In this study, lung cancer was identified using histological images. A convolutional neural network (CNN) was employed to classify a picture into the three categories of benign, adenocarcinoma, and squamous cell carcinoma.

For locating the cancerous spots and differentiating them into normal and abnormal, the CNN algorithm and Googlenet were selected. A deep convolution network design known as VGG-16 was chosen as the basis network for the CNN algorithm implementation. The suggested algorithm successfully detects lung cancer. This categorization is accomplished utilising transfer learning methods and a pre-trained convolutional neural network named AlexNet (Li, et al., 2013).

Sivanantham et al, (2022) explores Image processing is crucial for a variety of uses, such as content-based picture restoration, remote sensing, protected image data transmission, multimedia computing, biomedical imaging, pattern recognition, and image compression. Making photographs of the human body for use in inspecting, exposing, or diagnosing problems is known as medical image processing. The medical imaging approach displays the inside architecture of the human body without having to open it. Some of the most important methods used in medical imaging include: ultrasonic, gamma, CT, PET, X-rays, and more.

Monkam et al., (2019) delivers extensively used in a range of medical imaging tasks due to its extraordinary accomplishments. Applications for the study of lung, breast, prostate, and brain cancers have been the subject of several excellent assessments' recent publications. Other survey studies, on the other hand, have a main focus on CNNs and how they are used in radiological tasks. Despite the fact that these studies have covered artificial intelligence (with a focus on deep learning) applications to medical imaging tasks in great detail, there hasn't been any specialised analysis of the identification and classification of pulmonary nodules by CNNs. In this paper, we provide a number of cutting-edge deep learning methods that were recently used to analyse lung nodules.

According to Li et al., (2016), proposed a method for detecting lung nodules uses deep neural networks with convolutions. On the same LIDC dataset, it is impossible to directly compare our method to theirs because both the feature learning and nodule identification strategies in this research use deep convolutional neural networks. The LIDC database is used for the experimental evaluations. Two distinct strategies make up the test strategy. Training and testing data are separated from the dataset, with the 10-fold cross-validation (CF-test) test included (DD-test). Sivanantham (2022), explain about medical imaging technology makes it possible to see into the human body without cutting it open. Numerous noise-reduction techniques, segmentation of lung MRI images selected by a radiologist utilising straightforward thresholding, and using the query MRI slice, the slice is divided into two categories based on whether it has an illness or not.

From the reviews analysis, to detect and classifies the lung nodules with the help of neural networks are determined. The existing systems work under the different module in medical application. Also the classification of exact segment is slightly deviated from the particular part, and it takes slightly more number of time duration and produces less accuracy. So, this proposed system is designed to overcome the above limitations of time duration and accuracy. This system supports and working with the android application from the end user is efficiently and user friendly. Also it develops the overall better performance and delivers the better output result.

SYSTEM DESIGN

Implementation of the classification of lung cancer in this study using digital images of 20 images of benign cancer and 20 images of malignant cancer. Total digitals used are 40 images, each image measuring 512 X 512 pixels. The first stage in image processing is the initial processing stage. Each image sample is inputted into the application and saved with a 512 X 512 pixel size. The histogram equalisation procedure, picture enhancement, and intensity transformation are applied to the image, which has been stored at a resolution of 512 X 512 pixels. By using a procedure called histogram equalisation, the distribution of the value of the degree of grey in an image is evened out. By utilising median filters, hazy digital X-ray images can be improved.

Figure 1. Flow of proposed techniques

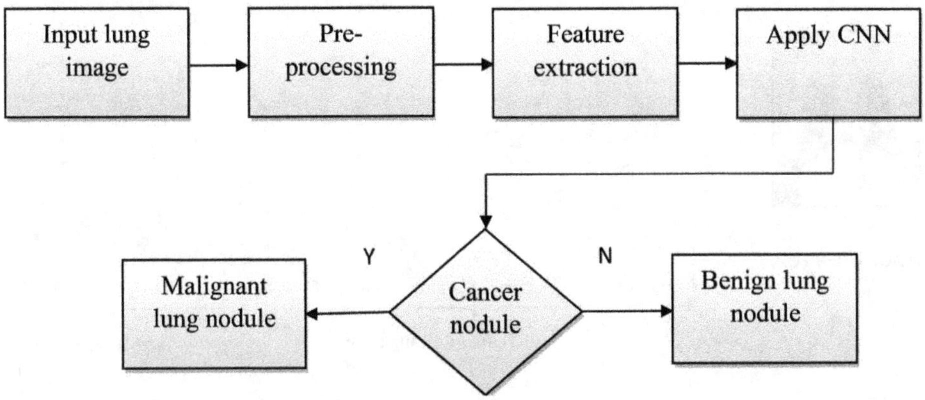

In the preceding figure 1, the proposed system design technique flow is displayed. After that, the intensity transformation of the gray scale image is carried out using the imadjust function. The imadjust function will map the intensity value in the gray scale image initial image with new intensity values in such a way. After the initial processing stage the next stage is segmentation stage. Goals to be achieved in segmentation are to define a specific object or an area in the image to match or identification. The X-ray image of the lung needs to be split into the right and left lobes. The scale values that correspond to the maximum and minimum values deduct from the overall scale value that corresponds to those values. The total number of scales that reach the maximum and minimum values is deducted from the values of m and n. The image was segmented using correlation values from Euler, and two dots representing the right side and left lung were added. Correlation is used in the segmentation result region.

To retrieve information from the image that has been successfully performed, the extraction feature is used, namely the texture feature. Based on the segmentation results' Gray Level Co-Occurrence Matrix (GLCM), these attributes were estimated. After searching the initial matrix's maximum value, the GLCM matrix is built with X as its maximum value. The study uses a distance of 1 and the average values of 0°, 45°, 90°, and 135° are used to compute the angle. The final image segmentation that has undergone a morphological procedure and been combined with the original image serves as the GLCM matrix for its values. The neural network uses the feature signals that were produced from the input image for classification. Back propagation is a technique that allows a convolutional neural network with several building blocks.

Figure 2. CNN Image classification

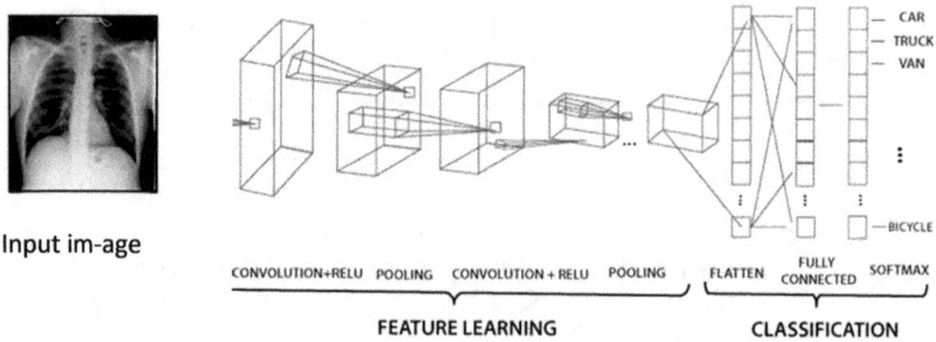

An image classification using the different layers is shown in above figure 2. A method for using relatively tiny arrays while allows CNN algorithms to function on huge images. We both contribute in different ways. An equation based on states and outputs governs how each cell in the array functions. The continuous time state and output equations were utilised to define how the cell would function because the first CNN architecture was based on analogue circuitry pseudo code in Algorithm in Figure 3.

Since CNN's runtime will at least be scaled by the ratio of the picture to CNN array size, comparing the total time directly between CNN and the ideal CNN (where the CNN array size equals the image size) is not really meaningful. One or more CNNs may be present in the CNN accelerator. The advantage of having several CNN is the ability to run multiple partitions simultaneously. Additionally, numerous CNN can more effectively conceal memory delay. However, too many CNN might result in increased memory bandwidth contention, which causes performance to scale less linearly. Additionally, there must be a trade-off between using a high number of small CNN arrays against a limited number of large arrays.

Architectures of CNN fall into two basic categories: segmentations the use of CNNs for the classification of image regions originating from one or more classes of semantically understandable objects. CNNs can assign each pixel to one or more different categories by using a set of real-world object categories. The main components of a convolutional neural network are input, output, and one or more hidden layers. These components are described in this session of the neurons in the layers of a convolutional network.

Figure 3. Pseudocode for CNN Algorithm

```
While change do
Change = false
Forp in partitions do
LoadState and input toCNN(p)
for n = 0; n < interval; n++, t++ do
Change |= parallel-for-i,j on p
Parallel-for-i,j on p
end for
SaveToNextStateFromCNN(p)
end for
swapState andNextStatePointers(p)
iter += 1, vt += interval
end while
```

Neural networks provide a lot of benefits, including as requiring less formal statistical training, having various training options, and being able to implicitly detect complicated nonlinear correlations between dependent and independent variables.

Step by step procedure for classifying an image in CNN.

1. First, pick a dataset.
2. Prepare the data set for training in step two.
3. Create training data in step three.
4. Shuffle the dataset in step four.
5. Choosing Labels and Features is Step 5.

6. X is normalised in step 6 and the labels are transformed into categorical data.
7. Divide X and Y for CNN use.

Figure 4. Neural Network model

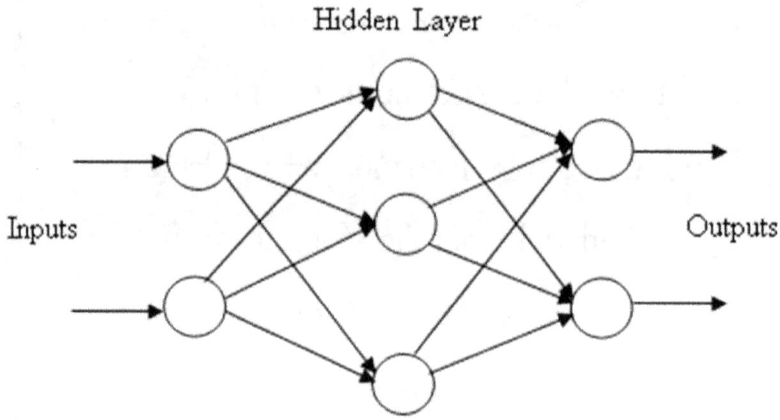

In figure 4 above, the generalised neural network model is displayed. Signals flow via the middle (hidden) layers to the output layer once stimulation is delivered to the first layer's inputs. There is a different weighting value for each link between the neurons. Assume that our convolutional layer comes before some N X N square neuron layers. Our convolutional layer output will be of size (N-m+1) X (N-m+1) if we employ a m X m filter. The contribution from the previous layer cell must be added together in order to calculate the pre-nonlinearity input to some unit our layer:

$$x_{ij}^{l} = \sum_{a=0}^{m-1}\sum_{b=0}^{m-1} w_{ab} y_{(i+a)(j+b)}^{l-1}$$

Then the convolutional layer applies its non-linearity:

$$y_{ij}^{l} = \sigma\left(x_{ij}^{l}\right)$$

Convolutional neural networks are an alternative method of processing dimensional and organised data from an architectural standpoint. Convolutional and max pooling layers impose weight sharing as opposed to fully connected layers, which assume that the location of the data in the input is immaterial. Applications for voice and picture identification typically use convolutional neural networks (CNN or ConvNet), a subclass of neural networks. Without sacrificing any information, its integrated convolutional layer minimises the huge dimensionality of images. Therefore, CNNs are ideal for this use case. Any image size can be entered, and these CNN feature maps will produce maps that are 32 times smaller. For instance, if you input 224x224, CNN will produce feature maps that are 7x7 in size. These CNN outputs include maps with a size of 16x16 if you input images with a resolution of 512x512. The area is referred to as the "local receptive field". If the training data is added consistently, the classification accuracy improves. The study's findings further demonstrate that the convolution neural network reliably handles output faults by adjusting the network weights for the subsequent input. The mobile application-based lung nodule classification techniques using CNN is very reliable to update the image and image classification performance results are good.

RESULT AND DISCUSSION

Utilizing the Android Studio environment with an Intel Core i5-2410M CPU running at 2.30GHz and 16GB of RAM, the suggested methodology is applied. The suggested CNN's performance is assessed based on a specific medical condition because it impacts a person's lifetime ability to move. The input image of lung and pre-processed image are shown in the figures 5(A) and 5(B) respectively.

Figure 5. (A) Lung Input Image; (B) Pre-processed image

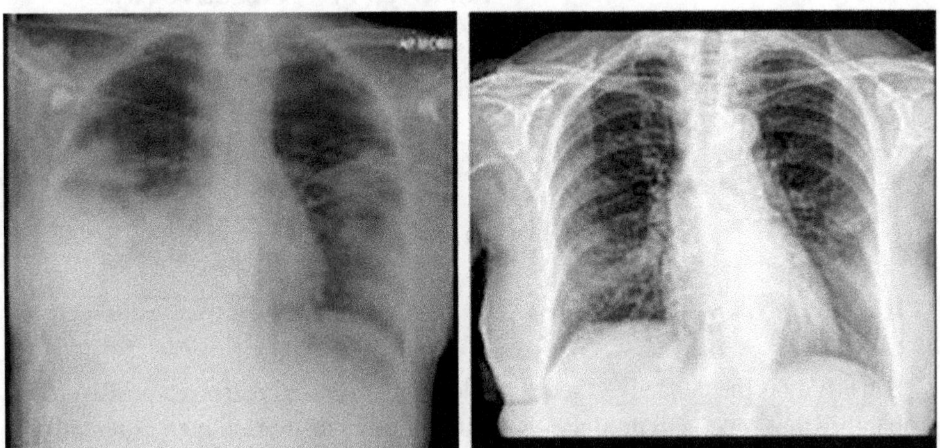

Figure 6. Lung diseases result of different stages

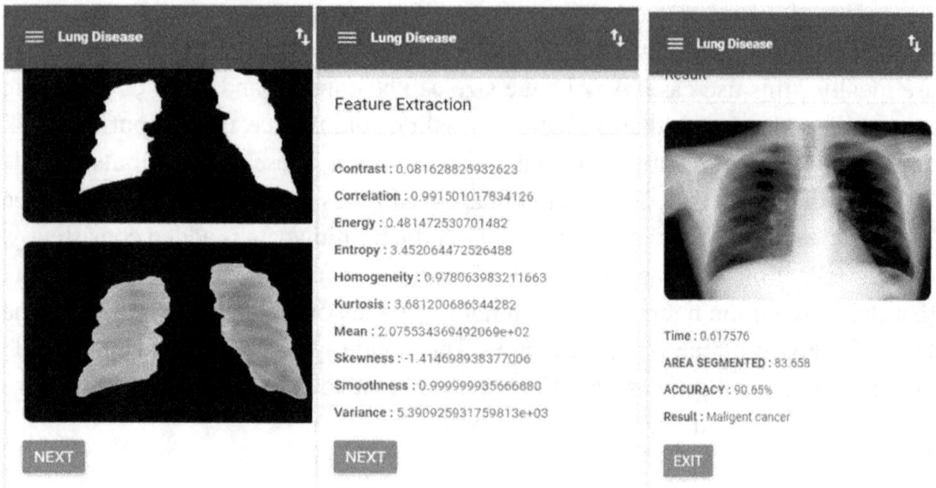

Figure 6 above displays the outcome in various lung disease stages. The matrix compares the machine learning model's predicted values to the actual target values. A confusion matrix is displayed in figure 7 below.

Figure 7. Confusion matrix

		Predicted Values	
		Positive (1)	Negative (0)
Actual Values	Positive (1)	TP	FN
	Negative (0)	FP	TN

Sensitivity is the capacity of a test to identify a diseased person as positive. The probability that a test will produce a genuine positive result is known as sensitivity (also known as true positive rate). Calculating sensitivity involves dividing the total number of accurate positive predictions (TP) by the entire number of positives (P). In Table 1, the sensitivity analysis is displayed.

$$\text{Sensitivity} = TP / (TP + FN)$$

Table 1. Sensitivity Analysis

Algorithm	Sensitivity (%)
Decision tree	78.33
Naive Bayes	81.5
SVM	84.7
CNN	91.1

The sensitivity analysis represents the different algorithm of decision tree, native bayes, SVM and CNN.

Figure 8. Sensitivity Graph

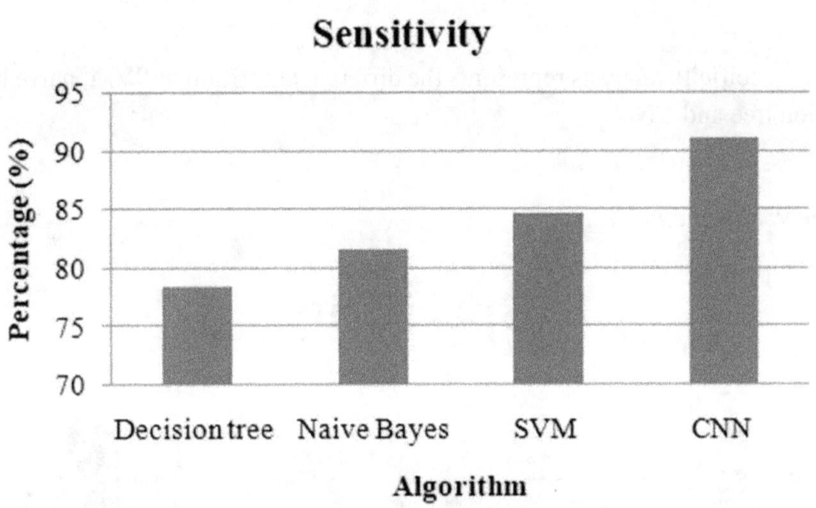

The sensitivity graph is shown in the above figure 8. In some previous calculations, the proposed CNN attains better sensitivity as 91.1% compared to SVM as 84.7% and naive bayes maintains 81.5 and also decision tree performs 78.33%. Thus, the proposed work maintains an overall 8% of better performance than the existing schemes.

Specificity is the proportion of people who test negative for a given disease among a sample of people who do not have the condition. Specificity is the likelihood that a test will be negative, assuming that it will be (true negative rate). Calculating specificity involves dividing the total number of negative outcomes (TN) by the number of negative outcomes that were accurately predicted (N).In tabulation 2, the specificity analysis is displayed.

$$\text{Specificity} = TN / (TN + FP)$$

Table 2. Specificity Analysis

Algorithm	Specificity (%)
SVM	83.33
Naive Bayes	88.12
Decision tree	86.45
CNN	90.65

The specificity analysis represents the different algorithm of SVM, naive bayes, decision tree and CNN.

Figure 9. Specificity Graph

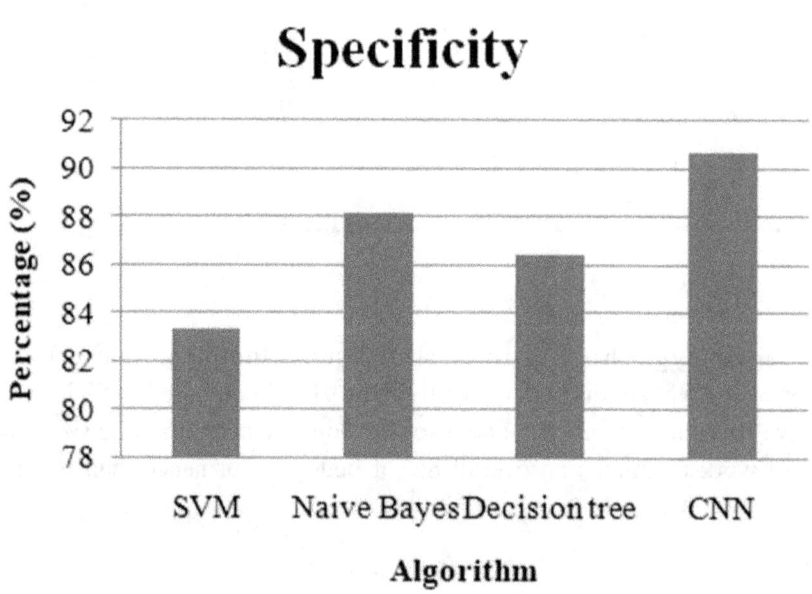

The specificity graph is shown in the above figure 9. For the above calculations, the proposed scheme of CNN attains better sensitivity as 90.65% compared to naivebayes as 88.12% and decision tree maintains 86.45% and also SVM performs 83.33%. Thus, the proposed work delivers an overall 14% of better performance than the existing schemes. Time Duration is the amount of time it takes for a process to complete. The time duration of the different algorithm is showed in below table and graph. The time duration is calculated and is listed in table 3.

$$\text{Time Period} = 2\pi\sqrt{(L/g)}$$

Table 3. Time Duration

Algorithm	Time Duration (m.sec)
SVM	16.85
Naive Bayes	13.85
Decision tree	14.25
CNN	11.12

Figure 10. Time Duration

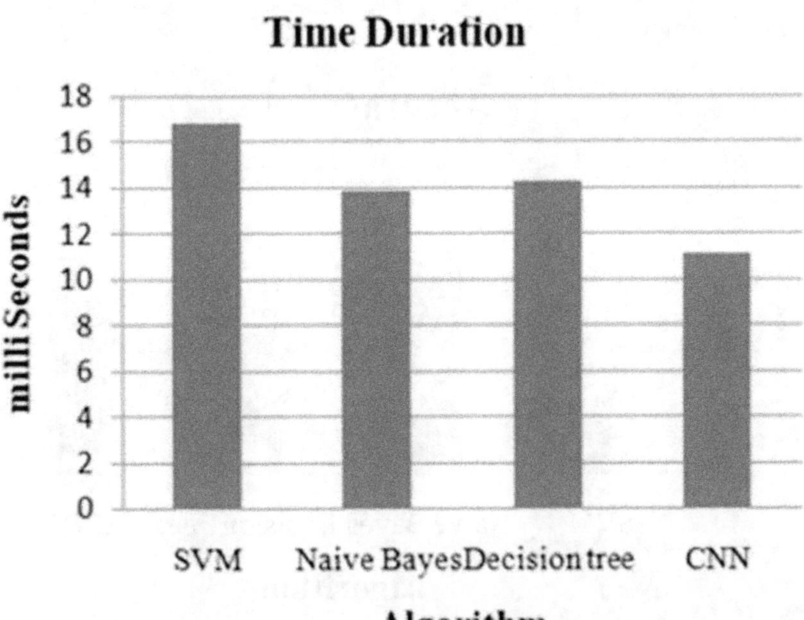

The time duration graph is shown in the above figure 10. The proposed scheme of CNN takes a smaller number of time duration as 11.12 milliseconds compared to naive bayes as 13.85 ms and decision tree maintains 14.25 ms and also SVM performs 16.85 milliseconds. Thus, the proposed work delivers an overall 7% of better performance than the existing schemes. Simply put, accuracy gauges how frequently the classifier predicts correctly. It is the ratio of occurrences that were accurately anticipated to all events that were expected. The accuracy of the above different algorithms are calculated and showed in below table and graph. The accuracy of different algorithm is calculate and shown in below tabulation 4.

$$Accuracy = (TP + TN)/(TP + TN + FP + FN)$$

Table 4. Accuracy

Algorithm	Accuracy (%)
SVM	78.95
Naive Bayes	87.56
Decision tree	83.65
CNN	90.65

Figure 11. Accuracy

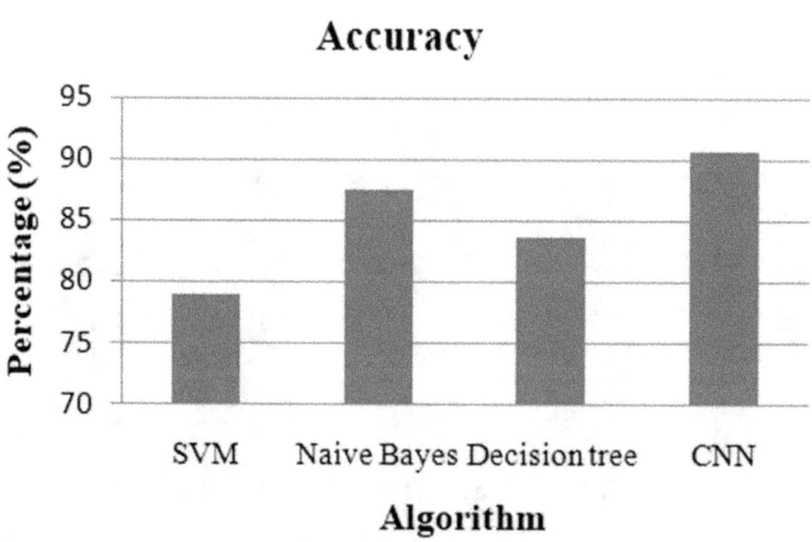

The accuracy graph is shown in the above figure 11. From the above calculations, the suggested proposed scheme of CNN attains better 90.65% accuracy compared to naive bayes as 87.56% and decision tree maintains 83.65% and also SVM performs 78.95% accuracy. So, the proposed work achieves an overall 13% of better accuracy performance than the other existing schemes.

CONCLUSION

The deep convolutional neural network (CNN) architecture described in this study uses U-Net architecture to identify interest spots and detect nodules in lung cancer patients. CNN pre-processes data at this step. The deep CNN models performed the best on the test set. The fact that the first layer functions as a pre-treatment layer for segmentation using different techniques is an exciting discovery. The algorithms Threshold, Watershed, and U-Net are used to determine the nodules of the patients. The network may be trained from the very beginning using raw picture patches. Its main need is the presence of a training database;aside from that, neither the objects of interest nor the underlying picture modality are assumed. Convolutional neural networks, which have several layers, are far better at extracting properties from data. It is effective with photos and requires little pre-processing. It can accurately recognise images by distilling an image down to its essential components using convolutions and pooling. Compared to other types of neural networks, CNN models can be trained with less initial parameters. It won't require many hidden layers because the convolutions can take care of the majority of the hidden layer detection. The variety of challenging issues to which CNNs can be applied is one of its best features. This type of data can be processed by convolutional neural networks, which can produce precise predictions. As a result of the foregoing discussions, the suggested system can quickly identify and categorise lung nodules utilising the neural network algorithm technique. Also it can be used in android applications, so it is suitable for user friendly and it gives the better performance of accuracy and the time consumption is less comparing than the other existing systems.

REFERENCES

Alakwaa, W., Nassef, M., & Badr, A. (2017). Lung cancer detection and classification with 3D convolutional neural network (3D-CNN). *International Journal of Advanced Computer Science and Applications*, *8*(8). doi:10.14569/IJACSA.2017.080853

Armato, S. G. II, Li, F., Giger, M. L., MacMahon, H., Sone, S., & Doi, K. (2002). Lung cancer: Performance of automated lung nodule detection applied to cancers missed in a CT screening program. *Radiology*, *225*(3), 685–692. doi:10.1148/radiol.2253011376 PMID:12461246

Brenner, D. R., Boffetta, P., Duell, E. J., Bickeboeller, H., Rosenberger, A., McCormack, V., & Hung, R. J. (2012). Previous lung diseases and lung cancer risk: A pooled analysis from the International Lung Cancer Consortium. *American Journal of Epidemiology*, *176*(7), 573–585. doi:10.1093/aje/kws151 PMID:22986146

Chhikara, B. S., & Parang, K. (2022). Global Cancer Statistics 2022: The trends projection analysis. *Chemical Biology Letters*, *10*(1), 451–451.

Chon, A., Balachandar, N., & Lu, P. (2017). Deep convolutional neural networks for lung cancer detection. Standford University, 1-9.

Hatuwal, B. K., & Thapa, H. C. (2020). Lung cancer detection using convolutional neural network on histopathological images. *International Journal of Computer Trends and Technology*, *68*(10), 21–24. doi:10.14445/22312803/IJCTT-V68I10P104

Kalimuthu, S. (2021). Sentiment Analysis on Social Media for Emotional Prediction During COVID-19 Pandemic Using Efficient Machine Learning Approach. *Computational Intelligence and Healthcare Informatics*, 215.

Kalimuthu, S., Nait-Abdesselam, F., & Jaishankar, B. (2021). Multimedia Data Protection Using Hybridized Crystal Payload Algorithm With Chicken Swarm Optimization. In Multidisciplinary Approach to Modern Digital Steganography (pp. 235-257). IGI Global. doi:10.4018/978-1-7998-7160-6.ch011

Li, Q., Cai, W., Wang, X., Zhou, Y., Feng, D. D., & Chen, M. (2014, December). Medical image classification with convolutional neural network. In *2014 13th international conference on control automation robotics & vision (ICARCV)* (pp. 844-848). IEEE. 10.1109/ICARCV.2014.7064414

Li, W., Cao, P., Zhao, D., & Wang, J. (2016). Pulmonary nodule classification with deep convolutional neural networks on computed tomography images. *Computational and Mathematical Methods in Medicine*, *2016*, 2016. doi:10.1155/2016/6215085 PMID:28070212

Masood, A., Sheng, B., Li, P., Hou, X., Wei, X., Qin, J., & Feng, D. (2018). Computer-assisted decision support system in pulmonary cancer detection and stage classification on CT images. *Journal of Biomedical Informatics*, *79*, 117–128. doi:10.1016/j.jbi.2018.01.005 PMID:29366586

Miah, M. B. A., & Yousuf, M. A. (2015, May). Detection of lung cancer from CT image using image processing and neural network. In *2015 International conference on electrical engineering and information communication technology (ICEEICT)* (pp. 1-6). IEEE. 10.1109/ICEEICT.2015.7307530

Monkam, P., Qi, S., Ma, H., Gao, W., Yao, Y., & Qian, W. (2019). Detection and classification of pulmonary nodules using convolutional neural networks: A survey. *IEEE Access: Practical Innovations, Open Solutions*, *7*, 78075–78091. doi:10.1109/ACCESS.2019.2920980

Sivanantham, K. (2022). Deep Learning-Based Convolutional Neural Network with Cuckoo Search Optimization for MRI Brain Tumour Segmentation. In *Computational Intelligence Techniques for Green Smart Cities* (pp. 149–168). Springer. doi:10.1007/978-3-030-96429-0_7

Sivanantham, K., Kalaiarasi, I., & Leena, B. (2022). Brain Tumor Classification Using Hybrid Artificial Neural Network with Chicken Swarm Optimization Algorithm in Digital Image Processing Application. *Advance Concepts of Image Processing and Pattern Recognition: Effective Solution for Global Challenges, 91.*

Thakur, S. K., Singh, D. P., & Choudhary, J. (2020). Lung cancer identification: A review on detection and classification. *Cancer and Metastasis Reviews*, *39*(3), 989–998. doi:10.100710555-020-09901-x PMID:32519151

Zhou, B., Zang, R., Zhang, M., Song, P., Liu, L., Bie, F., & Gao, S. (2022). Worldwide burden and epidemiological trends of tracheal, bronchus, and lung cancer: A population-based study. *EBioMedicine*, *78*, 103951. doi:10.1016/j.ebiom.2022.103951 PMID:35313216

Chapter 12
Class Model for Mobile Game Architecture

Leon Cana
Rochester Institute of Technology, Kosovo

Art Jashari
Rochester Institute of Technology, Kosovo

Rafaela Marku
Rochester Institute of Technology, Kosovo

ABSTRACT

Class models are an essential part of mobile game architecture, providing a framework for organizing and structuring the code. These models help developers to understand the relationships and dependencies between different objects in the game, allowing for more efficient and effective code development. The use of class models is likely to continue to evolve and improve in the future, as mobile games become more complex and sophisticated. As AI and ML technologies become more advanced, developers will need to create new classes and models to represent these technologies in mobile games. Additionally, the use of cloud computing and other distributed technologies is likely to become more prevalent, requiring the creation of new classes and models to represent these technologies in the game. Overall, class models will play a critical role in enabling developers to create more engaging and immersive mobile gameplay experiences.

DOI: 10.4018/978-1-6684-8582-8.ch012

INTRODUCTION

A class model is a type of diagram that is used in object-oriented modeling and design. It represents the classes and their relationships in a system, and it is used to visualize the structure and behavior of the system. Class models are an important tool in software engineering, and they are used to help developers understand and design the architecture of a software system. Class models usually include the following elements:

Classes: Classes are the fundamental building blocks of a class model. They represent the objects or entities in the system, and they are defined by their attributes, operations, and relationships with other classes.

Attributes: Attributes are the characteristics or properties of a class, and they describe the data or information that is associated with the class. Examples of attributes include name, age, gender, address, and phone number.

Operations: Operations are the actions or behaviors that a class can perform, and they define the functionality of the class. Examples of operations include calculate, display, update, and delete.

Relationships: Relationships are the connections or associations between classes, and they describe how the classes interact with each other. Examples of relationships include inheritance, aggregation, composition, and association.

Class models are often depicted using UML (Unified Modeling Language) notation, which is a standardized visual language for representing software systems. UML class diagrams are used to represent the structure of a system, and they show the classes, attributes, operations, and relationships of the system. Here is an example of a UML class diagram that represents a simple system of bank managing system information:

In this class diagram, the Customer class is the main class, and it has attributes such as name, address, and phone number. The Customer class also has operations such as OpenAccount, CloseAccount, and RequestCard. The Customer class has a relationship with the Loan class, which represents the type of loans that the customer has made. The Loan class has attributes such as type and accountId. Class models are an important tool in software engineering, and they are used to help developers understand and design the architecture of a software system. By representing the classes and their relationships in a visual diagram, class models provide a clear and concise representation of the system, and they help developers to identify and resolve potential issues or problems with the system.

The role of class models in software engineering is to represent the structure and behavior of a software system. Class models are a type of diagram that is used in object-oriented modeling and design, and they are an important tool for understanding and designing the architecture of a software system. The role of class models in software engineering is also to provide a clear and concise representation of the structure and behavior of a software system. By using class models, software developers can better understand the system, and they can identify and resolve potential issues or problems with the system. Class models are used throughout the software development process, from initial requirements gathering and analysis, to design and implementation, to testing and maintenance. (Sommerville, 2011). They are an essential tool for software developers, and they help to ensure that the system is well-designed, well-implemented, and easy to maintain.

Figure 1. UML for Bank Management System

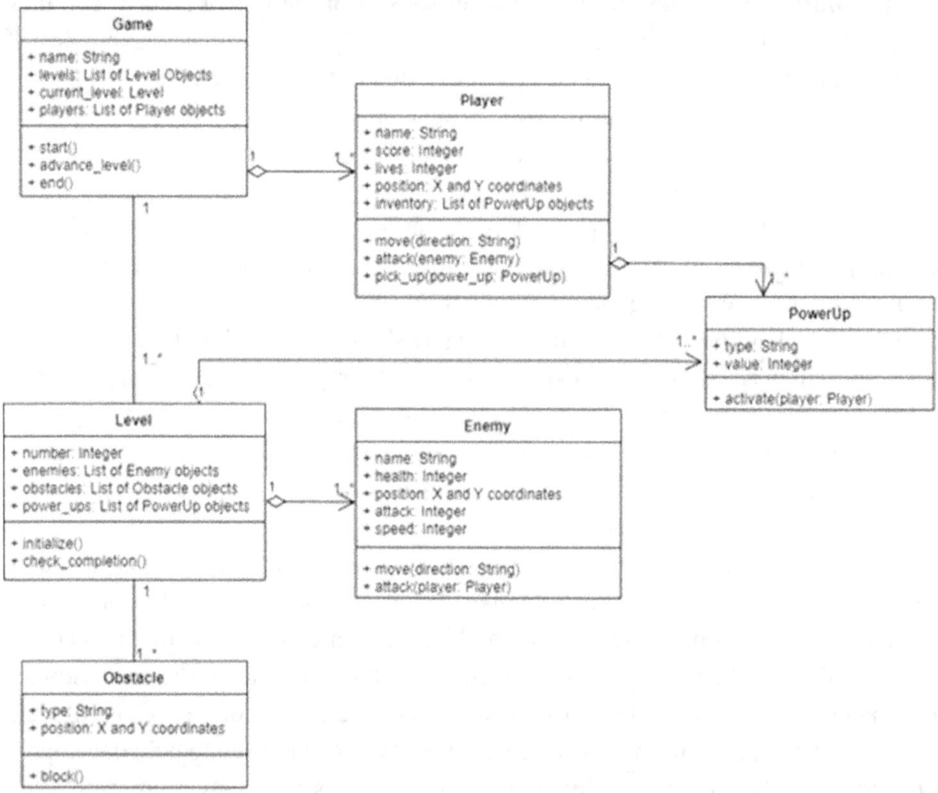

Class models are also used to communicate the design of a software system to other stakeholders, such as project managers, business analysts, and users. By using class models, developers can explain the structure and behavior of the system in a clear and concise manner, and they can get feedback and input from other stakeholders. We can say that class models are an essential part of mobile game architecture, as they provide a structured approach for organizing and implementing the various components of a game. Through the use of reusable code, object-oriented design principles, modular and scalable architectures, and maintainable and extensible designs, class models help to create high-quality mobile games that are both user-friendly and efficient. Class models are an integral part of mobile game architecture, as they provide a structured approach for organizing and implementing the various components of a game. By using class models, developers can create clear and concise designs that are easily understood by both themselves and other members of the development team.

One of the key advantages of using class models in mobile game architecture is their ability to support the creation of reusable code. By defining specific classes for different game components, such as characters, items, and levels, developers can easily reuse these classes in multiple games or even within the same game (Klint, 2013). This not only saves time and effort, but also ensures that the code is consistent and maintainable. Additionally, class models help to enforce object-oriented design principles, such as encapsulation and abstraction. Encapsulation refers to the practice of hiding the internal details of a class from the outside world, while abstraction involves the creation of simplified versions of complex concepts. By using class models, developers can create objects that are self-contained and easy to understand, making it easier to manage the complexity of a game. Furthermore, class models support the creation of modular and scalable game architectures. By defining well-structured classes and relationships between them, developers can create modular components that can be easily added, removed, or modified without affecting the overall structure of the game. This allows for the creation of flexible and scalable designs that can adapt to changing requirements or new features. Moreover, class models support the development of maintainable and extensible games. By using well-defined classes and relationships between them, developers can create designs that are easily understood and maintained by other members of the team. This not only reduces the risk of errors and bugs, but also allows for the easy addition of new features or functionality to the game. Therefore, we can say that developers find UML models a useful tool when building mobile game architecture.

MOBILE GAME

Mobile games have become increasingly popular in recent years, with a wide range of genres and styles to choose from. There are many different types of mobile games, including puzzle games, strategy games, action, adventure, sports, casino, gacha, etc., just to name a few. Some mobile games are designed to be played in short bursts, while others are more involved and require more time and commitment to play. Additionally, mobile games can be either single-player or multiplayer, and some are even augmented reality games that use the camera and other sensors on a mobile device to create an immersive gaming experience. The success of a mobile game often relies on its underlying architecture, which determines how well it performs and how engaging it is for players. The architecture of a mobile game's class model is an important aspect of the game's design, as it determines how the different components of the game interact with each other and how the game behaves.

MOBILE GAME LAYERS

A mobile game's class model consists of several layers, each of which serves a specific purpose and contributes to the overall functioning of the game. These layers can be broadly divided into three categories: data, presentation, and control.

The data layer is the foundation of the class model, and it is responsible for storing and managing the game's data. This layer is made up from several classes, such as the Player class, which stores information about the player, including their name, current level, and progress through the game; the Inventory class, which stores information about the items the player has collected; and the Level class, which stores information about the current level, including the layout of the level and the enemies and obstacles the player will encounter. The presentation layer is responsible for rendering the game's visuals and audio. This layer includes the Graphics class, responsible for rendering the game's graphics on the screen, and the Audio class, which is responsible for playing the game's sound effects and music. The control layer is responsible for managing the flow of the game and handling user input, consisting of several classes, such as the GameController class, which is responsible for managing the overall flow of the game and responding to user input; the InputController class, which is responsible for handling user input from the keyboard, touch screen, or other input devices; and the AIController class, which is responsible for managing the behavior of non-player characters in the game. Together, these layers form the architecture of a mobile game's class model, and they work together to provide the functionality and features that make the game enjoyable and engaging for players.

THE DATA LAYER

The data layer is responsible for managing and manipulating the game's data. Some common classes found in the data layer of a mobile game's architecture's class model include:

DataManager: This class is responsible for managing the game's data, including storing and retrieving data from a database or file system, as well as managing the game's data structures and objects.

GameData: This class represents the game's data, including the player's progress, high scores, and saved games.

Player: This class represents the player's in-game character and progress.

The associations between the classes in the data layer of a mobile game's architecture's class model are hierarchical, with the DataManager class at the top of the hierarchy, followed by the GameData class, and finally the Player class at the bottom. The DataManager class is responsible for managing the game's data, including the GameData and Player classes, while the GameData and Player classes are responsible for representing and storing the game's data and the player's progress, respectively.

The attributes of the classes in the data layer of a mobile game's architecture's class model are mostly focused on storing and managing the game's data. Some common operations found in the data layer of a mobile game's architecture's class model include:

Storing and retrieving data: The DataManager class includes operations for storing and retrieving data from a database or file system, such as operations for inserting, updating, and deleting data, as well as operations for querying and accessing data.

Managing Data Structures and Objects: The DataManager class includes operations for managing the game's data structures and objects, such as operations for creating and deleting data structures and objects, as well as operations for accessing and manipulating these structures and objects.

Managing Player Accounts: The DataManager class includes operations for managing player accounts, such as operations for creating and managing user profiles, as well as operations for managing the player's in-game progress and achievements.

Accessing and Manipulating Game Data: The GameData class includes operations for accessing and manipulating the game's data, such as operations for updating the player's progress, high scores, and saved games.

Accessing and Manipulating Player Data: The Player class includes operations for accessing and manipulating the player's stats, inventory, and other game-related data, such as operations for updating the player's stats, adding and removing items from the player's inventory, and managing the player's achievements.

These classes, associations, attributes, and operations work together to provide the game with a structured and organized approach to managing its data, ensuring that the game functions properly and provides a good user experience.

Figure 2. Operations in a data layer of a mobile game's class model

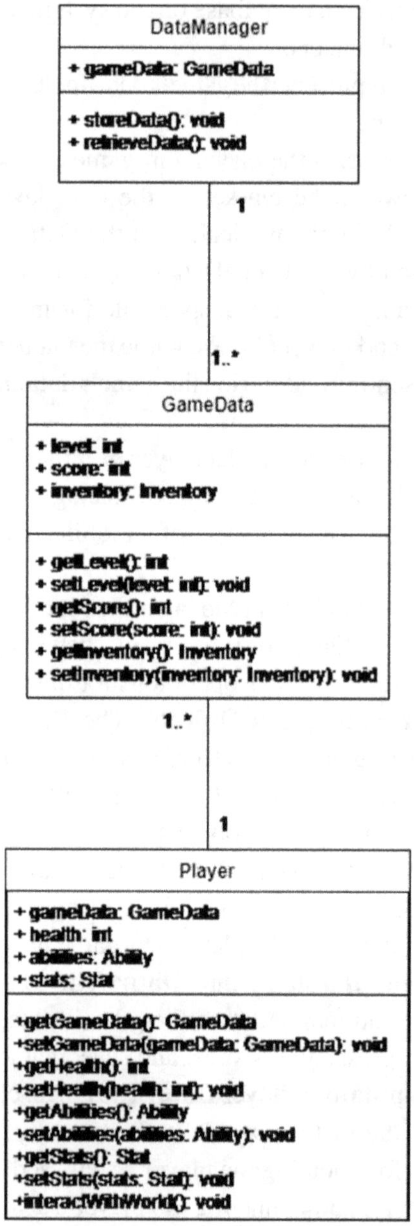

THE PRESENTATION LAYER

The presentation layer of a mobile game's architecture's class model is responsible for rendering the game's graphics and user interface. Some common classes found in the presentation layer of a mobile game's architecture's class model include:

GameRenderer: This class is responsible for rendering the game's graphics and user interface.

GameView: This class represents the game's view, including the game's graphics and user interface.

UserInput: This class represents the user's input, such as keyboard, mouse, and touch input.

The attributes of the classes in the presentation layer of a mobile game's architecture's class model are focused on rendering the game's graphics and user interface.

Some common operations found in the presentation layer of a mobile game's architecture's class model include:

Displaying Visual Elements: The GameRenderer class includes operations for displaying text, images, and other visual elements on the game's screen, such as operations for rendering text and images, as well as operations for managing the game's user interface and menus.

Handling user input and interaction: The GameRenderer class includes operations for handling user input and interaction, such as operations for processing keyboard, mouse, and touch input, as well as operations for managing the game's user interface and menus.

Implementing Animation and Special Effects: The GameRenderer class also includes operations for implementing animation and special effects, such as operations for animating objects and visual elements, as well as operations for applying filters and effects to the game's graphics.

Accessing And Manipulating Visual Elements: The GameView class may include operations for accessing and manipulating the game's visual elements, such as operations for updating text and images, as well as operations for managing the game's user interface and menus.

Accessing and manipulating user input: The UserInput class includes operations for accessing and manipulating the user's input data as well, such as operations for processing input events, as well as operations for managing the game's user interface and menus.

Overall, the presentation layer provides the game with a structured and organized approach to rendering its graphics and user interface, ensuring that the game functions properly and provides a good user experience.

Figure 3. Operations in a presentation layer of a mobile game's class model

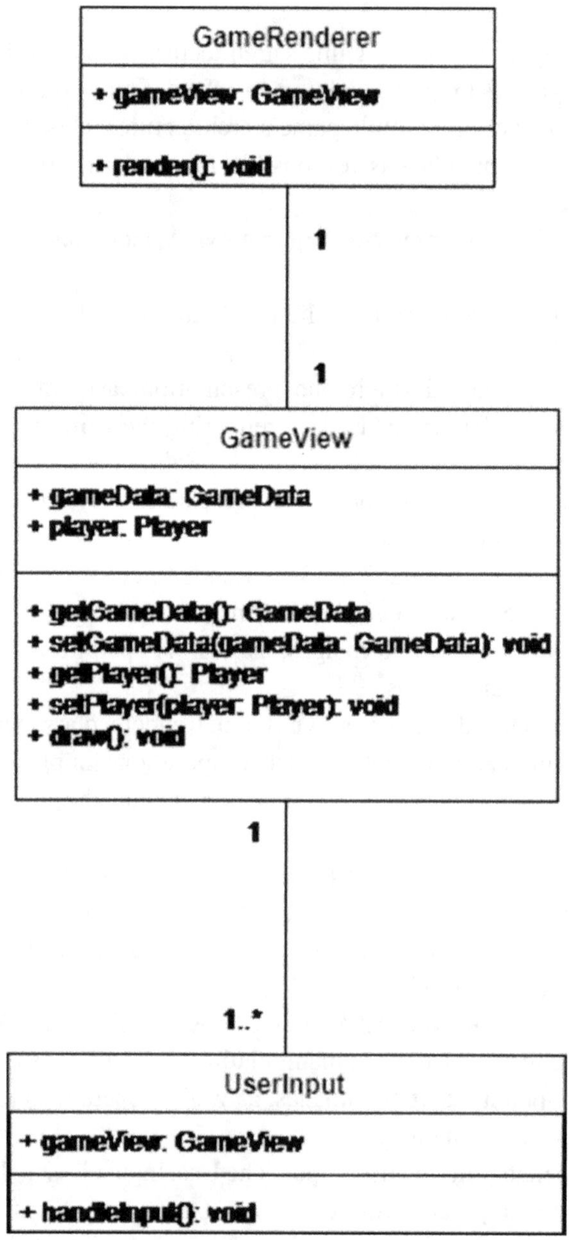

THE CONTROL LAYER

The control layer of a mobile game's architecture's class model is responsible for handling the game's logic and mechanics. Some common classes found in the control layer of a mobile game's architecture's class model include:

GameController: This class is responsible for handling the game's logic and mechanics.

GameState: This class represents the game's state, including the game's rules, mechanics, and progression.

GameObject: This class represents the game's objects, including the player's character, enemies, and other non-player characters.

The GameController class has attributes for implementing the game's rules and mechanics, as well as attributes for managing the game's flow and progression. Some common operations found in the control layer of a mobile game's architecture's class model include:

Implementing Game Rules And Mechanics: The GameController class includes operations for implementing the game's rules and mechanics, such as operations for determining how objects and characters move and interact, as well as operations for managing the game's flow and progression.

Handling User Input and Interaction: The GameController class handles user input and interaction, such as operations for processing keyboard, mouse, and touch input, as well as operations for managing the game's user interface and menus.

Managing AI And Non-Player Characters: The GameController class also includes operations for managing the game's AI and non-player characters, such as operations for determining how these characters move and act, as well as operations for managing their interactions with the player and other game objects.

Accessing and Manipulating Game State: The GameState class has operations for accessing and manipulating the game's state data, such as operations for updating the game's rules, mechanics, and progression.

Accessing and Manipulating Game Objects: The GameObject class may include operations for accessing and manipulating the game's objects, such as operations for moving, colliding, and interacting with objects, as well as operations for managing the properties and attributes of objects.

In addition to these core components, a mobile game's architecture may also include other components, such as social media integration, in-app purchasing, and analytics tracking. The specific architecture of a mobile game will depend on the requirements and features of the game, as well as the game engine and tools used to develop it.

Figure 4. Operations in a control layer of a mobile game's class model

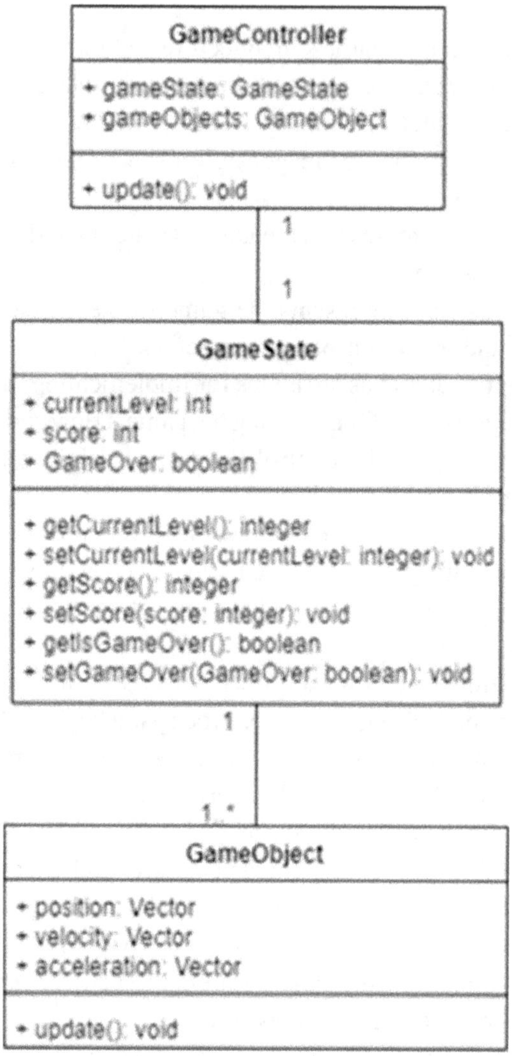

SPECIFIC INCLUSIVE GENRE

A mobile game's architecture can vary depending on the specific requirements and features of the game. For example, we wouldn't need a health bar when playing Candy Crush. But importantly, when a person says "mobile game," one of the first things they think of are games commonly known as RPGs, which is what we'll focus on for our next example. RPG stands for role-playing game, a game in which players don the persona of a character in a fictional setting. It is notably a term that covers a wide range of games as well. Inside the RPG game, so excluding the hardware side and the backend, the user's concerned with the following classes for a mobile RPG to work:

Classes:

- Game
- Level
- Player
- Enemy
- Obstacle/Terrain
- PowerUp

Attributes:

- Game:
- name: String
- levels: List of Level objects
- current_level: Level
- players: List of Player objects
- Level:
- number: Integer
- enemies: List of Enemy objects
- obstacles: List of Obstacle objects
- power_ups: List of PowerUp objects
- Player:
- name: String
- score: Integer
- lives: Integer
- position: X and Y coordinates
- inventory: List of PowerUp objects
- Enemy:
- name: String
- health: Integer
- position: X and Y coordinates
- attack: Integer
- speed: Integer
- Obstacle:
- type: String (e.g. "wall", "tree")
- position: Tuple of x and y coordinates
- PowerUp:

- type: String (e.g. "health", "attack")
- value: Integer

Operations:

- Game:
  start(): Initializes the game and sets the current level to the first level.
  advance_level(): Advances the game to the next level by updating the current_level attribute to the next level in the levels list.
  end(): Ends the game and displays the final scores of all players.
- Level:
  initialize(): Initializes the level by creating instances of Enemy and Obstacle objects and placing them in the level.
  check_completion(): Checks if the level has been completed by checking if all enemies have been defeated.
- Player:
  move(direction: String): Allows the player to move in a given direction (e.g. "up", "down", "left", "right").
  attack(enemy: Enemy): Allows the player to attack an enemy.
  pick_up(power_up: PowerUp): Allows the player to pick up a power up.
- Enemy:
  move(direction: String): Allows the enemy to move in a given direction.
  attack(player: Player): Allows the enemy to attack a player.
- Obstacle:
  block(): Prevents the player or enemies from moving through the obstacle.
- PowerUp:
  activate(player: Player): Activates the power up and applies its effects to the player (e.g. increasing health or attack).

RELATIONSHIPS

- A Game has many Levels.
- A Level belongs to a Game.
- A Level has many Enemies and Obstacles.
- A Player belongs to a Game.
- An Enemy belongs to a Level.
- An Obstacle belongs to a Level.
- A PowerUp can belong to a Level or a Player.

Figure 5. UML for an RPG game

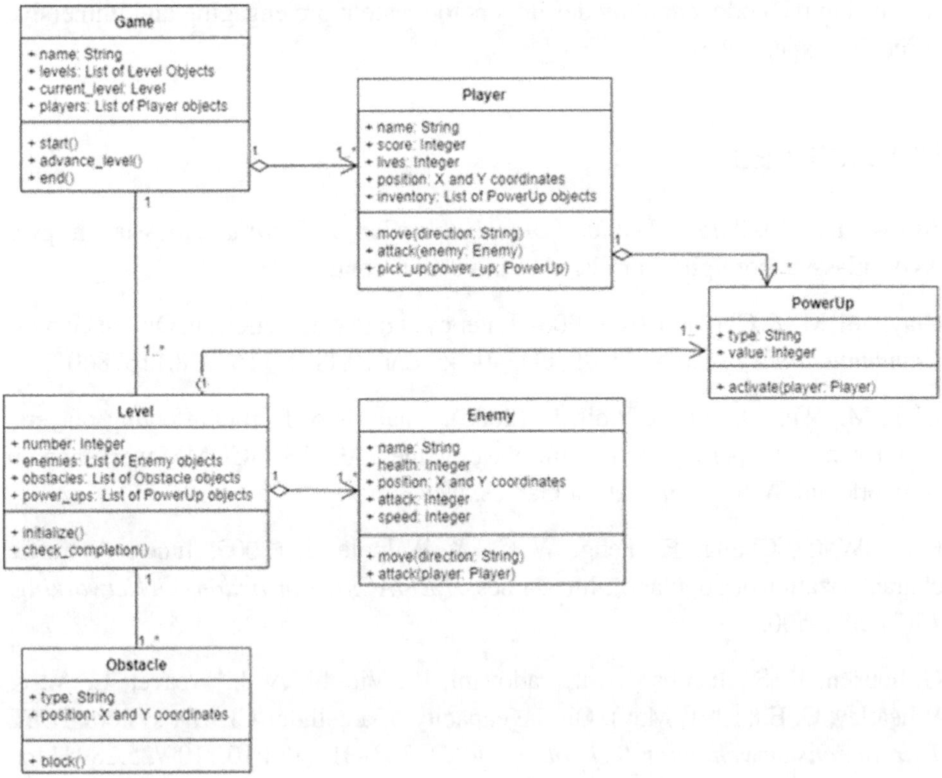

In conclusion, class models are an essential part of mobile game architecture. These models provide a framework for organizing and structuring the code, making it easier for developers to understand and maintain the game. The use of class models has several benefits, including providing a clear and consistent structure for the code, organizing the code into logical units, and identifying the relationships and dependencies between objects in the game.

CONCLUSION

As mobile games continue to evolve and become more complex, class models will play an increasingly important role in organizing and structuring the code. Future developments in AI and ML technologies, as well as the use of cloud computing and other distributed technologies, are likely to require the creation of new classes and models to represent these technologies in mobile games. Overall, the use of class models in mobile game architecture is likely to continue to be a critical part of the development process. These models provide a framework for organizing and structuring the code, enabling developers to create more engaging and immersive gameplay experiences.

REFERENCES

Atlassian. (n.d.). *What Is Version Control: Atlassian Git Tutorial.* Atlassian. https://www.atlassian.com/git/tutorials/what-is-version-control

Claypool, M., & Claypool, K. (2006). Latency and Player Actions in Online Games. *Communications of the ACM*, *49*(11), 40–45. doi:10.1145/1167838.1167860

Dick, M., Wellnitz, O., & Wolf, L. (2005). Analysis of factors affecting players' performance and perception in multiplayer games. ACM SIGCOMM workshop on Network and System Support for Games. NetGames.

Feng, W.-C., Chang, F., Feng, W.-C., & Walpole, J. (2005, June). A traffic characterization of popular on-line games. *IEEE/ACM Transactions on Networking*, *13*(3), 588–500.

Gilhousen, K. S., Jacobs, I. M., Padovani, R., Viterbi, A. J., Weaver, L. A., & Wheatley, C. E. (1991, May). On the capacity of a cellular CDMA system. *IEEE Transactions on Vehicular Technology*, *40*(2), 303–312. doi:10.1109/25.289411

Hands, D. S. (2004, December). A Basic Multimedia Quality Model. *IEEE Transactions on Multimedia*, *6*(6), 806–816. doi:10.1109/TMM.2004.837233

Honcharenko, W., Kruys, J. P., Lee, D. Y., & Shah, N. J. (1997, January). Broadband wireless access. *IEEE Communications Magazine*, *35*(1), 20–26. doi:10.1109/35.568192

Ranjan, R. (2022). The Mobile App Architecture Guide for 2022, Insights - Web and Mobile Development Services and Solutions. *Net Solutions.* https://www.netsolutions.com/insights/mobile-app-architecture-guide/

Ries, M., Nemethova, O., & Rupp, M. (2008). Video Quality Estimation for Mobile H.264/AVC Video Streaming. *Journal of Communication*, *3*, 41–50.

Schwartz, M. (1995).Network management and control issues in multimedia wireless networks. *IEEE Personal Communications, 2*(3), 8-16.

Tasaka, S., & Watanabe, Y. (2007, November). Real-Time Estimation of User-Level QoS in Audio-Video IP Transmission by Using Temporal and Spatial Quality. *IEEE Globecom Workshops*, 2661–2666. doi:10.1109/GLOCOM.2007.505

Winkler, S., & Dufaux, F. (2003). Video Quality Evaluation for Mobile Applications. *Proc. of SPIE Conference on Visual Communications and Image Processing*. IEEE.

Wong, W. W. S., & Sousa, E. S. (1999). A CDMA mixed cellular architecture with space division duplex (SDD) for symmetric/asymmetric wireless communications. WCNC *1999 IEEE Wireless Communications and Networking Conference (Cat. No.99TH8466),* (pp. 1338-1342). IEEE. 10.1109/WCNC.1999.796955

Wu, D. (2001). Streaming Video over the Internet: Approaches and Directions. *IEEE Transactions on Circuits and Systems for Video Technology*, (Mar), 282–300.

Yamagishi, K., & Hayashi, T. (2006, November). Opinion Model for Estimating Video Quality of Videophone Services. *IEEE Globecom Workshops*, 1–5. doi:10.1109/GLOCOM.2006.458

Yasui, T. (2005). Influences of network latency and packet loss on consistency in networked racing games. ACM SIGCOMM workshop on Network and System Support for Games. NetGames.

Chapter 13
SDLC Phases of a Mobile Application

Drin Hoti
Rochester Institute of Technology, Kosovo

Monika Maloku
Rochester Institute of Technology, Kosovo

Klinton Gashi
Rochester Institute of Technology, Kosovo

ABSTRACT

Class models are an essential part of mobile game architecture, providing a framework for organizing and structuring the code. These models help developers to understand the relationships and dependencies between different objects in the game, allowing for more efficient and effective code development. The use of class models is likely to continue to evolve and improve in the future, as mobile games become more complex and sophisticated. As AI and ML technologies become more advanced, developers will need to create new classes and models to represent these technologies in mobile games. Additionally, the use of cloud computing and other distributed technologies is likely to become more prevalent, requiring the creation of new classes and models to represent these technologies in the game. Overall, class models will play a critical role in enabling developers to create more engaging and immersive mobile gameplay experiences.

DOI: 10.4018/978-1-6684-8582-8.ch013

INTRODUCTION

The planning phase is the first phase of our Systems Development Life Cycle of the mobile application. Due to the complexity of mobile application development, preparation is crucial to project success. Failure to plan an application can result in an unclear concept and insufficient data (Frajtak et al., 2015)(Maqbool et al., 2019). Not to mention the potential delays and extra costs that can arise from poor organization, planning, and communication. More specifically, unforeseen changes, adjustments, and errors are costly and can lead to costly rework, delays, and even project failure (CodePlex Archive, Oct. 2019). For the planning phase, we need to make sure to complete certain tasks and objectives in order to have a complete SDLC project, later on, to be able to spot possible problems before they have an effect on the development process if they make a thorough plan for the upcoming development cycle (Nazir et al., 2017). A study and justification of the procurement management strategy are important in order to get the entire organization on board for the project. To read all the resources necessary (Ahmed et al., 2017) (Umuhoza et al., 2015).

Develop and refine the project's scope, timetable, risk profile, and costs. These ties into getting the resources ready for the project. Having completed the scope and timetable it will get into view if the project will be ready when needed and at minimum what resources will be needed to procure. The risk profile and costs will determine if the project is worth pursuing on the established constraints (Huang et al., 2018)(Roubi et al., 2015). Evaluate and explain procedures for coordinating related support plans. Establish common goals and objectives. Establishing common goals and objectives is the first step in coordinating related support plans. To do this, it is necessary to understand the wishes and needs of the project and to explain in detail what kind of support is needed. Creating an action plan is the next step. After establishing common goals and objectives, a strategy should be developed. This plan should include a schedule of when tasks will be completed, who will be responsible for them, what resources will be needed, etc. (Ambriola et al., 2006). Progress should be tracked regularly to ensure that the plan is being implemented as intended and that goals are being met. This should include feedback from the client of the project. We also will need to brainstorm for possible ideas ahead of time and select the best idea among the possible solutions.

Among other things, in order for the project to succeed we must make sure the following tasks are also considered, planned for, and completed:

- Review possible application development companies and their services, if we intend to get a third party to develop our application
- From that third party, discuss price and delivery with the selected vendor

- Ensure the development process is of high quality and in line with the project strategy
- Determine and purchase necessary hardware and software resources
- Oversee user testing procedures to ensure proper operation
- Develop quality control procedures and ensure that all defects and problems are accounted for and corrected in a timely manner
- Prepare project budgets and control expenses
- Develop strategies for post-launch maintenance and support

Gathering, researching, and documenting application needs and requirements are all part of mobile application requirements analysis. Understanding user requirements, application goals, competitive research, and technical requirements are also part of this process (M. D. Lozano et al., 2000). It also includes creating use cases, user stories, and wireframes to ensure that the application meets the needs of the users (Tiwari et al., 2019). Developing a strategy to test the application and ensure that it complies with the required specifications is the final step.

DEFINING REQUIREMENTS PHASE

In this phase of the Software Development Life Cycle, we will define the requirements of the mobile application. The requirements we have gathered for this project were done by gathering information to what the market needs (Rhazali et al., 2016) (Liu et al., 2004) (Escalona et al., 2003). We have concluded that the application with have the following functional requirements:

The program should provide the functions for group discussion to be held. When registering, all users must provide their name, username, a password, and their country of residence.

- It should provide the ability to search information
- Access to every function relevant to your device
- All users must be able to login with their username and password
- All users must be able to access their personal information and modify them
- Users can request to enter groups, if the groups are public
- Users can invite other users to the group
- Users are able to post images, videos, GIF's, and texts on the groups page
- Users are able to comment on every post
- Users should be able to react to every post with a like, dislike, or emote

NON-FUNCTIONAL REQUIREMENTS INCLUDE

The system stores all relevant user information. Users are notified of any changes to their accounts. Gives the user on an iOS or Android smartphone a complete experience (Bernaschina et al., 2018)(Zhong et al., 2018)(Sibarani et al., 2014). All user passwords and personal information will be stored, secured, and encrypted. The application will be easy to use with an accessible user interface. Images have a limited file size and dimensions in order to be easily implemented.

SEQUENCE DIAGRAM

Figure 1. Basic Sequence Diagram for a chat app

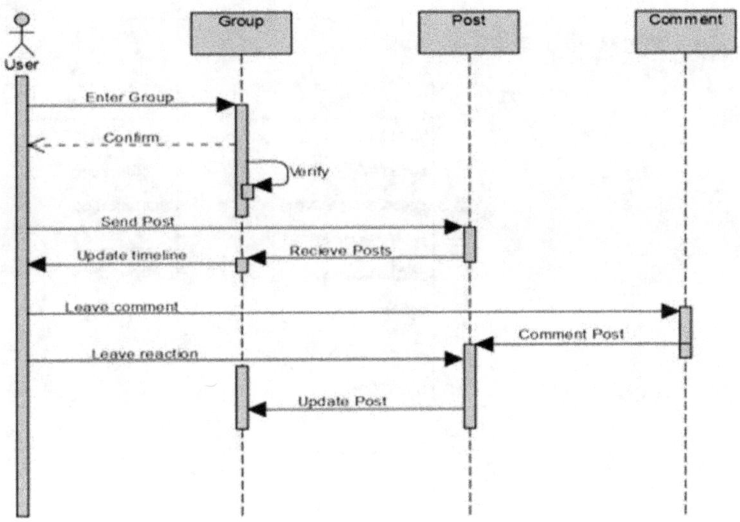

CLASS DIAGRAM

Figure 2. Class Diagram of a mobile chat app

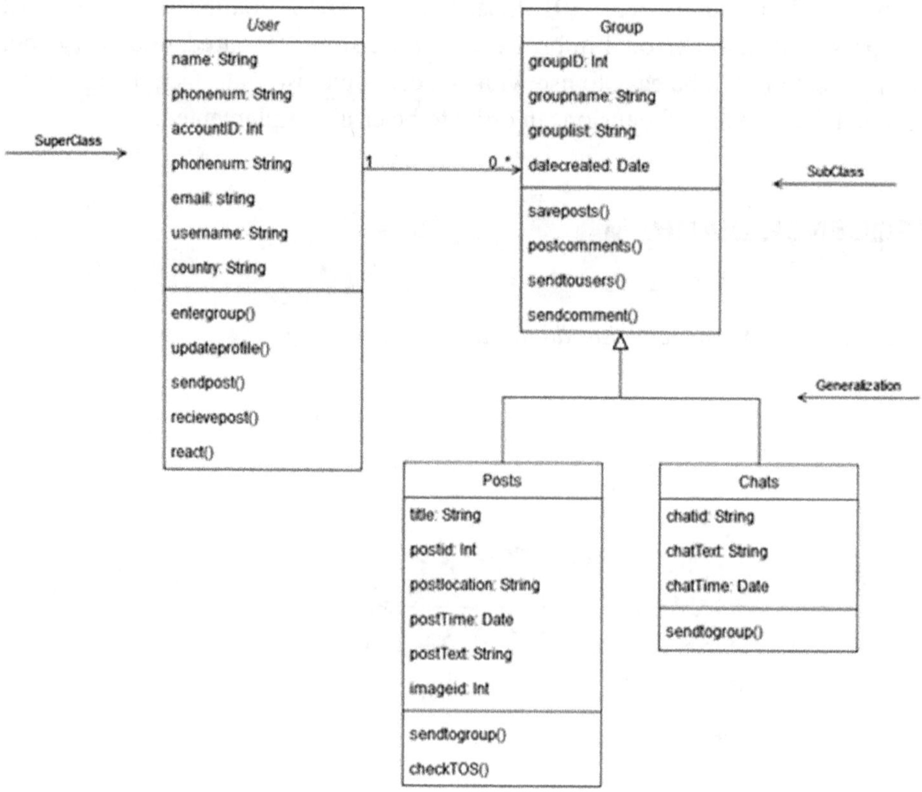

USE CASE DIAGRAM

Figure 3. Use Case of application

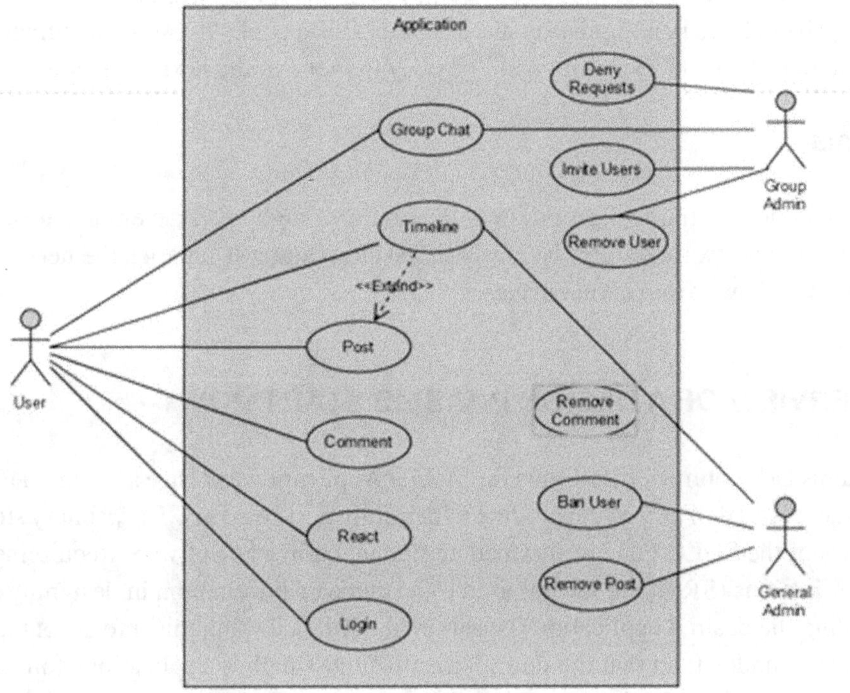

DESIGN PHASE

The design phase is the third phase of the Systems Development Life Cycle. This phase helps in determining the overall architecture of the system (He et al., 2016) (Vemuriet al., 2017). This phase's goal is to convert the business requirements established in earlier phases into a coherent system architecture that is practical, reliable, and beneficial to the business.

Objectives

In order to have a successful Design Phase there are a few objectives that need to be established:

Converting all logical requirements into a complete and thorough physical model where the specifications cover all the components and details of the system. Depending on the type of system that is intended to be developed, we put more specifications to the design phase (Delgado et al., 2016). In our case of a design phase in the Mobile System Development Life Cycle, there is a focus on the components that encompass screen design features for mobiles, data, UI components and prototypes, all these depending on the type of mobile device in which the system will be used in as well (Neyveli et al., 2019)(Yigitbas et al., 2016)(da Costa et al., 2014). A fulfillment of all the requirements so that it would be okay to move to the next sequential phase.

Goals

The intention of the Design phase is to achieve a high-level development of the system and software during the Design Phase, such that it satisfies the needs and objectives drawn from earlier stages.

OVERVIEW OF A DESIGN PHASE'S START POINT

Systems Development Life-Cycle for Mobile Applications all have a common goal; the successful delivery of the product with minimal to no errors. The primary steps/phases of the SDLC that are summed up in one, known as Software Requirements Specifications (SRS), are further used as a source of information in designing and building the desired application (Ansari et al., 2017). Taking this into account, we are led to understand that the development of most mobile applications follows a linear waterfall model, however, they do tend to overlap each other throughout the process, especially through the Design Phase where we see ourselves going back and forth to make changes so that the product comes out reliable, practical, and effective.

Figure 4. SDLC phases of chat application development

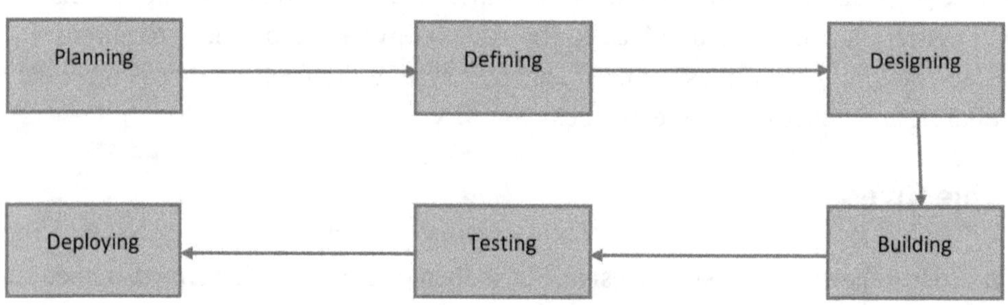

Figure 5. Design Documentation Specification

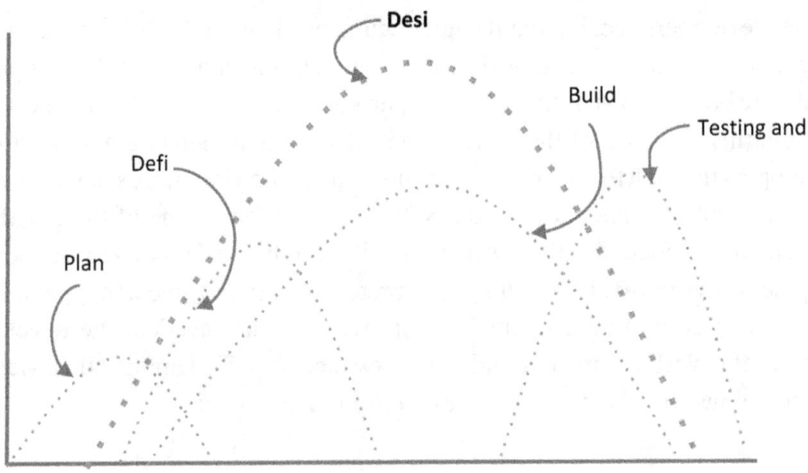

Throughout this phase, there is a recommendation of numerous designs that are usually documented in a Design Documentation Specification. The DDS can be a broad concept and it will be difficult to put it into use to create a system, particularly for beginners, if we don't focus on particular objectives. To make this easier to understand we take as an example the Instagram App for Mobile. Instagram is a platform for sharing photos and videos with other users. A user of this service can follow other users and share images. Each user will have their own newsfeed (Salini et al., 2016)(Harris et al., 2015). This platform needs to fulfill some functional requirements, so it will successfully deliver what the users are interested in. The users of Instagram should first be able to upload, download, and view images. Users can be followed and tagged by another user (Brambilla, Mauri et al., 2014). The system should also show a user's News Feed, which is a collection of images from everyone the user follows. All these design requirements and many more, are documented in the DDS as shown below:

The picture above is a very simple example of a DDS. That DDS is something that must be reviewed and especially approved by the stakeholders of the project (Laaz et al., 2016). Once those steps are completed, the work on the other phases can begin to happen. And so, we move on to the next phase: building phase.

BUILDING PHASE

Once the Developers receive the design document, then the building phase starts. This stage entails converting the design into a language understandable by computers. The tasks are broken down into sections or parts and assigned to different developers throughout this 4th stage of the SDLC. The software will then begin to be built by the developers through the formation of code in the coding languages that they select (Ilieva et al., 2005). This phase of the SDLC is regarded as one of the lengthiest. To implement the code, developers need specific established coding principles and development environments including interpreters, compilers, and debuggers. In the situation when certain adjustments or improvements are needed, the developers can present the work completed to the business analysts. To simplify this, we have portrayed below how the building phase pretty much works:

Figure 6. Build phase

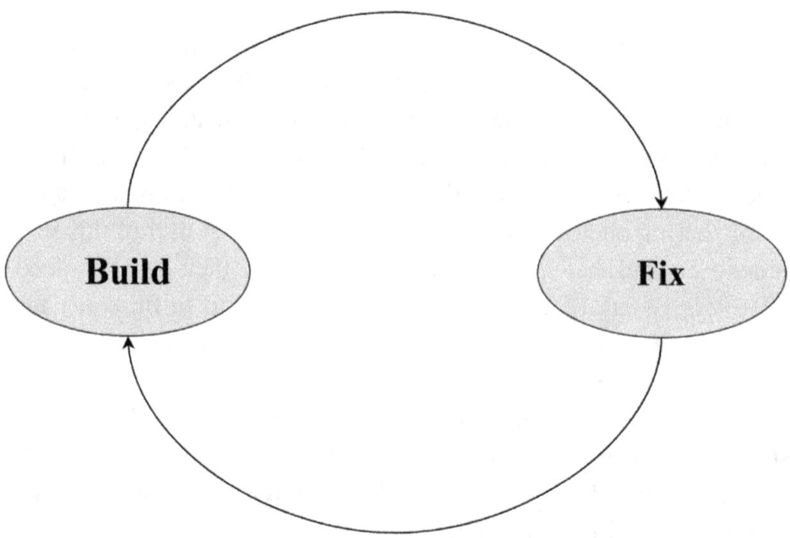

TESTING THE PRODUCT

The testing phase is also an important process in the SDLC. This phase is initiated after the initial development of the product or software (Nieto et al., 2016). The purpose of software testing is to focus on detailed investigation and discovery of empty spaces, logical errors in the code, obstacles in the user experience, and all the hidden things that our program can display.

Figure 7. Phases of testing

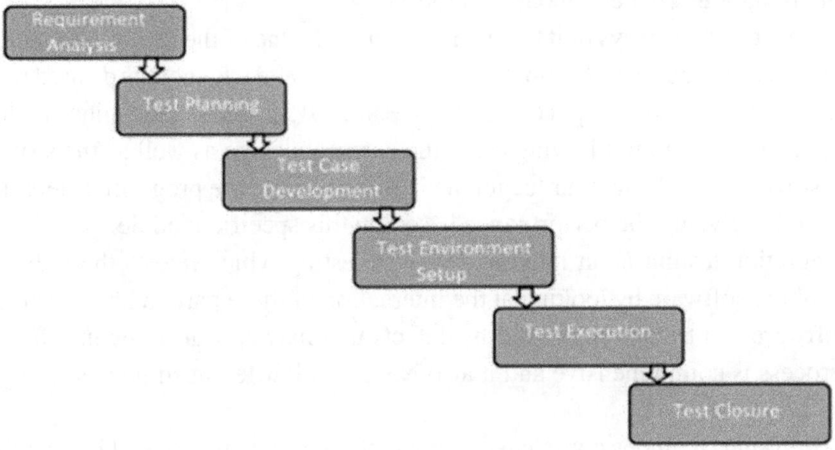

This phase is done by a group of software developers who are based on the data obtained from the previous phases of the SDLC. Various tests are performed on the software, assessing whether this software meets the requirements given by the clients and whether it has achieved the purpose for which this software is intended (Ahsan et al., 2017) (Brambilla et al., 2016). People who do the testing also make sure that the software contains all the functional and non-functional requirements received from the questionnaires from the clients. This phase, in addition to starting after the software construction phase, is also present at any time from the beginning of the SDLC. This process also can be divided into different phases in different methods (Leffingwell et al., 2003).

TESTING PHASES

Compliance testing: This type of testing, as I mentioned above, is present at each stage of the SDLC process for the application we want to build, in our case for a communication application. The purpose of this process is to check all the time whether we are following the procedures and policies established at the beginning by the company, us, or the client. Unit testing: This step is the first step that starts within the testing phase. Because it is the first step, it starts from the most basic things, dividing the code into individual parts and testing each part, verifying the functionalities one by one for each class, method, or function, procedure and fix or remove bugs or errors that can delay or hinder the process of creating the application (Bancora et al., 2015). Also, this part can be used by developers because it helps them to understand and make the necessary changes in the code to achieve specifically the functionalities created by our developers.

There are several ways unit testing is performed. One of the techniques includes part of the base code, tests the input and user interface and whether the desired results are being returned in the output (Meedeniya et al., 2015). Another technique includes testing the output but not leaving aside the external design as well as the structure of the software, while another technique includes giving the program different test cases, and analyzing the performance based on this specific module.

Integration testing is an internal phase of testing which retests the individual actors of the software by looking at the interaction of these parts with each other in the software and tries to highlight the defects that may appear in the interface. So this process is comprehensive and it also has several different strategies.

- Big Bang: Includes complete documentation of the process and is considered dangerous.
- Bottom-Up: Starts hierarchically, starting from low-level parts to high. This strategy is very efficient in the process of detecting errors.
- Top down: Tests the integrated high-level components first.

System testing is the main part of testing. This testing is done by evaluating the entire system as a whole, whether it is in coordination with the requirements obtained from the points above. This testing finds errors both within the internal units and the software as a whole. System testing evaluates functionality, scalability and design. It also tests the behavior towards different events and evaluates whether it is ready to meet the expectations of the customers. This test, like the entire testing process, is repeated endlessly until the results are successful and there are no other errors. One of the last stages in this process is performance testing. This test and the findings of this test are usually compared with non-functional requirements and should be adapted in such a way that there are no differences between expectation and reality (Antonelli et al., 2015) (Koren et al., 2018). During this test, it is evaluated how many one-time visits the client can accept, how long it takes to respond, when they are expected to be profitable, as well as some other proposals (Brambilla et al., 2017). We can do the testing of our application through these methods, where the first part will deal with the individual parts, then continuing with the other slightly more complicated steps. At this stage, all software installations, adjustments, and completions are completed. Our application will help you a lot in different aspects. This stage must be completed to enable the user's communication with the last device, be it a phone or a laptop (Brambilla et al., 2014).

DEPLOYMENT AND MAINTENANCE PHASE

Figure 8. Deployment inputs and deliverables
https://www.loginworks.com/blogs/categorize-sdlc-phases-development/

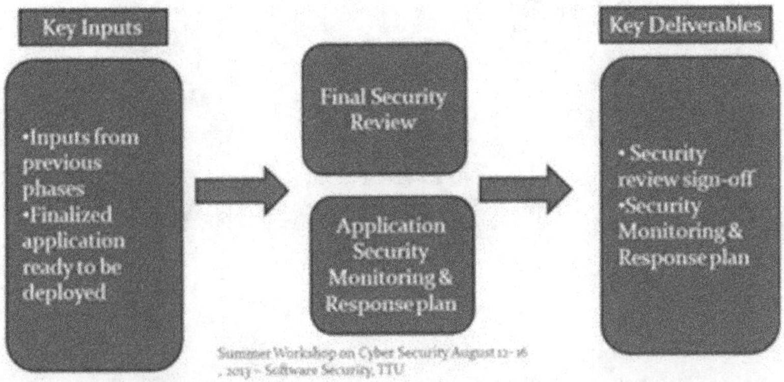

Deploying the new system in its intended environment is the main activity of the implementation phase. Training end users and getting ready to hand the system over to maintenance people are examples of supporting actions (Anwar et al., 2019). The system then moves into the Testing phase where the system is tested in different cases of users' actions interacting with it. Projects with numerous releases need to go through the implementation phase more than once if it has more than one version (M. Hamdani et al., 2018). The implementation phase is the longest phase in SDLC and it has to be revised quite well before it has ended.

DEVELOPING THE SYSTEM

A group of developers comes together or they work remotely on their private machines after they get the design documents and diagrams to implement the software application. This activity is made by the developer's team individually, guided by project managers. The software is separated into different modules, and each of them will have a project manager and developers' team (Acerbis et al., 2015). They will write the source code for their units, and get together to find solutions other than that. Deployment is the most complex process since the previous work comes into play again, but the process is not only deployment but also massage (Gotti et al., 2016).

Figure 9. SDLC diagram

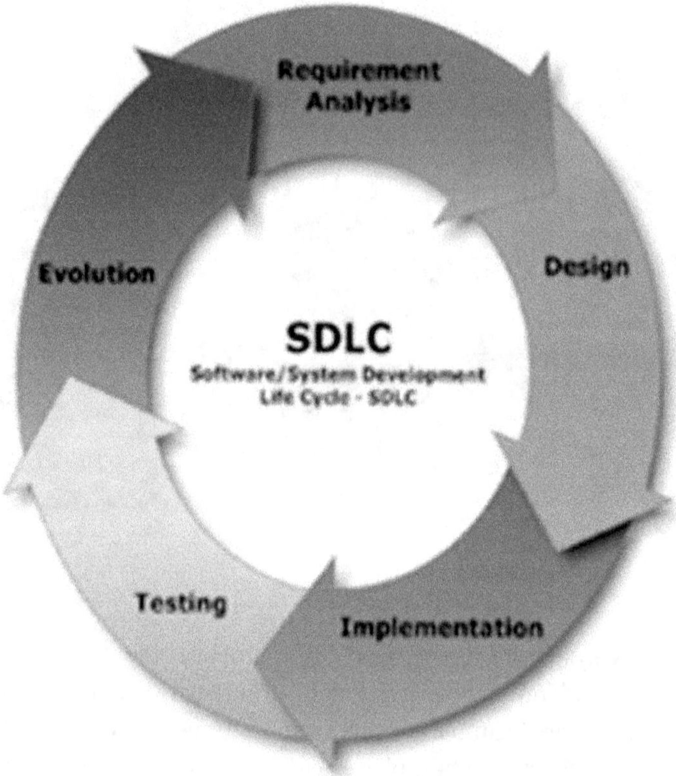

This stage is critical especially for us, to be able to complete the opportunity we have 20 times 80 (Yousaf et al., 2019). This also applies to our application. For example, to save the mayors of the communes for a while. So as a conclusion of SDLC, I help a lot in finding the training of system, also to increase the work in another aspect related to the aspect.

CONCLUSION

The architecture of mobile games must include class models since they give the code a framework for organization and structuring. These models enable more effective and efficient code development by assisting developers in understanding the connections and interdependencies between various game components. As mobile games get more sophisticated and complicated, the use of class models is anticipated to continue to develop and advance. To depict AI and ML in mobile games, developers will need to design new classes and models as these technologies improve. Additionally, the use of remote technologies such as cloud computing is projected to increase, necessitating the development of new classes and models to represent these technologies in the game. In general, class models will be essential for enabling developers to produce more deep and compelling mobile gameplay experiences.

REFERENCES

Acerbis, R., Bongio, A., Butti, S., & Brambilla, M. (2015). Model-driven development of cross-platform mobile applications with WebRatio and IFML. *Proc. MOBILESoft*, (pp. 170-171).

Ahmed, M. A., Butt, W. H., Ahsan, I., Anwar, M. W., Latif, M., & Azam, F. (2017). A novel natural language processing (NLP) approach to automatically generate conceptual class model from initial software requirements. Proc. 8th Int. Conf. Inf. Sci. Appl. (ICISA). IEEE.

Ahsan, I., Butt, W. H., Ahmed, M. A., & Anwar, M. W. (2017). A comprehensive investigation of natural language processing techniques and tools to generate automated test cases. *Proc. 2nd Int. Conf. Internet Things Data Cloud Comput. (ICC)*. IEEE.

Ambriola, V., & Gervasi, V. (2006). On the systematic analysis of natural language requirements with circe. *Automated Software Engineering*, *13*(1), 107–167.

Ansari, A., Shagufta, M. B., Fatima, A. S., & Tehreem, S. (2017). Constructing test cases using natural language processing. *Proc. IEEE 3rd Int. Conf. Adv. Elect. Electron. Inf. Commun. Bio-Inform. (AEEICB)*, (pp. 95-99). IEEE.

Antonelli, H., Silva, E., & Fortes, R. (2015). A model-driven development for creating accessible Web menus. Proc. DSAI Conf. ACM.

Anwar, M. W., Rashid, M., Azam, F., Kashif, M., & Butt, W. H. (2019, December). A model-driven framework for design and verification of embedded systems through SystemVerilog. *Design Automation for Embedded Systems*, *23*(4), 179.

Bancora, M., Brambilla, M., Ripamonti, D., & Vaccarella, A. (2015). Personal social and event organization through Web and mobile apps: The fluxedo case. *Proc. ICWE Eng. Web Big Data Era*, (pp. 609-612). IEEE.

Bernaschina, C. (2018). *ALMOsT.js: An agile model to model and model to text transformation framework*. Proc. Web Eng. ICWE.

Brambilla, M., & Fraternali, P. (2014, September). Large-scale model-driven engineering of Web user interaction: The WebML and WebRatio experience. *Science of Computer Programming*, *89*, 71–87.

Brambilla, M., Koka, T., Mauri, A., & Umuhoza, E. (2017). Integrating modeling languages and Web logs for enhanced user behavior analytics. Proc. 26th Int. Conf. World Wide Web Companion, (pp. 171-175). IEEE.

Brambilla, M., Mauri, A., Franzago, M., & Muccini, H. (2016). A model-based method for seamless Web and mobile experience. *Proc. 1st Int. Workshop Mobile Develop.*, (pp. 33-40). IEEE.

Brambilla, M., Mauri, A., & Umuhoza, E. (2014). Extending the interaction flow modeling language (IFML) for model driven development of mobile applications front end. Proc. MobiWIS, (pp. 176-191). IEEE.

CodePlex Archive. (2019). CodePlex Archive. *Codeplex*. https://sharpnlp.codeplex.com/

da Costa, S. S. L., Neto, V. V. G., & de Oliveira, J. L. (2014). A user interface stereotype to build Web portals. *Proc. 9th Latin Amer. Web Congr.* IEEE.

Delgado, A., Calegari, D., & Arrigoni, A. (2016, December). Towards a generic BPMS user portal definition for the execution of business processes. *Electronic Notes in Theoretical Computer Science*, *329*, 39–59.

Escalona, M. J., & Koch, N. (2003, February). Requirements engineering for Web applications—A comparative study. *Journal of Web Engineering*, *2*, 193–212.

Frajtak, K., Bures, M., & Jelınek, I. (2015). Transformation of IFML schemas to automated tests. Proc. Conf. Res. Adapt. Convergent Syst. (RACS). IEEE.

Gotti, Z., & Mbarki, S. (2016). GUI structure and behavior from Java source code analysis. *Proc. IEEE Int. Colloq. Inf. Sci. Technol. (CiSt),* (pp. 251-256). IEEE.

Hamdani, M. (2018). A systematic literature review on interaction flow modeling language (IFML). *Proc. Int. Conf. Manage. Eng. Softw. Eng. Service Sci.*, (pp. 134-138). IEEE.

Harris, C. B., & Harris, I. G. (2015). Generating formal hardware verification properties from natural language documentation. *Proc. 9th Int. Conf. Semantic Comput. (IEEE ICSC),* (pp. 49-56). IEEE.

He, C., & Mussbacher, G. (2016). Model-driven engineering and elicitation techniques: A systematic literature review. Proc. REW. IEEE.

Huang, A., Pan, M., Zhang, T., & Li, X. (2018). Static extraction of IFML models for Android apps. *Proc. 21st ACM/IEEE Int. Conf. Model Driven Eng. Lang. Syst. Companion (MODELS),* (pp. 53-54). ACM/IEEE.

Ilieva, M. G., & Ormandjieva, O. (2005). Automatic transition of natural language software requirements specification into formal presentation. *Proc. Int. Conf. Appl. Natural Lang. Inf. Syst. (NLDB)*, (vol. 3513, pp. 392-397). IEEE.

Koren, I., & Klamma, R. (2018). The exploitation of OpenAPI documentation. Proc. Companion Web Conf. (WWW), (pp. 781-787). IEEE.

Laaz, N., & Mbarki, S. (2016). Integrating IFML models and owl ontologies to derive UIs Web-Apps. *Proc. Int. Conf. Inf. Technol. Org. Develop. (IT4OD).* IEEE.

Leffingwell, D., & Widrig, D. (2003). *Managing Software Requirements: A Use Case Approach.* Addison-Wesley.

Liu, D., Subramaniam, K., Eberlein, A., & Far, B. H. (2004). Natural language requirements analysis and class model generation using UCDA. *Proc. Int. Conf. Ind. Eng. Appl. Appl. Intell. Syst.*, (pp. 295-304).

Lozano, M. D., Gonzalez, P., & Ramos, I. "User interface specification and modelling in an object-oriented environment for automatic software development", Proc. Int. Conf. Technol. Object-Oriented Lang. Syst. (TOOLS), pp. 373-381, 2000.

Maqbool, B., Azam, F., Anwar, M. W., Butt, W. H., Zeb, J., Zafar, I. (2019). A Comprehensive investigation of BPMN models generation from textual requirements—Techniques tools and trends. *Proc. ICISA*, (vol. 514). IEEE.

Meedeniya, D., Perera, I., & Bowles, J. (2015). Transformation and composition of software design models for Model Driven Development. Proc. Int. Conf. Ind. Inf. Syst. (ICIIS). IEEE.

Nazir, F., Butt, W. H., Anwar, M. W., & Khattak, M. A. K. (2017). The applications of natural language processing (NLP) for software requirement engineering—A systematic literature review. *Proc. 8th Int. Conf. Inf. Sci. Appl. (ICISA)*,(vol. 425, pp. 485-493). IEEE.

Neyveli, V. R. N., Sivakumar, S. S., Arunagiri, D., Arumugam, C., & Veeramani, A. M. (2019). An approach to estimate the size of Web application using IFML user interface model. *Proc. Amity Int. Conf. Artif. Intell. (AICAI)*, (pp. 292-295). IEEE.

Nieto, V., Castro, V., Lopez, F., Ferro, R., & Gonzalez, C. (2016, January). Model driven architecture software and interaction flow modelling language for tourism data acquisition in Colombia. *Communications in Computer and Information Science, 657*, 368–379.

Object Management Group IFML Specifications Version 1. (2019). OMG.org. http://www.omg.org/spec/IFML/1.0/

Object Management Group Portals. (2019). OMG Wiki. https://www.omgwiki.org/

Rhazali, Y., Hadi, Y., & Mouloudi, A. (2016). A model transformation in MDA from CIM to PIM represented by Web models through SoaML and IFML. *Proc. 4th IEEE Int. Colloq. Inf. Sci. Technol. (CiSt),* (pp. 116-121). IEEE.

Roubi, S., Erramdani, M., & Mbarki, S. (2015). A model driven approach to generate graphical user interfaces for rich Internet applications using interaction flow modeling language. Proc. ISDA. IEEE.

Salini, A., Malavolta, I., & Rossi, F. (2016). Leveraging Web analytics for automatically generating mobile navigation models. *Proc. IEEE Int. Conf. Mobile Services (MS),* (pp. 103-110). IEEE.

Yigitbas, E. & S. Sauer, S. (2016). Engineering context-adaptive UIs for task-continuous cross-channel applications. In *Human-Centered and Error-Resilient Systems Development*. Springer.

Sibarani, E. M., Hutagaol, A., Simarmata, D., & Manihuruk, J. (2014). Actor and use case extraction from text based requirement specification. Proc. Int. Conf. Image Process. Comput. Ind. Eng. (ICICIE). IEEE.

Tiwari, S., Ameta, D., & Banerjee, A. (2019). An approach to identify use case scenarios from textual requirements specification. *Proc. ISEC*. IEEE.

Umuhoza, E., Brambilla, M., Cabot, J., & Bongio, A. (2015). Automatic code generation for cross-platform multi-device mobile apps: Some reflections from an industrial experience. Proc. 3rd Int. Workshop Mobile Develop. Lifecycle, (pp. 37-44). IEEE.

Vemuri, S., Chala, S., & Fathi, M. (2017). Automated use case diagram generation from textual user requirement documents. *Proc. IEEE 30th Can. Conf. Electr. Comput. Eng. (CCECE),* (pp. 1-4). IEEE.

Yousaf, N., Azam, F., Butt, W. H., Anwar, M. W., & Rashid, M. (2019). Automated model-based test case generation for Web user interfaces (WUI) from interaction flow modeling language (IFML) models. *IEEE Access: Practical Innovations, Open Solutions*, *7*, 67331–67354.

Zhong, Z., Guo, J., Yang, W., Xie, T., Lou, J.-G., & Liu, T. (2018). Generating regular expressions from natural language specifications: Are we there yet? *Proc. Workshop NLP Softw. Eng. (NL4SE)*. IEEE.

Chapter 14
Model View Controller Frameworks for Mobile View Website

Jyoti Khandelwal
University of Engineering and Management, Jaipur, India

ABSTRACT

Websites have evolved over the past few years utilizing a variety of programming languages and frameworks. There were other developers working on them concurrently. Developers deploy their projects utilizing a number of design patterns to handle these sophisticated web apps, which makes the code simpler and easier to work with. They are well known for using the model view controller approach. Model, view, and controller, or MVC, are the three components that make up an application in accordance with the MVC architectural pattern. Each of these parts is made to address particular facets of application development. This chapter focuses on MVC for mobile device website viewing. The various sections of this chapter address the introduction to MVC, MVC components, the architecture pattern, several techniques used to implement the pattern in the project, and an example of MVC architecture. Towards the end of the chapter, the advantages and disadvantages of MVC are examined.

INTRODUCTION

The creation of web applications has advanced significantly since the world wide web's inception. Today, a variety of technologies and programming languages can be used to build web applications, but due to the world wide web's explosive growth,

DOI: 10.4018/978-1-6684-8582-8.ch014

these technologies were unable to keep up. Many players tried to develop novel and unusual technologies, mostly to enhance user experience and aid web application developers in creating quicker and more potent websites. Some of these technologies, such as Microsoft Silverlight and Java Applets, have played significant roles in web development but have suffered significant usage declines in recent years. On the other hand, numerous technologies, like JavaScript, Flash, and XML, have developed from being insignificant toys to being strong and vital components of the modern web ecosystem. Today's web environment uses HTML and CSS to convey data to consumers, and JavaScript is used for interactivity. These systems are referred to as "front-end" or "client-side" systems. Contrarily, "back-end" or "server-side" technologies refer to methods for processing and storing data (Subari et al.,2021). The fact that web technologies are being employed more frequently to create various complicated applications has led to another issue. The critical point to take away from this is that an application must clearly distinguish between appearance, logic, and data storage. The model-view-controller (MVC) design is presently receiving the most attention, even though there are a few application design paradigms and patterns that can help with this problem. MVC programming implements this three-way combining to organize tasks relating to the application domain (its model), the display of the application's state (its view), and the user's interaction with the model as well as the view (the controller).

THE MVC PATTERN AND LITERATURE OVERVIEW

The MVC was invented by Trygve Reenskaug in 1978 or 1979, when he was visiting a scientist at Xerox Palo Alto Research Center (PARC). According to the first reports on MVC, it is named "Thing Model View Editor," but it was quickly modified to "Model View Controller."

The primary reason for creating MVC was to provide user control over complex data sets. The MVC practice evolved in response to time and demand. Because it was created before web browsers, the MVC pattern was initially used as an architectural pattern for graphical user interfaces (GUI). MVC is now used to create web applications. Ruby on Rails, Laravel, Zend Framework, CherryPy, Symphony, etc. are a few web frameworks that leverage the MVC idea (Dey, 2011).

The architecture that offers several views of the same data is known as the MVC design pattern. It splits the application objects into three classes: the Model class, the View class, and the Control class. It also distinguishes between the data layer and the expression layer. The business and data logic are handled by the Model class, the display logic by the View class, and the control processing by the Control class. Three different types of logic are used by MVC to create control and display logic-oriented applications, as well as business and data logic-oriented applications. The business and data logic are unaffected by the change in business processes. The model class is the only thing that changes when business concepts and algorithms change (Leff & Rayfield, 2001). In order to ensure the independence of the modules, MVC separates the data access and display.

A dynamic e-business system or an Internet application can be created and implemented using the MVC design pattern. The e-commerce market is expanding quickly, and the enterprise-level application is emerging quickly. As a result, an increasing number of initiatives at the enterprise level are following the trend of embracing Web technology as their primary technology. Middleware and other server-end technologies are becoming more and more important at the same time. Therefore, the enterprise's information technology department needs a workable strategy for creating the application and connecting it to that middleware, which is adaptable and portable.

UNDERSTANDING MODEL-VIEW-CONTROLLER

We may not even be aware of the MVC examples we see every day. Let's examine an illustration: In a restaurant, the waiter takes your order, goes to the kitchen, and tells the chef what you ordered. To prepare your food, the chef pulls the items from the refrigerator. The waiter goes to the kitchen to retrieve the dish and presents it to you. In the example above, "you" serves as the "view," "waiter" serves as the "controller," "chef" serves as the "model," and "fridge" serves as the "data." You, the view in this example, are unable to speak with the "chef," the model in this instance, directly. The controller, "Waiter," does not need to understand how to carry out your instructions; instead, it simply passed them on to the model, "Chef." In this scenario, the data "fridge" can only be accessed by the model, who is also the "cook."

It is quite simple to maintain even a large database because MVC is essentially a means to arrange and organize your code. MVC architecture is used by many web frameworks. Examples of such web frameworks are shown below:

- Django
- Laravel
- CakePHP
- CodeIgniter
- Catalyst
- Ruby on Rails

MVC (MODEL-VIEW-CONTROLLER) ARCHITECTURE

The model, view, and controller (MVC) pattern architecture emphasize the idea of separation of concerns and facilitates its use among these classes in programs (Majeed & Rauf, 2018). Separation of concerns makes it easier for us to test our program since the relationships between the different application components are more logical and consistent. With the use of MVC, a test-driven development technique is made possible, in which automated test cases are developed before actual code is written. The requirements of new code are defined and verified using these unit test cases before the code is created.

In essence, the MVC pattern layout has three layers. There are different application-specific characteristics. User input logic is implemented in the first layer, business logic in the second, and user interface logic in the third. The connectivity between these three layers is somewhat lax with MVC. The MVC design dictates where each piece of application logic should be placed. By using MVC patterns, parallel programming is made possible. Three developers can work simultaneously on a single application thanks to the fact that each application layer is independent of the others (Kumar et al., 2016).

COMPONENTS INTERACTIONS

The model-view-controller design specifies the interactions between the three sorts of components that make up the application. The application's data management is the responsibility of the template. The controller provides a user input to it. View refers to displaying the model in a specific format. The controller reacts to user input and engages with data model components. The controller receives the data and verifies it before sending it to the model. The MVC cooperation diagram, which shows the connections and interdependence between the figures, is shown in Fig. 1 (a–b) below.

MODEL

The model layer of an MVC application is crucial. In order to portray the output with the aid of views, it maintains the information in the form of data. The database records are shown. The application data are managed by it. The application data, logic definition, function specification, and business rule participation are generally all included. A model could be a single thing or a group of objects put together(Jailia et al., 2016). This layer allows the database to communicate, allowing users to insert,

Figure 1. MVC Collaboration Diagram (a) Communication in MVC (b) 3-Tier Architecture of MVC

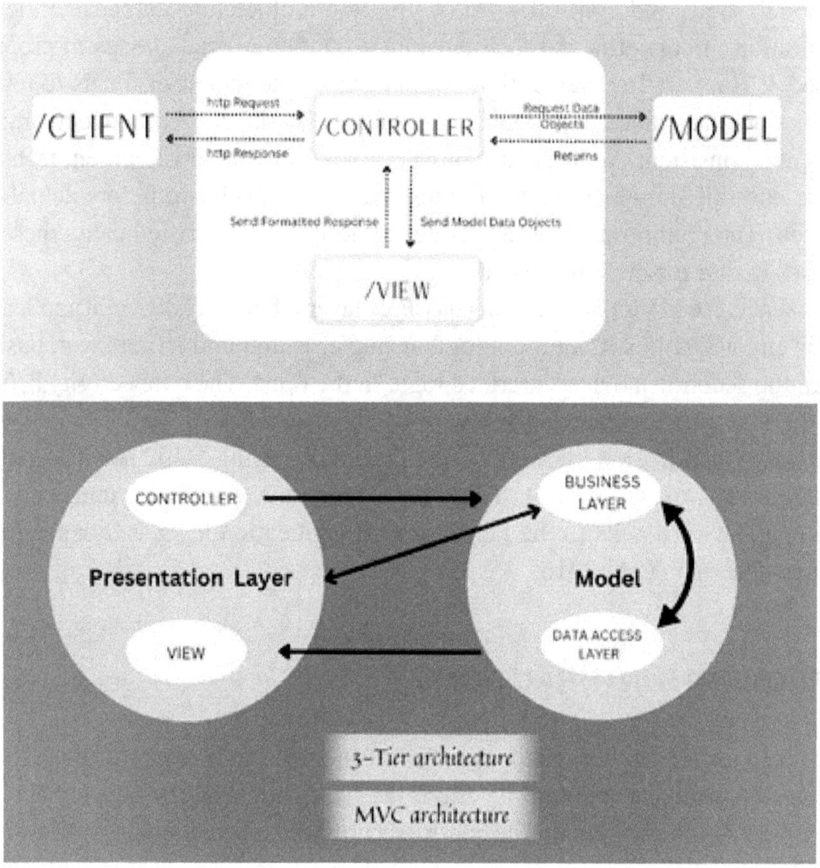

retrieve, and change data in the database. The application's domain model needs to abide by the following guidelines:

- Hold the domain information.
- Include the logic for generating and changing domain data.
- Offers a suitable API that makes the model data and operations on it available.
- Domain data shouldn't abide by the following guidelines:
- Provide specifics on the collection and management of model data.
- Include logic that manages user interaction-based requests to call models.
- Include logic for the user's data display.

VIEW

Views are used to create our program's user interface. Users interact with our web pages using the user interface. The results of the data in the model are shown in view. Through a view, all of the model's data is shown in its totality. Only the essential properties are visible; the unnecessary ones are hidden. It provides us with the advantage of presentation encapsulation as a consequence. Like JSP, ASP, and PHP, it is script-based, making it easy to incorporate AJAX technologies. The guidelines below should be adhered to:

- Contain the markup and logic necessary to deliver data to the user
- The following rules should not be used by a view:
- Have intricate logic
- Include logic for building, storing, or modifying the model.

CONTROLLER

The controller layer is utilized to react to user requests. Controller, carry out the user's requested activities. A controller serves as the link between a user and a machine. The driver controls the model and the view. Every time the data are altered, it modifies the model's data flow and modifies the view. The problem is the separation between the Model and the View (Singh, 2020). The command that changes the state of the data model is executed by the controller after it has processed the data and received it. These components collaborate with the model layer to choose the appropriate view for the user to see based on their requests. The guidelines that the controller abides by are:

- Include the reasoning necessary to set up the model's data.
- Include the logic necessary for the view to display data from the model.
- Include the logic necessary to modify the model in response to user interactions.
- Following rules should not be used controller:
- Alter data outside of the intended use.
- Be composed of logic that controls data persistence.
- Include logic that shows users info.

In essence, the MVC pattern architecture has three layers. It divides the qualities of application. User interaction logic relates to its third layer, while business logic relates to its second layer and user input logic to its first layer (Pop, & Altar, 2014).

Extremely loose connectivity between these three layers is provided by MVC. The MVC pattern is used to specify where all application logic should be placed. The capability of concurrent development is provided by MVC patterns. In other words, three developers will work on one layer of the program since none of the layers are dependent on one another. There will be three engineers working on user input logic (controller logic), user interface logic (view), and business logic at the same time (model). User input processing will be handled by one programmer (controller logic).

WHEN TO USE MVC PATTERN ARCHITECTURE

If we're developing an application with sufficiently compelling client-side functionality to reject using JavaScript exclusively. We shouldn't utilize the MVC pattern design for creating applications that require a lot of server-side processing and little client-side communication. Instead, we should use a straightforward configuration like a web-based form model. The following traits will help us decide whether or not to adopt MVC design in our application (Majeed & Rauf, 2018):

Asynchronous backend communication is required for our application.

Our application contains a feature that prevents whole pages from reloading, such as scrolling endlessly or commenting on a post when using Facebook.

The majority of data manipulation occurs on the client side (browser), as opposed to the server.

On a same page, several methods of delivering the same sort of information are used (navigation).

When our application uses numerous pointless connections to change data (button, switches).

FEATURES OF MVC FRAMEWORK

1. More Rapid Development

MVC enables quick and concurrent development. When a particular web application is created using an MVC model, it is possible for one programmer to work on the view while the other creates the business logic for the web application in the controller. Therefore, compared to an application built using another development model, an MVC-based application can be finished three times quicker (Kalelkar et al., 2014).

2. Ability to Offer Multiple Perspectives

There are various layouts you can create for a model in the MVC Model. New application access points are increasingly needed, and MVC development is without a question the ideal solution. Additionally, this method significantly reduces code duplication by separating data and business logic from the display.

3. Asynchronous Technique Support

The MVC architecture can be combined with the JavaScript Framework. As a result, MVC applications can be set up to work with desktop widgets, browsers that are unique to certain websites, and PDF files. MVC also supports the use of an asynchronous approach, allowing programmers to build apps that load incredibly quickly.

4. The Modification Has No Effect on The Model as a Whole

The user interface of any online application will probably alter more frequently than even the business's operational procedures. The frequent changes show how frequently you update your web application, changing things like colors, fonts, screen layouts, and adding support for new mobile or tablet devices. Additionally, the MVC methodology makes it relatively simple to build a new type of view because the Model component is independent of the Views portion. Therefore, any modifications to the Model won't have an impact on the design as a whole.

5. Data Is Returned Without Formatting by MVC Model

Without any formatting, data is returned using the MVC pattern. As a result, the same components can be utilized by all interfaces. For instance, any type of data can be formatted with HTML, Dream viewer, or Macromedia Flash.

6. SEO Friendly Development Platform

The MVC platform allows for the creation of SEO-friendly web applications and pages. This tool makes it incredibly easy to create URLs that will increase traffic to a particular application. This architecture for development is commonly used in applications that use test-driven development. To create feature-rich online apps, MVC can also be combined with scripting languages like JavaScript and jQuery.

ADVANTAGES AND DISADVANTAGES OF MVC ARCHITECTURE

Advantage

- The MVC architectural style divides an app into three parts: a model, a view, and a controller. This makes it easier for all of us to handle the complexity of an app.
- The MVC front controller design. A controller is used in the front controller pattern to manage the many inbound requests. The front controller provides centralized management. We only need to set up one device, not several.
- MVC is ideal for developers who want full control over the behavior of their applications because it does not require server-based forms.
- The front controller aids in the communication routing needed to build our web application.

Disadvantage

- It's challenging to comprehend an application's movement.
- It is not appropriate for low-level applications and is somewhat difficult to execute.
- The design of the MVC is challenging to comprehend.
- Its implementation is a little challenging.
- Methods must be governed by strict guidelines.
- The software is complicated because jQuery, AJAX, and javascript are used so frequently.
- MVC is challenging to master because a programmer needs to have a thorough understanding of patterns.
- MVC does not include any built-in user controls for web sites.
- Model View Controller for Mobile View Website
- Web Development Using a Non-MVC Approach

Initially, HTML output for dynamic Web sites was sent to a browser using instructions like out. println (" h2>Class List/h2>"); in programs that operated on a web server. Technologies that employ this strategy include Java Servlets and Common Gateway Interface (CGI). With this strategy, the program commands and HTML commands must be tightly interspersed.

New methods that start with the HTML code and then enable server-side programming commands to be inserted inside the HTML commands have made it possible for web designers to work from an HTML perspective rather than a programmer perspective. PHP, Active Server Pages (ASP), and Java Server Pages (JSP) are manifestos that use this strategy (Hypertext Pre-processor). These tools enable web designers to build static pages, which they can then deliver to programmers for any server-side programming that may be required. Because HTML code and program code are closely linked, much like CGI and servlets, designers and programmers must collaborate on the same files (Patel & Pan-choli, 2018). Unwanted pages that are badly coherent and intricately woven result from this. In addition, both web designers and coders are needed for the creation and maintenance of a website. Inter-crosscutting results in chaotic code because it employs criteria that depend on a supplementary resource (a database, other pages, etc.). This makes software management challenging because all pertinent Web pages must refresh in response to changes made to the database or other websites. When web applications have a lot of interdependent pages, it becomes challenging to keep but manageable in small systems. As a result of distributed code and tangled code issues, the MVC architecture for dynamic Web apps was created.

FRAMEWORK OF MVC

PHP MVC Framework

PHP Application data and business logic (model) are separated from the presentation by the MVC application design pattern (view). Model, View, and Controller, or MVC. Between models and views, the controller acts as a middleman. Consider the MVC design model to be a car with a driver. The driver (controller) utilizes the car's windscreens (view) to keep an eye on the road ahead before deciding whether to speed up or brake.

PHP MVC Feature

MVC By hiding all the specifics of the difficult implementation, PHP frameworks make working with complex technology easier. It offers standardized techniques for us to use when creating our applications.

Enhanced developer productivity due to partially completed basic implementation of tasks like database logging and user input sanitation.

It adheres to industry coding best practices.

PHP-based system architecture. The MVC design pattern was used in the development of the PHP framework. Using PHP classes and web pages, the creators of the controllers and views (.php). The models were made using Entity Classes, an additional collection of PHP classes (.php).

Figure 2. System Architecture of PHP

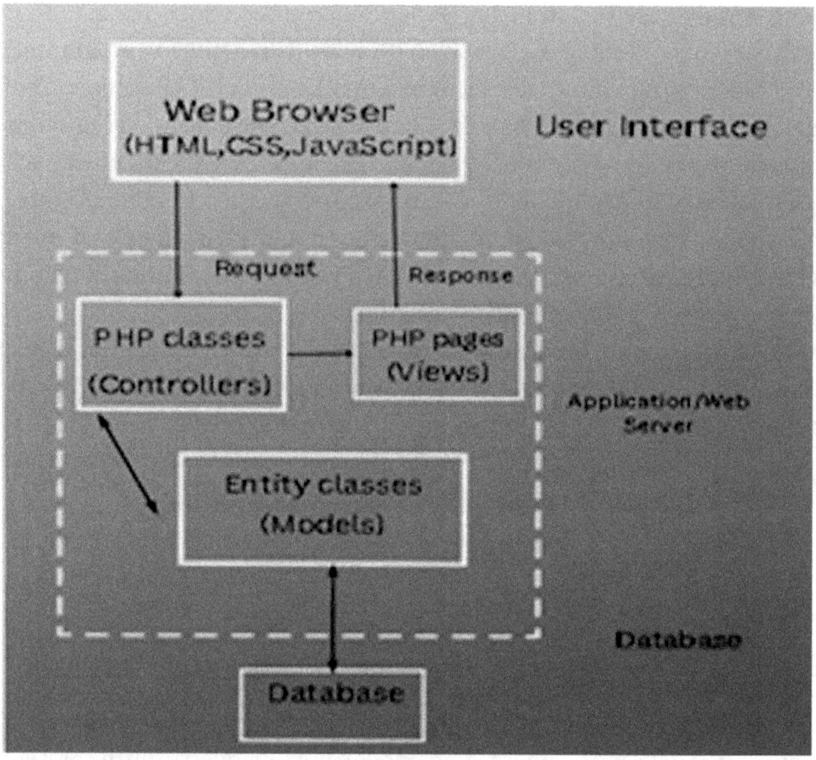

Figure 2, represent the system architecture of PHP. The user interact with the front end techniques like HTML, CSS, and JavaScript. The user request is forward to the controller class and forward to suitable models. The model contains the entity classes which help system to access the data(Liu & Cheng, 2013). The response forward back to the controller and user able to the response using different views.

PHP Implement MVC. The top server-side scripting language is PHP. Efficiency, stability, maintainability, and scalability are in great demand due to the Internet's rapid development for the creation of Web applications. PHP is presently one of the most significant and well-liked Web development languages because of its intuitive, user-friendly, quick, cross-platform, open source, and other characteristics. By utilizing the MVC architecture, it completes the separation of business logic from user interface logic. Fig. 3 depicts the PHP data flow for the MVC design.

The php page can establish the fundamental resource required as a front controller based on Fig. 3. To decide who will handle the current request, the router examines the http request. If the cache file is present, it will be transmitted straight to the browser rather than following the standard system execution sequence. The http request and the user-submitted data will be screened before the program is loaded. The model is loaded by the controller to handle a particular request. The view then displays the content in the web browser. The view will be cached first if the cache is open so that it can be used for later queries.

Figure 3. PHP Data Stream for MVC

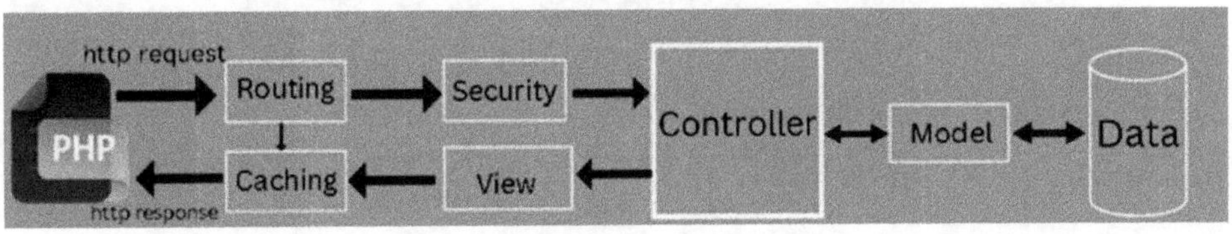

ASP.NET MVC FRAMEWORK

Microsoft is responsible for developing the ASP.NET MVC Web application infrastructure. This infrastructure utilizes the MVC (model-view-controller) paradigm. It is open-source software with the exception of the ASP.NET Web Forms component, which is private. Aspects of ASP.NET MVC. It offers an extendable and pluggable frame that is simple to swap out and modify, making it ideal for developing lightweight but complex apps. It makes use of component-based application architecture by logically dividing it into Model, View, and Controller components.

Use ASP.NET when constructing a system. Model-View-Controller (MVC) architecture was used to create the program in ASP.NET, as shown in Fig. 4. With the exception that the view was created using Active Server Pages (ASP) and the controller was written in C#, this is comparable to the MVC design model for Java. ASP.NET HTML pages were used to create the views. For creating views, Razor (*. cshtml) and Aspx (*. aspx) are both supplied. Every view was built using the Razor engine, while the handlers were built using C# classes (*SP.NET). With the exception of the fact that the controller was written in C# and the view was created using Active Server Pages (ASP), this is identical to the MVC design paradigm for Java. ASP.NET HTML pages were used to create the views. For creating views, Razor (*. cshtml) and Aspx (*. aspx) are both supplied. The Razor engine was used to create the models with entities, which are also C# classes (*.cs), as well as the controllers (Xuebin, 2017).

Figure 4. System Architecture of ASP.Net

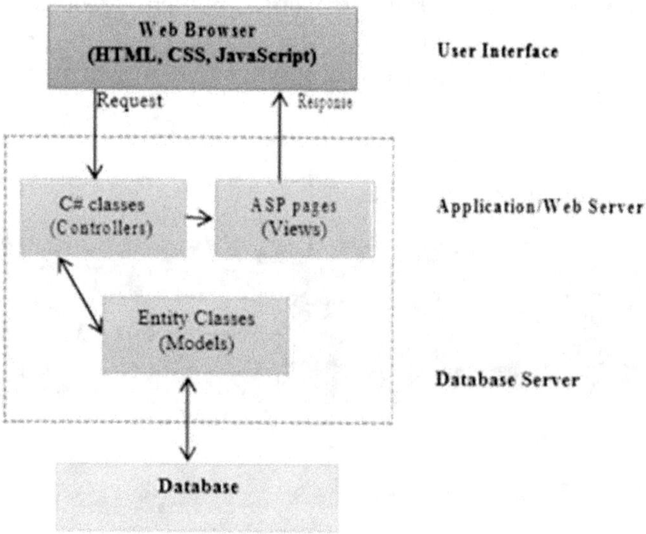

ASP.Net uses MVC. Microsoft unveiled a new Web development architecture called ASP.Net MVC in April 2009. Model, View, and Controller are kept apart, and standard MVC development principles are combined with ASP.Net. Here, we demonstrate how ASP.Net implements the MVC architecture with a straightforward one-page solution that is a login screen. Using the login screen application, a user can view a login page where they can enter their identify and click the login button to log in to the system. The user can also enable the checkbox to recall them. In order to maintain the user's activities while he uses the Web application, a fresh session is also created on the server side when the user logs in. Actually, because the architecture is service-oriented, you need a separate controller to function as a business services controller. We have incorporated a security services controller into this application. Fig. 5 depicts the ASP.Net MVC programming model.

The system was created in ASP.NET using the ASP.NET MVC 4 architecture. ASP.NET MVC integrates the Model View Controller paradigm, which is used to build online applications generally, into the ASP.NET framework. Models were created using C# feature classes, while controllers and views were built using the Razor view engine. The Razor view engine uses a combination of HTML tags and C# code to produce dynamic web sites. The controllers invented the C# language to handle inbound requests, execute model actions, and select which views to display to the user. Entity classes, which also act as representations of database objects, carry out the business functionality.

Figure 5. The ASP.NET Programming Model for MVC

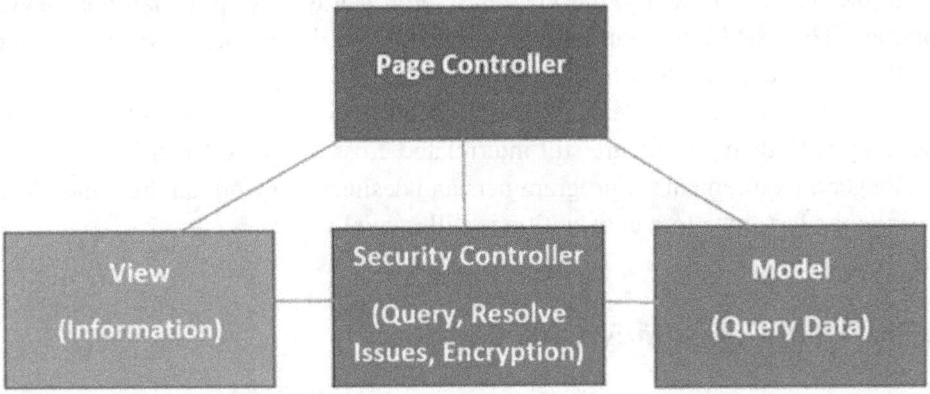

MVC ARCHITECTURE USING JAVA

An MVC design developed using Java-based technologies is shown in Fig. 6. User HTTP requests are routed through a controller in this architecture, which is frequently implemented as a servlet.

Figure 6. System Architecture of JAVA

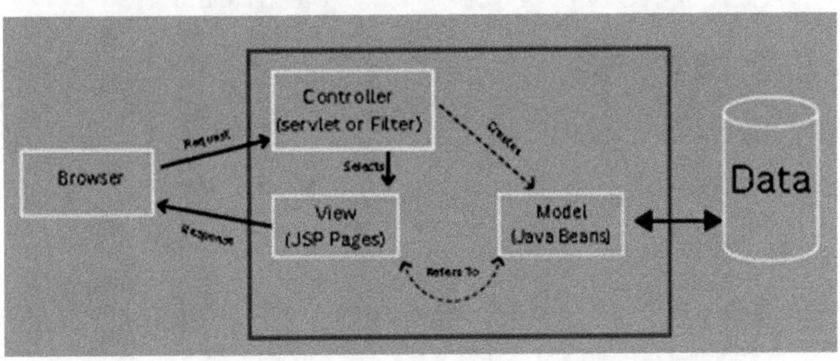

It is up to the controller to manage the request, create any required model classes, and choose which response should be sent to the browser. When JSP pages are used as the display, the JSP page can access data stored in the models-JavaBeans using the JSP Expression Language (Thakur & Pandey, 2019).

This approach effectively separates the HTML code from the program code, allowing programmers and designers to work much more freely together. The controller and model code can be written by coders without incorporating HTML from the display. Similarly, the view has significantly less code than non-MVC options. The JSP Expression language and tag libraries are less invasive coding methods than adding script lets and directives to a JSP page.

Thus, while this strategy lessens intra-crosscutting problems, it does not completely do away with them. There are still interrelated cross-cutting difficulties. There is no longer a requirement for programmers and designers to work on the same files, despite the fact that close cooperation is still crucial.

STRUTS MVC FRAMEWORK

The MVC framework Struts provides a pre-written controller and an XML-based configuration file to match URLs with actions (Singh & Iyer, 2016). In the example above, Struts automatically builds the action class (Product) and populates its properties with user inputs. As a result, the MVC controller coding from the prior example is greatly simplified.

Figure 7. Struts architecture

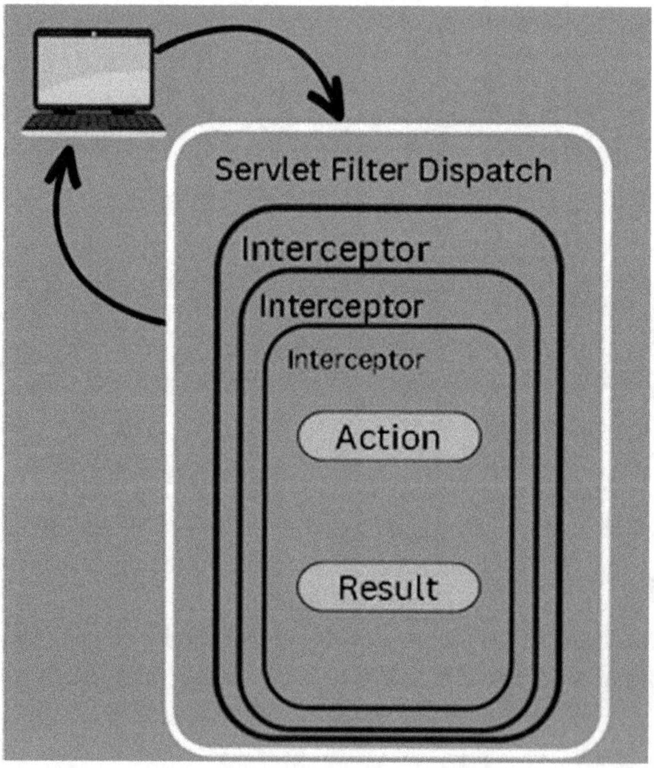

The Struts design is displayed in Fig. 7. The process begins with the browser making a request for a resource, the controller determining the best course of action, the interceptors automatically applying common processing to the request (such as validation, authorization, etc.), the action method specified being executed (typically storing or retrieving data from a database), and the output being rendered to the browser, typically as a JSP page.

CONCLUSION

The MVC architecture allows for the creation of scalable, transparent, and portable web-based applications. Since the MVC architecture divides the application's functionality into three layers, it enables multiple developers to work on the web application's three tiers concurrently. There is often high setup expenses associated with the MVC design. Large MVC projects process faster than small MVC projects, with small MVC projects and non-MVC projects frequently processing in the same length of time.

REFERENCES

Dey, T. (2011). A Comparative Analysis on Modeling and Implementing with MVC Architecture. *International Journal of Computers and Applications*, *1*(1), 44–49.

Jailia, M., Kumar, A., Agarwal, M., & Sinha, I. (2016). Behavior of MVC (Model View Controller) based web application developed in PHP and. NET framework. *2016 International Conference on ICT in Business Industry & Government (ICTBIG)*, 1–5.

Kalelkar, M., Churi, P., & Kalelkar, D. (2014). Implementation of model-view-controller architecture pattern for Business Intelligence Architecture. *International Journal of Computers and Applications*, *102*(12), 16–21.

Kumar, V., Kumar, A., Sharma, A. K., & Singh, D. (2016). Implementation of MVC (Model-View-Controller) design architecture to develop web based Institutional repositories: A tool for Information and knowledge sharing. *Indian Research Journal of Extension Education*, *16*(3), 1–9.

Leff, A., & Rayfield, J. T. (2001). Web-application development using the model/view/controller design pattern. *Proceedings Fifth IEEE International Enterprise Distributed Object Computing Conference,* 118-127.

Liu, X. K., & Cheng, G. G. (2013). Analysis and implementation of ASP.NET and PHP frameworks based on MVC Architecture. *Advanced Materials Research, 798-799*, 749–752.

Majeed, A., & Rauf, I. (2018). MVC Architecture: A Detailed Insight to the Modern Web Applications Development. *Peer Review Journal of Solar & Photoenergy Systems, 1*(1), 1–7.

Patel, S. J., & Pancholi, P. D. (2018). Implementation and Comparison of MVC Model in ASP.net Framework and PHP Framework. *International Journal of Research and Analytical Reviews, 5*(4), 827–835.

Pop, D.-P., & Altar, A. (2014). Designing an MVC model for rapid web application development. *Procedia Engineering, 69*, 1172–1179.

Singh, S. (2020). MVC Framework: A Modern Web Application Development Approach and Working. *International Research Journal of Engineering and Technology, 77*(1), 51–55.

Singh, S., & Iyer, J. (2016). Comparative Study of MVC (Model View Controller) Architecture with respect to Struts Framework and PHP. *International Journal on Computer Science and Engineering, 5*(3), 142–150.

Subari, A., Manan, S., & Ariyanto, E. (2021). Implementation of MVC (model-view-controller) architecture in online submission and reporting process at Official Travel Warrant Information System based on web application. *Journal of Physics: Conference Series, 1918*(4), 042145.

Thakur, R. N., & Pandey, U. S. (2019). A Study Focused on Web Application Development using MVC Design Pattern. *International Research Journal of Engineering and Technology, 6*(8), 237–244.

Xuebin, J. (2017). Research on Computer Network Virtual Laboratory based on asp. net. *2017 International Conference on Computer Systems, Electronics and Control (ICCSEC)*, 464–467.

Chapter 15
Design for Mobile View Website Using Model View Controller

Veton Klinaku
Rochester Institute of Technology, Kosovo

Deni Turku
Rochester Institute of Technology, Kosovo

Alisa Qatipi
Rochester Institute of Technology, Kosovo

Lumi Zhubi
Rochester Institute of Technology, Kosovo

ABSTRACT

This chapter gives a broad outline of machine learning and artificial intelligence and introduces the reader to many novel and most recent developments in the field of machine learning. The first half of this compilation provides an all-round view of the classical concepts of machine learning, namely: ensemble learning, concept of big data, handling of big data, and predictive data analytics using big data. Examples of machine learning (ML) frameworks are discussed, which are computer vision (CV), swarm algorithm, network science/graph theory and applications in machine learning, Bioinformatics using machine learning, and internet of things (IoT). A side note—R language is added as is the second most common language used worldwide for machine learning and this chapter spotlights mostly on Python language for ML. Deep learning, concepts, models, types, and algorithms in machine learning are elaborated in the subsequent section, followed by a detailed introduction to Neural networks, concepts of weight initialization, propagation, and vanishing gradient problem.

DOI: 10.4018/978-1-6684-8582-8.ch015

INTRODUCTION

The front end of this website is built with React.js. Bootstrap was utilized in the development of the design. Bootstrap also helps with the responsiveness of the webpage and we can analyze the mobile view of the website. The client folder organizes the various components of each page into their own distinct folders, making it simpler for the developer to access the information they need. The images that are used throughout the entirety of the website each have their own folder.

Figure 1. Dependencies used in the project

```
{
  "name": "client",
  "version": "0.1.0",
  "private": true,
  "dependencies": {
    "@fortawesome/free-solid-svg-icons": "^6.2.0",
    "@fortawesome/react-fontawesome": "^0.2.0",
    "@testing-library/jest-dom": "^5.16.5",
    "@testing-library/react": "^13.4.0",
    "@testing-library/user-event": "^13.5.0",
    "axios": "^1.0.0",
    "bootstrap": "^5.2.2",
    "node-sass": "^7.0.3",
    "react": "^18.2.0",
    "react-bootstrap": "^2.5.0",
    "react-datetime": "^3.2.0",
    "react-dom": "^18.2.0",
    "react-export-table-to-excel": "^1.0.6",
    "react-router-dom": "^6.4.1",
    "react-scripts": "5.0.1",
    "react-social-icons": "^5.15.0",
    "react-toastify": "^9.0.8",
    "web-vitals": "^2.1.4"
  },
```

The image above shows the dependencies of the project: bootstrap, router-dom, react-dom, toastify, and fontawesome.

Below, we can see the App page which holds the links to each page of the website.

Figure 2. Paths that take you to each page

```
function App() {
  return (
    <>
      <BrowserRouter>
      <Header/>
      <Routes>
        <Route path="/" element={<Home />}/>
        <Route path="/about" element={<About />}/>
        <Route path="/cars" element={<Cars />}/>
        <Route path="/reservation" element={<Reservation />}/>
        <Route path="/contact" element={<Contact />}/>
      </Routes>
    </BrowserRouter>
    <Footer/>
    </>
  );
}
```

HOME PAGE

The website that we have created is completely responsive and optimized for use on mobile devices. The landing page is what you see as soon as you open it for the first time. There is a bar for navigation at the very top, and because we are on mobile view, the menu is in the form of a hamburger. If we select it, we are presented with the following options: Home, About Us, Cars, Reservation, and Contact. Let's proceed with the landing page that we began with, Home. At the very top, we have a banner with a slogan and a button that takes you to the Cars page, from there you can view the available options. After that, the website displays the reasons why the company is the best by displaying three cards with the qualities of Special Prices, Safest Option, and Time Efficiency respectively. After that, there is a segment on the website that displays a portion of the car collection that is available. Both a button that will take you to the Cars page and a carousel that will show some of the cars are included in this section. The buttons on the page that take you to the Cars page are there to encourage the user to view the collection and rent one of the options. Putting it at the beginning and the end of the page was a strategic idea since we considered users that are already convinced that they want to rent from us but also the users who may need a little convincing regarding our credibility. The footer can be found at the very bottom of the website. The user can obtain contact information for the company, navigate to other pages on the website, access the company's social media accounts, and read a brief description of the company all from this part of the webpage. The navigation bar and the footer are present on all pages.

THE NAVIGATION BAR

Figure 3. The navigation bar

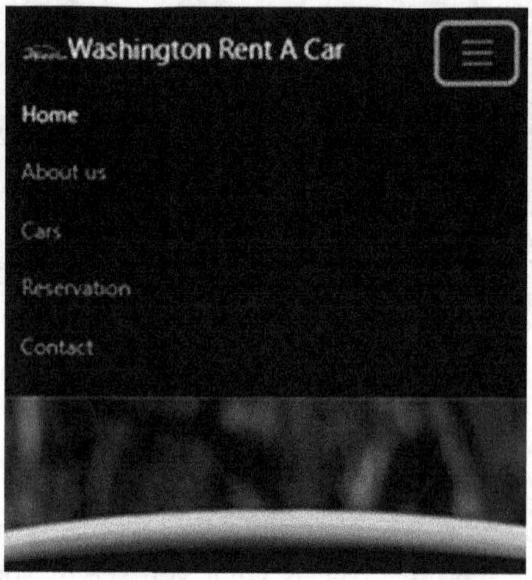

The home page:

Figure 4. The home page

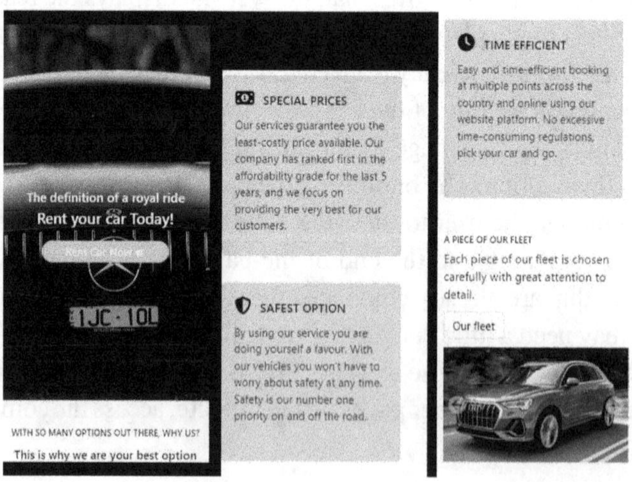

THE FOOTER

Figure 5. The footer of the website

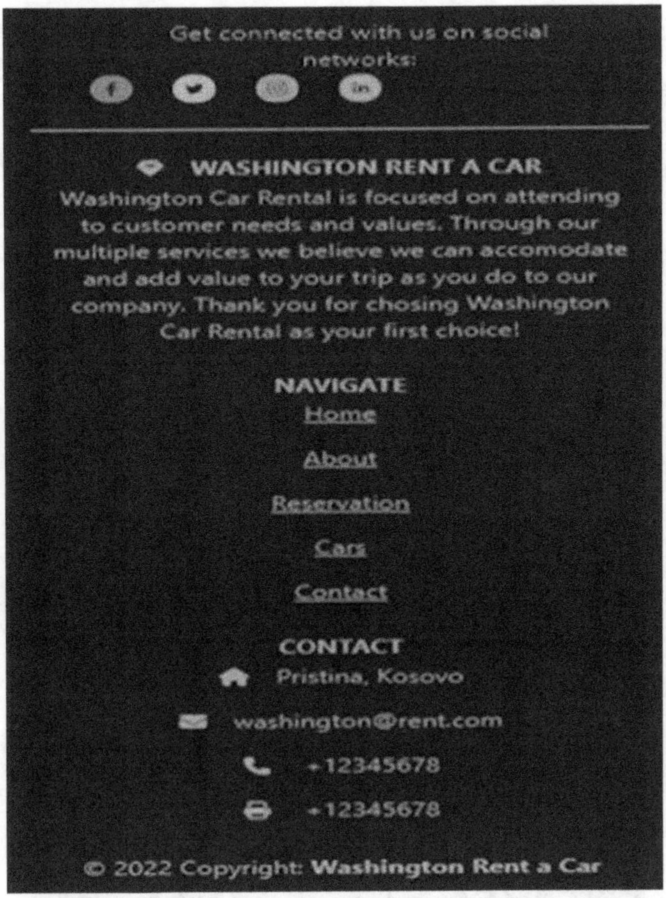

ABOUT US PAGE

The page retains the same structure as the landing page, with the navigation bar located at the top and the footer located at the bottom. The user is presented with a banner that reads "Who we are" at the very top of the page. This provides the user with an improved comprehension that the page will provide necessary information about the company. We immediately continue with reviews from customers; these reviews are presented in the form of cards that show an image of the user, and when the card is hovered over – or clicked when we are using a mobile device – it "turns around" and displays what the user has to say about their experience with us. After that, there are three cards on the page that each display a different article related to automobiles that is pertinent at the present time. The purpose of this page is to provide the user with the opportunity to read about the positive experiences of other clients who have rented cars from the company in the past and gain the knowledge that these clients have been satisfied with the service. The purpose of these articles is to make the content more relevant to individuals who have a passion for automobiles. This is accomplished by demonstrating to readers that we are aware of and keep up with current industry developments.

The cards and their messages:

Figure 6. The banner of the About page

Figure 7. The messages of the card

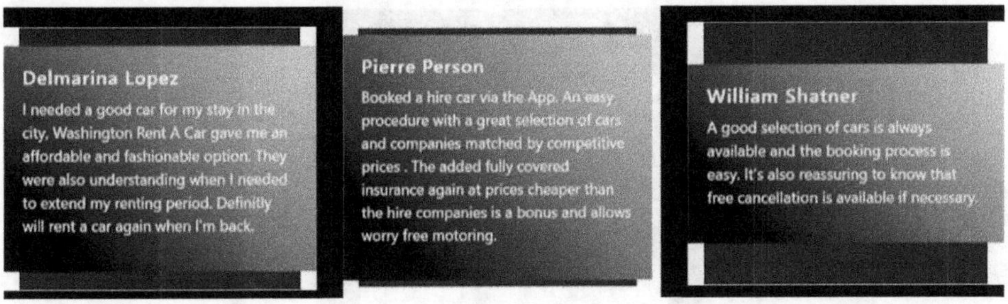

Figure 8. Articles about cars

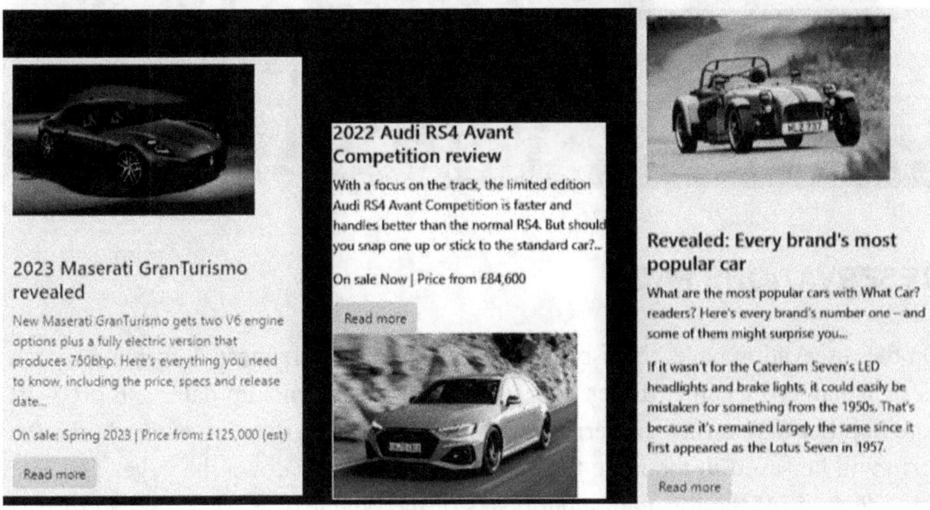

CARS PAGE

The Cars page displays a collection of cards that contain photographs of the automobiles that are available from the company. On the cards is also printed the make and model of the vehicle, as

well as information about the number of people it can seat, whether or not it has air conditioning, the kind of fuel it requires, and how much it costs to rent for a day. Last but not least, each card features a button that, when clicked, takes you to the page where you can make reservations. After the cards are over, the page will display a list of guidelines that they are expected to be familiar with before renting a vehicle.

Figure 9. Cards about the car options offered

RESERVATIONS PAGE

On the reservation page, the user can fill out a form that requests information such as their name, surname, driver license ID, chosen car, pickup date, and drop-off date in order to rent their preferred automobile. The data that you entered onto the form will be saved in the table that can be found further down the page. The first three sections of the form are input boxes of the type text. Next, we select the car from a select box, and finally, we select the dates of the renting period from an input box of the type date. After we have clicked the button, all of the information will be added to the table that is located beneath the form, and you will receive a toast notification at the very top of the page to confirm that your submission has been received. If you try to make the reservation before you have completely filled out the form, you will also receive a toast notification asking you to complete the form. The table is responsive, which means that you can scroll through it horizontally to see all of the information within it without having to make the table extremely small or go outside the dimensions of the website. The Delete and Edit buttons can be found in the very last column of the table. The row from the table as well as the information in the database can be removed by using the Delete button. On the other hand, clicking the Edit button will open a modal window. This window will have a form that is already filled out with all of the information that can be retrieved from the database when accessing the id of the row we clicked the button for. Each of the form's inputs can have its value altered, and the resulting changes will be reflected both in the table and in the corresponding database fields. The Reservation page is where the crud operation happens. Firstly, through the post method, all the information that the user inserts in the form goes in the database.

Figure 10. Code about the post method

```
    axios
      .post(
        "http://localhost:8080/api/insertreservations/insertreservation",
        reservationdata
      )
      .then((data) => {
        console.log("data", data.data);
        showReservations();
      });
  } catch (error) {
    console.log(`Posti nuk u  krijua ${error}`);
    toast.error("");
  }
```

Through the get method, all the information from the database is taken and mapped in a table.

Figure 11. Code about the get method

```
  const showReservations = () => {
    axios
      .get("http://localhost:8080/api/insertreservations/read")
      .then((data) => {
        setReservationInformation(data?.data);
      })
      .catch((err) => {
        console.log("err", err);
      });
  };
```

Through the delete method, the information of the certain ID is deleted from the database and also with the delete function, it deletes from the table as well.

Figure 12. Code about the delete method

```
const handleDelete = (id) => {
  axios
    .delete(`http://localhost:8080/api/insertreservations/delete/` + id)
    .then((data) => {
      console.log("deleted ajshe", data.data);
      toast.success("Deleted successfully!");
      showReservations();
    })
    .catch((err) => {
      console.log("err", err);
      toast.error("Something went WRONG!");
    });
};
```

The update of the information in the modal happens through the put method, where we take the ID to retrieve the information which the user can later modify and fix according to their need.

Figure 13. Code about the put method to edit the reservation

```
const handleEditReservation = (e) => {
  console.log("Reservimi", e);
  e.preventDefault();
  const editReservationData = {
    name: e.target.name.value,
    surname: e.target.surname.value,
    driversLicenceId: e.target.driversLicenceId.value,
    car: e.target.car.value,
    pickupDate: new Date(e.target.pickupDate.value),
    dropoffDate: new Date(e.target.dropoffDate.value),
  };
  console.log("dataaaa", props.data.pickupDate);

  axios
    .put(
      `http://localhost:8080/api/insertreservations/update/` + props.data._id,
      editReservationData
    )
    .then((data) => {
      console.log("data", data.data);
      window.location.replace("/reservation");
```

The table uses the map method to go through each database entry and display its information in its cells. If there is no information is the database or there is no access to it for whatever reason, a message “No Users Found” will show.

The modal to edit a reservation:

Figure 14. Reservation modal

CONTACT PAGE

There are two primary components to the contact page. The first option is a form that you can fill out with your information and a message that you want to send to the company. The message can be written in the text area provided, in addition to the two text inputs that are available on the form. The information will be saved in the database once you have clicked the Send button. In the second section, there is a map that displays the location of the business; for the purpose of this example, we used the city of Pristina; however, this location can, of course, be altered. The contact information can also be found in the footer of the website, so the purpose of the page is sending emails without having to discuss with someone physically. The contact page also uses the post method in order to get the user input from the form and insert it in the database of the application.

Figure 15. Post method of the contact form

```
axios.post("http://localhost:8080/api/posts/insert", data);
// console.log("hi")
toast.success('Message sent!', {
  position: "top-right",
  autoClose: 2000,
  hideProgressBar: true,
  closeOnClick: true,
  pauseOnHover: true,
  draggable: true,
  progress: undefined,
});
```

THE CONTACT PAGE

Figure 16. Contact form

Figure 17. Maps

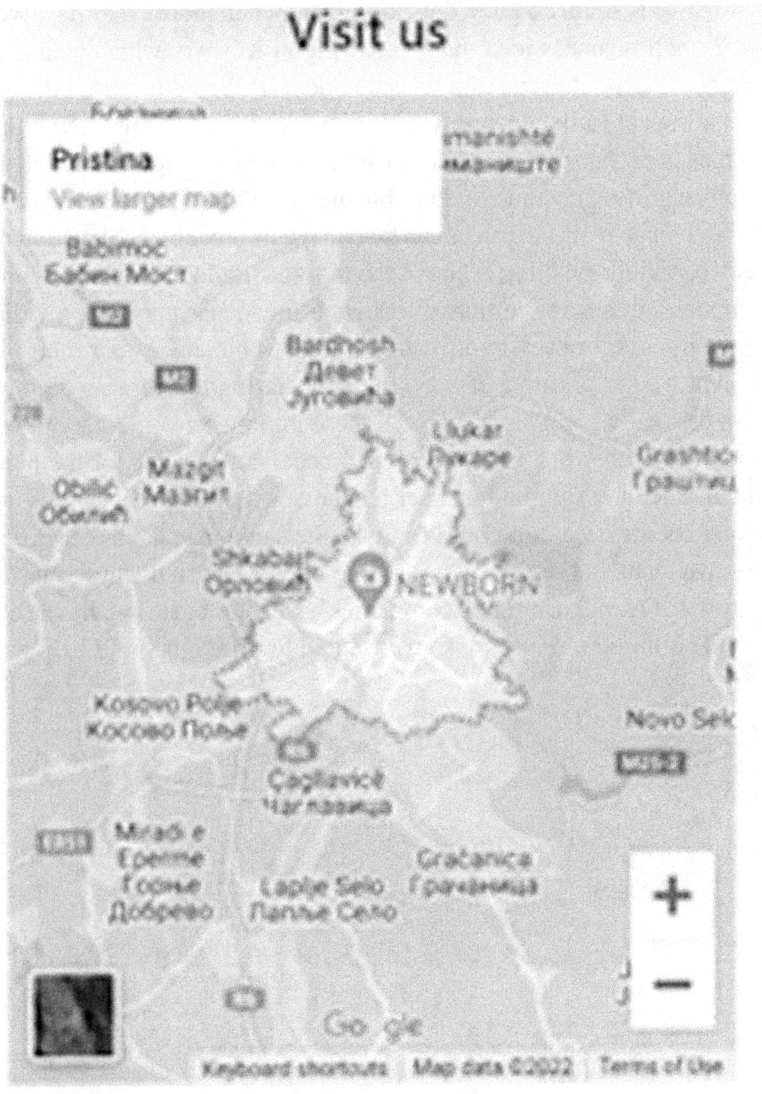

CONCLUSION

This chapter serves as a primer to young student of data science or computer science, introduces the concept of Machine learning, Artificial Intelligence, Neural networks, and their functioning. The second half highlights the latest developments in the field of Machine learning, recent and interesting advances in the application of Machine learning, like ChatGPT and Stable diffusion. The chapter focuses on simplicity, and lucid language which should be easy to grasp for anyone who is interested in data sciences. Kosovo is small country, developing more and more every day. We wanted to choose a small business idea that is growing in Kosovo, which also caught our interest and create a prototype of a website about it. We believe that the car rental business can impact in the economy of Kosovo, especially considering that it can become a means of transportation for tourists. It is a way to open more job positions and also to bring more revenue to other businesses. Other than that, it is a business that with the right growth and prices can help in the improvement of our environment since Kosovo, mainly Pristina, is one of the most air-polluted places in Europe. Car rental companies can bring initiative to switch to electric cars by providing them for renting purposes, or even by investing in renewable energy for their offices. It is also a great way to promote car pooling and sharing to decrease the number of vehicles on the road as well as traffic. All these initiatives would not only decrease their carbon footprint while still providing a necessary service for the population, but would also set an example for other businesses to follow. Since Kosovo is still a developing country, we don't have to make the same mistake as other developed countries have made in their developing time but we can find ways to have a more positive growth. Overall, car rental companies have the chance to help the economy, tourism, and pollution of Kosovo, so we wanted to offer our small help to it.

ADDITIONAL READING

Montenegro, A. A., Carvalho, P. C. P., Velho, L., & Gattass, M. (2004). Space carving with a hand-held camera. *Proceedings of the XVII Brazilian Symposium on Computer Graphics and Image Processing (SIBGRAPI'04)*, (pp. 396-403). 10.1109/SIBGRA.2004.1352986

Chapter 16
Machine Learning-Based Mobile Applications Using Python and Scikit-Learn

Santhosh Kumar Rajamani
https://orcid.org/0000-0001-6552-5578
Maharashtra Institute of Medical Education and Research, India & Dr. Bhausaheb Sardesai Talegoan Rural Hospital, India

Radha Srinivasan Iyer
https://orcid.org/0000-0001-7387-4401
SEC Centre for Independent Living, India

ABSTRACT

This chapter gives a broad outline of machine learning on Android mobile phones using the Scikit-learn module. The first section introduces the reader to Python language; next, Python on Android is introduced with a brief historical note on implementations of Python on Android mobile phones. Pydroid3 is introduced in the subsequent section. This is followed by instructions on setting up an Android phone for machine learning. This is followed by a description of supportive modules for machine learning that are available for Pydroid3, and some example codes, namely: os, pathlib, Pandas, NumPy, SciPy, Matplotlib, Seaborn, PySimpleGUI, NetworkX, Biopython, WordCloud, Kivy, and Jupyter Notebook. The last section of this compilation describes the Scikit-learn library, basic concepts of the Scikit-learn module, and algorithms available with this module, namely: Linear Regression, Logistic Regression, Principal Component Analysis (PCA), XGBoost, K-nearest neighbors, and support vector machine.

DOI: 10.4018/978-1-6684-8582-8.ch016

INTRODUCTION

A robust and user-friendly toolkit for the Python programming language, *Scikit-learn* offers several machine learning methods. Because of its extensive toolkit, which includes modules for data pre-processing, model creation, model evaluation, and model selection, it is one of the most popular libraries among developers. Along with factor analysis, ability to extricate features, and scale dimensions on complex datasets. *Scikit-learn* also supports many supervised and unsupervised learning techniques like random forests, k-nearest neighbors, and support vector machine SVM. It is based on the well-known Python libraries *NumPy* and *SciPy*. Additionally, its algorithms are simple to use because of the detailed documentation and tutorials that provide step-by-step explanations, making it the perfect tool for anyone perusing machine learning. Finally, *Scikit-learn* module is the best choice for many machine learning applications thanks to its user-friendly API and strong emphasis on automation (Lutz, 2013).

PYTHON

Python is a popular, easy to learn, simple, object-oriented, interpreted, high-level programming language, based on more abstract, difficult, and symbolic C language. Python was created by Guido van Rossum in 1990s. It has become known for its ease of use, readability, low-level syntax, and ability to rapid develop ideas. Python also has a wealth of robust machine learning libraries that make it easy to implement complex algorithms and platforms in Python. This makes Python an attractive choice when working with machine learning problems. Python and its source code are released under liberal GNU General Public License (GPL) which makes it attractive option even for commercial projects (Lutz, 2006).

Currently, Python is the preferred programming language of more than 80% of the developers. One key advantage of using Python is that it is generally easier to get started with than many other languages promoted as suitable for machine learning. This is partly due to Python's readability and partly due to the abundance of free resources available online. Additionally, Python has been widely adopted by data scientists, making it a familiar toolkit that can be easily integrated into existing neural networks and business applications. For example, Flutter (Dart)is currently the de facto language of Android development is a bracket hell, brackets to enclose chunks of code, brackets to enclose those brackets, *ad infinitum.*, making code confusing and incomprehensible to the reader (Lutz, 2013 and https://neurodeb.pirsquared.org)

THE POWER OF PYTHON FROM THE SMALLEST TO THE LARGEST

In contrast despite its obvious simplicity Python is a powerful language capable of implementing many complex ML algorithms using vast collection of additional libraries. On the other hand Python can also be used to automate common menial tasks like for instance, document conversions (*docx2pdf, pdfplumber,PyPDF2*), creating PowerPoints (*pptx*), taking screenshot, copy and paste text from the system clipboard (*pyperclip*), merging word documents (*docx*), manipulate Excel sheets (*openpyxl*), image background removal (*rembg*)or lookup a stock price(*yfinance)* or lookup currency exchange rate (*forex_python*) or changing colour images to greyscale (*pillow*) or spell-check a text (*textblob*) or download a file (*urlib*). Office automation task to complex Machine learning algorithms like for instance, Distributed Evolutionary Algorithms in Python (DEAP) (*deap* module) and Symbolic mathematics (*sympy*). DEAP gives the power of mathematical modelling, optimization, genetic programing, particle swarm optimization, and evolution strategies. Many of these modules are also available via pip on Pydroid3 and code conveniently tested on Android itself. Even the complex module *deap* is obtainable on Android. Python is indeed the Swiss army knife of programming language.

LIMITATION OF PYTHON

Python interpreter itself is written in C and converted into computer-executable machine code. When Python code is written and executed, it is first parsed and converted by the interpreter into bytecode. The interpreter then runs this bytecode, carrying out the operations dictated by the code. Thus, every instruction in Python must be interpreted in symbolic byte language, this makes Python much slower than other languages (Kane, 2017).

MACHINE LEARNING PARADIGMS IN PYTHON

Machine learning implementation in Python is the most popular tool used by data scientists to solve complex problems and improve existing models (http://github.com). By leveraging powerful libraries such as *Scikit-learn*, *NumPy*, and *Tensor Flow*, users can build sophisticated models that can make decisions based on large amounts of data and can even improve the accuracy of predictions over time. With python's ease of use and versatile syntax, developers of all levels can quickly access powerful machine learning algorithms when developing new applications or looking to apply existing ML models. Python also offers a robust suite of graphing libraries that allow users to visualize their data to better inform model building or interpret results from existing ML models. The first part of this section focuses on adjunct software on available Android for Machine learning along with some usage examples (Apeltsin, 2021).

PYTHON ON ANDROID

Python on Android is an increasingly popular platform for developers who want to create applications on mobile devices. It works by providing a virtual machine (VM) layer which sits between the operating system and the application, allowing users to develop apps with Python code. The VM optimizes the performance of an app, as well as simplifying development tasks such as porting existing apps from other platforms. Additionally, it also allows developers to use common libraries like TensorFlow which makes it easier to create machine learning models. With the rise of 5G networks making streaming video more feasible and enabling faster file transfers, Python on Android could become even more popular with developers looking for a powerful platform for their projects.

VARIOUS PYTHON DISTRIBUTIONS PORTED TO ANDROID OS

Although Python has not always been supported on the Android operating system, there are currently several choices available. The QPython app, which was published in 2012, was one of the first ways to execute Python on Android. This app included a Python interpreter built-in, as well as several libraries and tools for working with Python on Android, allowing users to run Python scripts and projects on their Android devices. Other Python on Android possibilities, such the *Kivy* framework, which makes it simple to develop cross-platform mobile apps using Python, have developed in recent years. Additionally, Python code may now be run on Android using a web browser thanks to the popularity of Jupyter notebooks. Additionally, due to the popularity of Jupyter notebooks, applications like JupyterLab and Colab may now be used to run Python code on Android online using a web browser (Mirjalili, 2019).

PYDROID V. 3.9.7

Pydroid3 is an Android app that allows users to write and run Python scripts on their Android devices. It provides a Python development environment for Android, allowing users to create and run Python scripts and programs directly on their phones or tablets. Pydroid3 includes a built-in text editor, interpreter, and debugger, making it a powerful and convenient tool for learning and experimenting with Python on Android devices. The current stable version of Pydroid3 available on Google Play store is 3.9.7, that was released on Oct 6, 2021 (running GCC on Linux 11.1.0). GNU Compiler Collection is known as GCC, which is the default compiler on many Linux distributions including Android GCC offers high-quality, effective, and portable code creation for a variety of applications.

Figure 1. Features of Pydroid3 that can be utilized for Machine Learning

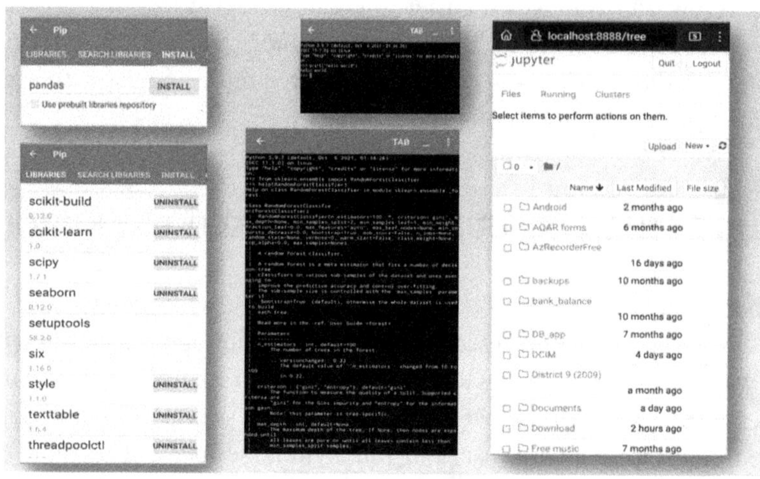

Pip is available as a convenient GUI program, B. Installed packages can be seen in a single scroll-view C. Python interpreter mode is also available D. Scikit-Learn installed on phone showing implementation of algorithm Random Forrest Classifier (via. *RandomForrestClassifier*) E. Jupyter notebooks can be installed locally on Android phone (http://localhost:8888/tree) via Pip, does not need an internet connection and data stored on SD card. Users can conveniently execute Python scripts on their Android devices with the Pydroid3 software. It is made for people who wish to learn or use Python on the move. Overall, Python support for Android has developed significantly in recent years, and it is now a viable option for many types of platform development. Users of Pydroid3 may create and execute

Python scripts directly on their Android smartphones, thanks to the inclusion of a built-in Python interpreter, a code editor, and several libraries and modules which are discussed individually in the subsequent sections.

A NOTE ON PACKAGE VERSION SYSTEM MENTIONED IN THIS PAPER

In this compilation, package names are mentioned in Italics to help user recognize these as Python addon packages or modules that can be install via Pip. Major version number of package, along with minor version, is mentioned in this work as these remain stable and unchanged for many years, even decades. This follows Python version identification and dependency specification scheme (PEP 440) for modules, which follows Semantic versioning scheme which is depicted as follows: "Major.Minor.Patch", for example, latest release of *SciPy* package on Pydroid3 is "1.7.1". The patch versions are changed frequently as software is updated and minor version numbers gradually change with timeframe of months and years, for instance Python 3.10 has been released as this work is being compiled. On the other hand, many packages do not follow such clear-cut scheme of release and versioning (7. Releasing and Versioning — Python Packages, n.d.).

SETTING UP PYDROID3 (PYTHON V. 3.9) ON ANDROID MOBILE PHONE FOR MACHINE LEARNING

Setup the Android Permissions for Python Development

Android permissions provide access control for applications to access system services and parts of the operating system. Permissions are broken into two major categories: Normal and Dangerous. Dangerous permissions require user approval, as they could potentially cause malicious or privacy-violating effects on the device if granted. Applications must request permission when they communicate with other data sources; however, users can also modify application permissions in their settings menu to approve or deny requests made by apps or carriers to access resources on their device. Android's permission system provides an additional layer of security that helps protecting data and sensitive information stored on mobile devices from unauthorized access.

Android permissions are important security measures that allow users to control which functions, data and resources third-party apps can access on their devices. When the user downloads a new app, they have the option of granting or denying individual permissions for specific features or data. For example, a user may be asked to grant access to their camera and contacts list for a photo-sharing app. These settings help protect personal information from being shared or transferred without explicit permission from the user, as well as preventing malicious apps from accessing private data and system resources. The Android permissions system is an important tool for protecting user privacy and security on mobile devices (Cohen et al., 2014).

SETUP ANDROID SD CARD PERMISSIONS FOR DATA ACCESS

Android devices use the SD Card permissions system to control access to certain areas of the internal storage. This means that applications must be granted specific permission from the user before they can read, edit, or delete files stored on external SD cards. The Android manifest must also provide such permissions within its code for an application to fully leverage SD card functionality. Furthermore, this permission system helps ensure data security by preventing apps from accessing sensitive information that is stored on external media and granting users increased control over which app can write data to their SD cards. All in all, SD Card permissions are a simple but effective way of controlling file access and protecting users' data (Cohen et al., 2014).

As a part of setting up ML on Android phone, user must grant SD card read-write permission for Pydroid3 to be able to access the SD card and load the datasets and assets. This is accomplished in ***GUI by Settings ® Apps®Your apps® Pydroid3 ®Permissions® Files and media access***. This only solves half the problem of access as described in the following section.

Due to inherent restriction on the system, user cannot directly access the vital operating system files which reside in the system folder like: ***"data/system",*** unless the phone is rooted. Download folders at ***"/storage/emulated/0/Download"*** is a fixed path folder which user can both read and write.***"/data/user/0/ru.iiec.pydroid3/ app_HOME***" (https://linuxhint.com). is the default folder of Pydroid3 app, is inaccessible by default except when read or written by host Python program. For example, if we save an output plot *matplotlib* via the code:

```
# Save the rendered figure to disk
plot.savefig('trends_in_hospital_admission.png', dpi=300)
plot.show()
# (OR) saving the data frame
dataframe.to_csv(r'my_data.csv', index=False)
```

Please note, that plt.show() which displays the graph is placed after save, logic dictates that plot be shown then saved. This 'trends_in_hospital_admission.png', PNG image file or data_file.csv, CSV data file will be saved in ***"app_HOME"*** folder and no file manager can directly access this folder for editing or visualization. This creates a unique, Android-specific access problem for loading datasets and other assets for Machine learning.

MY WORKAROUND FOR THIS PROBLEM

For instance, the code *df = pd.readcsv("data.csv")* will load the dataset data.csv only from app_HOME folder. User themselves have little access to this directory which is hidden deep in *Android* system directory. The Python program host have access to the folder, so the workaround would be to copy this output file from the ***'app_HOME'*** folder to any other folder which we have full access, like the *'Downloads'* folder.

Another simpler solution is to use OS module to change current working directory to one of those folders that the user has full access. This is accomplished by code on my phone (this maybe different on other phones, but will be something in similar lines, use ES-Explorer/MiX file manager from Play store to find out the correct system specific path).

```
# Remove this line in Windows and Linux, causes program to crash by null error
    as no such directory exists on Windows and Linux, use some file manager to
    find the exact path to our media storage folder in Android
import os
os.chdir('/storage/emulated/0')
```

Pathlib purists immediately would frown upon me for my approach of hard coding the path. But this approach gets the work done, keeps the code clean, makes loading, and saving data, much easier and faster. Just load and save directly, no need to bother with the path or access.

The code *pd.read_csv("/storage/emulated/0/data.csv")* assumes that the data.csv file is in path /storage/emulated/0 is fixed, full path location needs to be specified every time one reads or writes the file.

Please note for both the above codes to work without a null error, we must enable Files and media access permission (SD card access) via the Android System GUI.

PIP Python Package v. 21.2 Management Systems on Pydroid3

Pip is a powerful and versatile package management system for easily installation, upgrading, dependency management of additional Python modules. Pip is preinstalled onsite with App installation of Pydrioid3 (Python 3.9) from Google Play Store. In addition to the ability to manage packages, it also provides support for multiple version control systems, can handle complex dependencies, provides a secure authentication system, and more. Furthermore, Pip supports several platforms, such as Linux distributions and Windows PCs as well as Mac OS X devices. Pip simplifies the workflow of software engineers by providing an automated way of keeping track of dependencies, enabling easier development and testing of applications. This makes it one of the most important tools for modern software development teams looking to streamline their process in creating high quality products (Vanderplas, 2017).

PIP is the default package management system which is supplied with every installation of Python. This package management system is used to install additional packages and add-on functionality to the existing Python installation. For example, *Pyinstaller* compiler add-on which can compile Python code to independent single EXE files for porting to other Windows computers. *Kiwi* is a similar compiler add-on available on Linux system, which can be used to create standalone APK files to port Python programs to other Android mobile phones (Vanderplas, 2017).

OS vs. on Pydroid3 for Setting Up Machine Learning on Android Phone

The Python OS module is an incredibly powerful tool that provides numerous ways to interface with the underlying operating system. It offers basic functionality such as manipulating files and directories, managing environment variables, and executing external programs. More advanced features include working with process IDs, creating sockets for network communication, and even interacting with process signals.

The Python OS module provides access to the operating system's functionality such as creating directories, deleting files and directories, interacting with environment variables, and others. It is a built-in library of modules that contains functions related to the operating system's different resources such as hardware and memory. Additionally, it can be used to access command line arguments, set file permissions, stat info (file size, creation time etc.) As its operations may involve dealing with sensitive information or operations that could affect an entire system's stability, it is important for developers to become familiar with proper use of the OS module before performing any tasks related to it.

Table 1. Housekeeping commands of Python PIP package manager, which is the most popular package manager and comes bundled with all distributions of Python, including Pydroid 3 distribution which runs on Android mobile phone where it is a GUI application

Command (Windows, Linux, Mac)	Purpose	Equivalent command on Android Pydroid3 PIP
>pip install package_name Example: >pip install pandas	This command installs package_ name	Installation is via GUI under left drawer menu- PIP-Install- Enter library name- Click on install button
>pip uninstall pandas	This command uninstalls pandas package	List of installed packages can be accessed via GUI under left drawer menu PIP-Scroll header LIBRARIES
>pip install –upgrade pip	This command upgrades self, PIP	PIP is upgraded automatically with App upgrade from Google Play Store
> pip install –upgradepackage_name	This command upgrades the package_name	None, need to install and reinstall hoping that they upgraded for Android
>pip show package_name	This command displays information on package	Installed List can be accessed via GUI under left drawer menu- PIP-Scroll header LIBRARIES
>pip check	This command verifies that there are no broken packages and dependency issues.	None
>pip list	This command lists all the packages installed on the system	Installed List can be accessed via GUI under left drawer menu PIP-Scroll header LIBRARIES

Additionally, this module also provides a wide variety of convenience functions for accessing operating system specific data such as user information, current date & time information, and more. For these reasons, the OS module can prove invaluable for developers who frequently deal with system internals or need to perform more complex tasks than typical everyday operations. OS module is preinstalled onsite with every installation of Pydroid3 (Python 3.9) like Pip. This module is very important in setting up our phone for machine learning.

Pathlib v. 1.0 on Pydroid3

Python *pathlib* ver. 1.0 module is a new, succinct, and clearer implementation of OS module is also available for Pydroid3. Pathlib code is easier to understand and debug than OS module. The main concept behind pathlib package is to do away with hard-coded fixed path and keep a flexible path independent implementation of the program. Example flexible path code sample is *my_home=pathlib.Path.home*(). This *home()* method always points to home path irrespective of the system, in Windows it is user folder and ~ in Linux/Posix systems. The equivalent in Android system is the default path of execution of Pydroid3, which is mostly *"/data/user/0/ru.iiec.pydroid3/app_HOME"*(https://linuxhint.com). Unfortunately, due to the restrictive director permissions and fixed path in Android, this flexibility is of little utility in Android OS.

Pandas v. 1.3 on Pydroid3

Pandas is an essential tool for manipulating, organizing, and analyzing data in the Python programming language. It provides powerful capabilities for working with tabular data, such as intuitive manipulation, sorting and filtering functions, grouping operations, and efficient hierarchical indexing. Additionally, *Pandas* makes robust use of the *NumPy* library for optimized numerical computing and statistical operations. This combination makes it a suitable choice for dealing with complex datasets requiring sophisticated operations at scale. With features that allow forward-filling missing values in temporal series, merging multiple data sources into one coherent dataset, automatic inference, and detection of data types across columns, as well as customized access commands allowing users to quickly pivot and reshape data to suit their needs. *Pandas* is the most popular choice amongst analysts and data scientists (Chen, 2018).

NumPy v. 1.2 on Pydroid3 (Python 3.9)

NumPy is a powerful open-source numerical computing library for Python. It is used by scientists, engineers, and data analysts alike to perform array operations such as calculations, manipulations, and transformations. *NumPy* features a powerful array object which makes it suitable for high-performance operations on multi-dimensional arrays and provides a versatile set of linear algebra functions. Furthermore, *NumPy* provides tools to read and write n-dimensional arrays from/to external files as well as integrate with existing data analysis tools like *SciPy*, Pandas, and TensorFlow. Through these capabilities, *NumPy* allows smooth interoperability between different libraries in Python's scientific computational ecosystem enabling users to quickly implement complex algorithms in their applications (Bressert, 2013).

NumPy has support for large, multi-dimensional arrays and matrices along with a collection of high-level mathematical functions to operate on these objects. *NumPy*'s fundamental data structure is a multidimensional array, which enables it to efficiently manipulate large datasets. In addition, it provides an interface to many other popular libraries such as *Pandas* and *Matplotlib*. With *NumPy*, linear algebra operations such as matrix inversion and singular value decomposition can be performed quickly and efficiently. Furthermore, its built-in random number generator makes it the ideal choice for random simulations or stochastic optimization problems. For many machine learning projects involving artificial intelligence or data mining tasks, *NumPy* is key to achieving reliable results in terms of performance and accuracy (Idris, 2011).

SciPy v. 1.7 on Pydroid3 (Python 3.9)

SciPy is a powerful open-source scientific computing library for Python, built on the *NumPy* extension. *SciPy* version 1.71 is latest module that is available for Pydroid3.*SciPy* is designed to provide fast access to a collection of specialized mathematical routines such as integration, optimization, statistics, and linear algebra (Hans Petter Langtangen, 2016). It also provides an extensive set of capabilities such as visualization and plotting functions that allow developers to demonstrate their results in real time. *SciPy* simplifies the development process by allowing users to create complex models quickly and efficiently. Furthermore, it has strong external linking ability which allows researchers to easily link with other software packages for data analysis tasks or numerical simulation. This makes *SciPy* highly efficient for collaborative projects as it enables multiple scientists to work together without significant overlap or risk. All in all, *SciPy* is an ideal toolset for both newcomers and experts alike due to its broad feature support and wide range of usage options (Blanco-Silva, 2015).

SciPy is a powerful open-source scientific computing library, based on *NumPy* and providing a wide variety of advanced mathematics solutions for industry standard applications. It features functions for data analysis, image processing, linear algebra and probability statistics offering comprehensive performance in these areas. *SciPy* has a high level of compatibility with other scientific computing libraries such as MATLAB or R, so users can easily leverage previously established code in the existing environment. The library also provides an easy-to-use interface for developing self-contained proprietary functionality (Bressert, 2013).

Additionally, it is designed to efficiently combine multiple existing scientific libraries in complex setups across multiple hardware platforms. By creating both simple and complex software solutions together with free access to updates and community support, *SciPy* makes scientific development more accessible than ever before. *SciPy* is a collection of algorithms that can be utilized in various machine learning tasks. It provides easy-to-use and efficient implementations of standard routines in supervised learning, unsupervised learning, and semi-supervised learning. It provides tools for data mining and data analysis. It also has various command line tools to perform specific tasks such as preprocessing data, transforming it into a different representation, performing model selection, or making predictions (Bruce & Bruce, 2017).

Matplotlib v. 3.4 and Seaborn v. 3.4.3 on Pydroid3 (Python 3.9)

Matplotlib provides comprehensive 2D plotting and high-quality publication-ready graphics. It supports both interactive graphical options as well as scripting with the powerful *PyLab* interface. Specifically, *Matplotlib* includes several unique features to aid in creating graphs through a user-friendly interface, including a variety of different color palettes that can be used to customize plots; customization of legend labels, tick marks, line styles and other details; built-in support for plotting arbitrary datasets alongside one another; integration with *NumPy* and *SciPy* packages for statistical datasets; resolution aware display formats plus panning and zooming tools. With its ability to generate high quality vector graphics pertaining to publications or web service applications, it also serves as an excellent choice to represent data-intensive scientific problems (Chi et al., 2017).

Figure 2. Output plot of matplotlib and seaborn 10 countries with their population, generated on Pydroid3. Such good quality data visualization is possible on mobile devices.

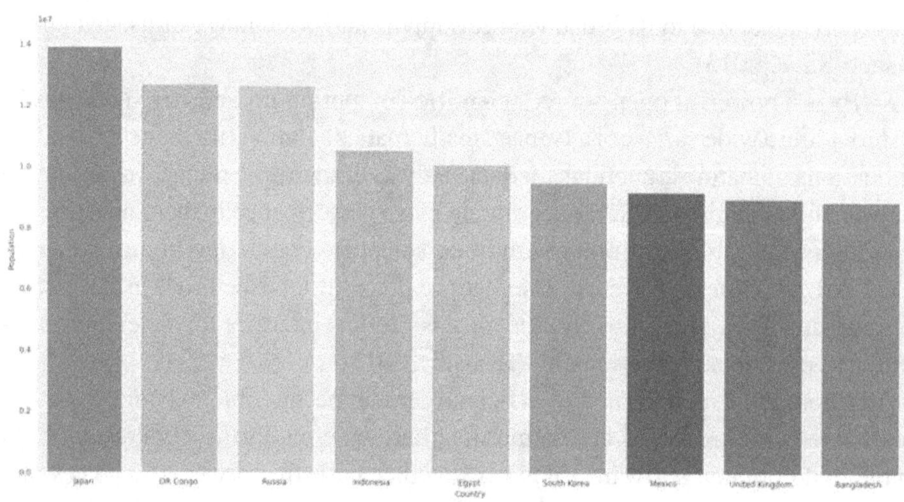

Seaborn is a graphing module used to visualize data, Seaborn a wrapper module of higher API function calls to *Matplotlib* which allow users to quickly generate attractive, articulated plots. By incorporating features like color-coded matrices and hierarchical clustering, Seaborn takes advantage of intuitive and visually appealing concepts to improve the understanding of complex datasets. These capabilities make

Seaborn ideal for exploratory data analysis, offering visualizations such as line graphs, bar plots, heat maps, violin plots and joint plots that all aim to improve the workflow in a quantitative analysis environment (Chi et al., 2017).

PySimpleGUI v. 4.6 on Pydroid3 (Python 3.9)

The *PySimpleGUI* module for Python is an easy-to-use, robust graphical user interface (GUI) library optimized around the *Tkinter* framework. It provides a platform-independent wrapper that makes creating custom GUI programs in Python both simple and easy. This module makes developing cross-platform applications with a wide range of functionality a breeze, it works on all common computer systems including Mac OS X, as well as popular Android OS. Additionally, its drag-and-drop feature allows one to quickly create widgets without having to deal with any complicated coding. The module also features built-in themes for further customization and offers several tools such as row/column creation wizards, text boxes and labels that are sure to provide users with the perfect GUI choice for their application.

NetworkX v. 3.0 (Released on Jan 08 2023) on Pydroid3 (Python 3.9)

Networkx allows users to deploy traditional network analytical tools as well as generate graphical visualizations of networks. Networkx offers several tools that make it an excellent choice for network applications including structured graph types such as directed graphs, convenience methods for node and edge manipulations.

There is support for network algorithms for structural characteristics like clustering coefficient, betweenness centrality, closeness centrality etc., writers that support exporting to standard file formats such as GML, GraphML and Pajek (Platt, 2019).

The package is easily extendable with user-defined plugins which makes it highly versatile and adaptable to various domain tasks. Additionally, its features strong interoperability with other Python packages such as *NumPy/NumPy* or *Matplotlib* which makes it ideal for performing sophisticated analysis tasks within data science pipelines (Goerzen, 2004). NetworkX library has been ported for Machine learning and Network analysis on Android mobile phones. The graph above shows output of NetworkX generated using sample code on Pydroid3.

Figure 3. Output graph produced by NetworkX on Pydroid3 (Py3.9) using following options: "node_size": 1500, "alpha": 0.2 and "node_color": "blue"in NetworkX to render the graph on Mobile device

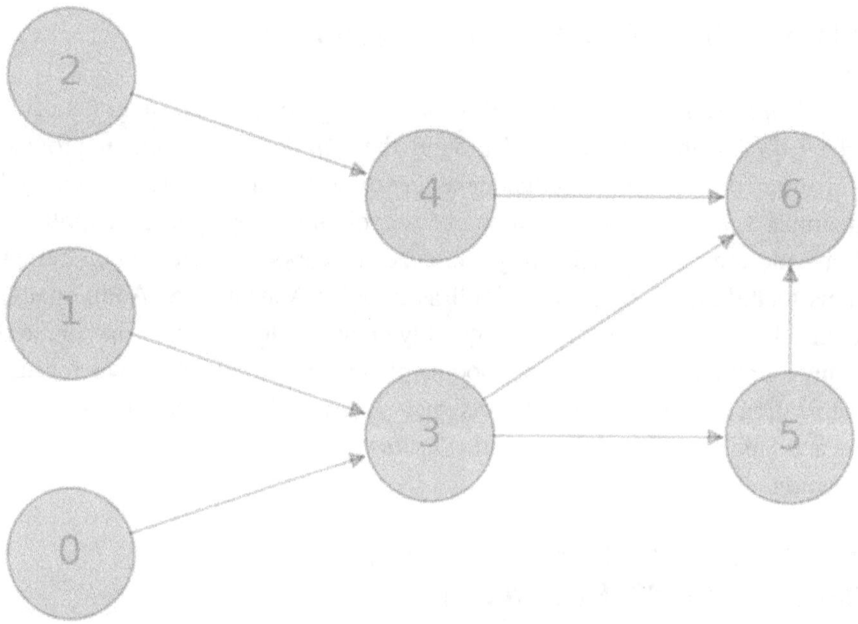

Biopython v. on Pydroid3 (Python 3.9) for Bioinformatics Research

Biopython (*biopython*) is an open source, collaborative Python project that provides a comprehensive set of tools for biological computation. It offers modules for a variety of bioinformatics tasks and supports Python 2, 3 as well as Java and C code. *Biopython* provides interfaces to useful bioinformatics programs and libraries, allowing users to leverage the benefits of both third-party code and existing user expertise without having to learn all the details. Furthermore, the project includes a large collection of modules for dealing with the different types of data generated by high-throughput biology experiments. This makes it easy for bioinformaticians to quickly develop powerful analysis pipelines in their chosen programming environment. *Biopython* is also available on Pydroid3, bioinformatics research can be carried out on Android Mobile phone with ease and portability (Zhang & Rajapakse, 2009).

Word Cloud in Pydroid Using "wordcloud" v. 1.8 Module

Word clouds are a key tool used in visualization techniques, allowing data and text to be shown graphically. Word clouds provide an excellent way to quickly identify words that are most relevant to a set of data. This type of analysis is helpful for visually summarizing topics from surveys, interviews, or documents, as the size, color, orientation, and font can all be adjusted to emphasize the keywords or phrases within the collection of text. Additionally, word clouds give users an opportunity to explore sentiment analyses, trends and correlations that can be discovered by analyzing different variations within word clouds. Thus, word clouds have become commonplace in market research as well as various other forms of visualization.

The *wordcloud* module for Python is a versatile and powerful library for creating word clouds. It enables users to produce visually attractive diagrams, comprised of words from text-based data sources. The frequency of each word used in the diagram is represented by its size in the drawing. This module facilitates a range of options for representing different fonts and color palettes, as well as adjusting image sizes and orientations. In addition, it supports rendering of complex shapes in the final drawing through support for custom masks (David Paper, 2019).ig4. Demonstration of "*wordcloud*" module of Python for generating Word clouds which show words that are most relevant to a set of data. This Word Cloud is generated for main keywords for this article pertaining to Machine Learning.

Figure 4. Showing the sample output of wordcloud module, this example uses few common Machine learning terms, more complex examples can be implemented by this module

KIVY FOR COMPILING PYTHON TO ANDROID APPS APKS

A free Python framework library for creating cross-platform graphical user interfaces is called Kivy (GUIs). It can make use of multitouch events to make apps that are more interactive and user-friendly, and it is designed for use across all platforms, including mobile ones. Applications for Linux, Windows, MacOS, Android, and iOS can be made using the Python and Cython programming language known as Kivy. Kivy apps maybe deployed on common platforms, like Mac, Android OS. It's made to make it simple for users to develop cross-platform applications that work on a range of gadgets, including smartphones, tablets, PCs, and Raspberry Pi. Python programs can be converted into APK files that can be installed and used on Android devices thanks to Kivy's provision of the required tools and features for creating and packaging Android applications. Kivy's packaging tools maybe utilized to turn the program into an Android device installable APK file. Kivy documentation can provide further details on building Android applications (Cohen et al., 2014).

Jupyter Notebook v. 1.0.0 on Pydroid3

Users can integrate text, visuals, and programming code in a single document using Jupyter notebooks, which are web-based interactive computing platforms. Jupyter notebook ver. 1.0 is available for Pydroid3 which users can utilize to harness the full power of Jupyter notebooks. They are a common tool among data scientists and academics and are frequently used for data analysis and scientific computing (Gupta, 2021).

For instance, Jupyter notebooks offers a more user-friendly and intuitive design, complete with tab completion and syntax highlighting. Additionally, it has sophisticated capabilities like magic commands that make it simple to complete complicated tasks without having to write a lot of code. An improved interactive Python interpreter is called IPython. Jupyter notebooks, allows one to produce, share and collaborate on notebooks containing python programs, latex code, pictures, graphs, interactive maps, and formatted textual matter (like headings, justification), are just one of the tools that IPython offers for collaborating with different programming languages and technologies. Overall, IPython is a useful tool for anyone who works with Python and can significantly increase your productivity and efficiency (Gupta, 2021).

SCIKIT-LEARN LIBRARY ONPYDROID 3

There are several popular Python libraries for machine learning including RapidMiner, MXNet, TensorFlow and *Scikit-Learn.* Each library has its own set of features and advantages, but they all share a common goal: to enable us to develop efficient M.L models.

Scikit-learn machine learning module implemented for Python and distributed under the 3-clause BSD license. The module is open-source, and anyone can suggest improvements, corrections, and even new algorithms. The liberal BSD license makes it free to use for both commercial and non-commercial purposes. It is a powerful tool that facilitates the development of classification, regression, and clustering algorithms using pre-made decision tree implementations. With *Scikit-learn*, it is possible to build complex models with minimal effort and fine tune them as needed. The library can generate more advanced prediction results by incorporating additional features into existing decision trees to make more accurate predictions than individual decision trees alone (Mirjalili, 2019).

In addition, *Scikit-learn* provides tools for feature selection, feature extraction, cross validation, model evaluation metrics (e.g., accuracy score), and visualization capabilities. By leveraging these capabilities, data scientists can customize their analyses and target specific goals like prediction accuracy or computational efficiency, opening new possibilities for sophisticated data analysis techniques with decision tree methods (Mirjalili, 2019).

Scikit-learn has several utilities for preprocessing data, model selection, and validation. *Scikit-learn* makes it easy to quickly prototype and develop an initial model with minimal code. Its built-in cross-validation tools simplify assessing the accuracy of various models while its efficient components can be combined into complex pipelines that leverage the power of parallel computing architectures. In addition, detailed online documentation combined with an active community of users make *Scikit-learn* an ideal platform for professional machine learning tasks (Collins, 2018).

Scikit-learn is a powerful and versatile machine learning library which allows users to create, test and apply regression models. Specifically, *Scikit-learn* includes various linear and non-linear regression algorithms such as Linear Regression, Ridge Regression, Lasso Regularization and Elastic Net Regularization. These methods allow us to fit a predictive model according to given parameters thereby obtaining an optimal set of coefficients that best explain the relationship between the target variable and its predictors. Moreover, *Scikit-learn* also offers various sorting methods like random forests that are best suited for datasets with multiple correlated parameters. It also provides cross validation methods which enable us to assess the quality of our model's predictions by keeping a separate test set aside from training data. Therefore, beginners and experts alike highly appreciate this Python library for its clear interface and ease of use in doing simple or advanced regression tasks. *Scikit-learn* is widely used in academic and commercial research, as well as in data science competitions and hackathons (Collins, 2018).

BASIC SCIKIT-LEARN CONCEPTS OF ESTIMATORS, PREDICTORS, AND TRANSFORMERS

An estimator is an object that is used to fit a model to data in the context of machine learning. Any object that implements the fit () function, which trains the model on a training data, and the predict outcomes using the predict () method, which uses the trained model to make predictions on new data, is stated to be an *estimator*. Linear Regression and Logistic Regression classes for modelling continuous and binary data, respectively, as well as the Random Forest Classifier for training decision tree ensembles, are a few examples of estimators in *Scikit-learn*. *Scikit-learn*'s estimators are implemented in Python and made to be quick and effective while also offering a robust and adaptable user interface for working with data (Kane, 2017).

A *predictor* in machine learning is a model that has been trained to forecast upcoming events or results based on data in the setting of *Scikit-learn*. For creating and training predictors, the *Scikit-learn* library has several methods, including Linear regression modelling for the forecasting of continuous values, Logistic regression modelling that forecast binary outcomes as zero and one, models for making predictions via a succession of decisions based on decision trees, and Support vector machines for applications involving regression and classification (Kane, 2017).

A transformer is a sort of estimator that is employed in the setting preprocess the data by changing its representation. Any object in the *Scikit-learn* package that employed the fit and transform methods, is referred to as a transformer. Examples of transformers in *Scikit-learn* include the StandardScaler, which normalizes data by scaling to unit variance and subtracting the mean, and the PCA (principal component analysis) transformer, which maps data onto a lower-dimensional space by exploiting directions of maximum variance. In a machine learning pipeline, transformers are frequently employed to prepare the data for modelling (Kane, 2017).

LINEAR REGRESSION ALGORITHM ON SCIKIT-LEARN LIBRARY OF PYDROID3

A statistical technique for simulating the relationship between a dependent variable and one or more independent variables is linear regression. This means that the change in the dependent variable is inversely proportionate to the changes in the independent variables because it is assumed that the dependent variable and the independent variables have a linear connection. In other words, linear regression seeks to identify the line of greatest fit between the independent and dependent variables. Based on the known values of the independent variables, this line can then be used to forecast the value of the dependent variable(Weisberg, 2013). The coefficient of determination (R-squared) and other metrics are used to assess the accuracy of the predictions(Bruce & Bruce, 2017). Logistic regression is employed to predict a variable's outcome using single or more independent parameter that can only have one of two possible values, such as "success" or "failure" (Hackeling, 2017).

LOGISTIC REGRESSION ALGORITHM ON SCIKIT-LEARN LIBRARY OF PYDROID3

Logistic regression uses a non-linear function to represent the non-linear association between the dependent parameter and an autonomous variable as opposed to linear regression, which assumes a linear relationship between the two parameters.

Performance of a classification model is pictorially assessed using AUC-ROC curve. The true positive rate are plotted on ordinant against the false positives are enlisted on the abscissae, resulting in score which can range from 0 to 1. A score nearer to 1 indicates better predictive accuracy of the model. Additionally, it is useful for establishing thresholds at which models adjust their decision boundaries between classes; these points may be determined by comparing different ROC curves. As with all statistical tests, however, results should be interpreted with high caution and multiple cross-validation experiments should be conducted to establish reliable evaluation metrics (Collins, 2018).

POLYNOMIAL REGRESSION

Polynomial regression is a method of modeling nonlinear relationships between predictors and outcomes in linear regression. It applies a nonlinear transformation to the features before calculating a linear relationship between them, allowing for the creation of predictive models with greater complexity than those described by standard linear equations. Polynomial regression can be used to address more complex issues, such as multiple interactions and higher order terms, as well as situations where a nonlinear effect appears in data; however, it can also result in overfitting if not applied correctly. This practice should be avoided by using cross-validation techniques whenever possible to ensure accurate prediction models are created (Giancarlo Zaccone, 2017).

PRINCIPAL COMPONENT ANALYSIS ALGORITHM ON SCIKIT-LEARN LIBRARY OF PYDROID3

Principal Component Analysis (PCA) scales the dimensionality of large data sets, by uncovering patterns and correlations within the data. *Scikit-learn*, provides an effective application of PCA to analyze large datasets searching for meaningful patterns. *Scikit-learn*'s PCA functionality utilizes a flexible API which allows users to control parameters such as the count of components, to retain and nature of transformation applied on the data (whether linear or kernel). Furthermore, it provides users with visualization tools such as scree plots and biplots to gain further insight into their data. Ultimately, *Scikit-learn*'s PCA module is robust enough to be used on large datasets while still yielding high performance results, making it a valuable tool in the exploratory phase of any effective Machine Learning workflow (Cross, 2015).

XGBOOST ALGORITHM ON SCIKIT-LEARN LIBRARY OF PYDROID3

XGBoost is a powerful and efficient machine learning algorithm. It is highly scalable implementation of the gradient boosting algorithm. It has been used in many different domains to solve many problems. XGBoost has been developed by Tianqi Chen and company. XGBoost has gained much popularity in applied machine learning field due to its remarkable performance gains over traditional algorithms, ability to scale across multi-dimensional data sets, and easy integration with existing frameworks such as *Scikit-learn* (Wade, 2020).

XGBoost library offers fast, accurate and efficient solution for many problems including regression and classification tasks. XGBoost leverages several techniques including regularization techniques such as L1 and L2 penalty, tree pruning techniques which helps reduce overfitting and various optimization techniques such as coordinate descent making it easier to tune hyperparameters. XGBoost is increasingly being used by data scientists in real world applications thanks to its speed, scalability, and accuracy (Wade, 2020).

K-MEANS CLUSTERING ALGORITHM ON SCIKIT-LEARN LIBRARY OF PYDROID3

K-nearest neighbors, or KNN machine learning algorithm is usually deployed for cataloguing and regression tasks on datasets. This technique searches the training dataset to find the most similar instances to each test instance and classifies it based on its neighbor's class labels. It works by finding the Euclidean distance between each point and its nearest neighbors, who are then assigned votes equal to their relative distance from the query data point. The label of the majority group of data points is then assigned as the same label for that query data point. This ensures robust accuracy, limited computational time and an easy-to-interpret output since it relies less on mathematical functions compared to other algorithms. Overall, KNN can be effectively applied when working with large datasets because of its ability to quickly group similar features together in unseen datasets (Vanderplas, 2017).

KNN is a form of unsupervised type of algorithm that divides the fed dataset into collections or clusters. The procedure divides the data points into clusters so that the totality of distance square between each datapoint and the center of the cluster is effectively minimalized. This clustering method is popular because it's easy to understand and implement, and it can be parallelized to speed up computation time(Vanderplas, 2017).

Inputs comprises of the Data set, Number of clusters denoted by K, and a distance function to measure the distance between two observations (e.g., Euclidean distance). The K-mean algorithm produces output inform of clusters of input data, where each cluster contains observations with similar characteristics, and each observation belongs to only one cluster

SVM OR SUPPORT VECTOR MACHINE

Support Vector Machine is machine learning procedure that is used for organization of data and can be employed to solve any data problem which comprises of two categories. The algorithm, in general, finds the hyperplane that divides the data into two classes such that the distance of each point from this plane is as small as possible. This approach is rooted in statistical learning theory (Taweh Beysolow, 2018).

Finding the optimum hyperplane that can distinguish between the various classes in the data is the aim of an SVM. The nearest points to this hyperplane are referred to as support vectors, and this boundary is known as the decision boundary. The objective is to clearly differentiate the various classes by placing the decision border as far from any class's closest data points as is practical. Due to its ability to frequently locate a decision boundary that is more effective than those found by other algorithms, SVM is excellent when dealing with high-dimensional data (Pardalos et al., 2016) Hao & Ho, 2019.

CONCLUSION

Scikit-learn is a powerful and user-friendly library for the Python programming language. It is one of the most used libraries among developers because of its comprehensive set of tools, which includes modules for data pre-processing, model building, model evaluation and selection. *Scikit-learn* also supports supervised and unsupervised learning methods like SVM, as well as factor analysis, feature extraction and reduction of dimensionality techniques such as PCA. Furthermore, its algorithms are easy to use thanks to clear documentation and tutorials providing step by step explanations which makes it ideal for learning machine learning.

The benefits of implementing Machine learning mobile devices are now abundantly evident as an increasingly mobile workforce changes its focus to remote work. With the aid of smartphones and tablets, people can perform data cleansing, model training, model fitting, optimization, and prediction—all while on the go. This promises greater independence and higher productivity because they can better manage their time, keep organized, manage projects digitally, and start activities more quickly. Developers have unprecedented access to sample code and the opportunity for real-time cooperation when linked via a mobile device.

REFERENCES

Apeltsin, L. (2021). *Data Science Bookcamp*. Simon and Schuster.

Beysolow, T. (2018). *Applied natural language processing with Python: Implementing Machine Learning and Deep Learning Algorithms for Natural Language Processing*. Ca Apress. doi:10.1007/978-1-4842-3733-5

Blanco-Silva, F. J. (2015). *Mastering SciPy*. Packt Publishing Ltd.

Bressert, E. (2013). *SciPy and NumPy*. O'reilly.

Bruce, P., & Bruce, A. (2017). *Practical Statistics for Data Scientists*. O'Reilly Media, Inc.

Chen, D. Y. (2018). *Pandas for everyone: Python data analysis*. Addison-Wesley.

Chi, A., Yik, C., & Yuen, K. (2017). *Matplotlib 2.x by example: multi-dimensional charts, graphs, and plots*. Packt Publishing Ltd.

Cohen, R., Tao Wang, & Springerlink. (2014). *Android Application Development for the Intel Platform*. Apress.

Collins, R. (2018). *Scikit-Learn in Details*. Independently Published.

Cross, R. (2015). *Principal component analysis*. Clanrye Intl.

Garreta, R. & Moncecchi, G. (2013). *Learning scikit-learn: Machine learning in Python: Experience the benefits of machine learning techniques by applying them to real-world problems using Python and the open source scikit-learn library*. Packt Publishing Ltd.

GitHub. (n.d.). [Code Sharing]. GitHub. http://github.com

Goerzen, J. (2004). *Foundations of Python Network Programming*. Apress. doi:10.1007/978-1-4302-0752-8

Gupta, P. (2021). *Practical data science with Jupyter: Explore data cleaning, pre-processing, data wrangling, feature engineering and machine learning using Python and Jupyter*. Bpb Publications.

Hackeling, G. (2017). *Mastering machine learning with scikit-learn: Learning to implement and evaluate machine learning solutions with scikit-learn*. Packt Publishing Ltd.

Hagberg, A. A., Schult, D. A., & Swart, P. J. (2008). Exploring Network Structure, Dynamics, and Function using NetworkX. In G. Varoquaux, T. Vaught, & J. Millman (Eds.), *Proceedings of the 7th Python in Science Conference* (pp. 11–15).

Hao, J., & Ho, T. K. (2019). Machine Learning Made Easy: A Review of Scikit-learn Package in Python Programming Language. *Journal of Educational and Behavioral Statistics, 107699861983224*(3), 348–361. doi:10.3102/1076998619832248

Idris, I. (2011). *NumPy 1.5: Beginner's guide*. Packt Publishing.

Kane, F. (2017). *Hands-On Data Science and Python Machine Learning*. Birmingham Packt Publishing.

Langtangen, H. P. (2016). *A Primer on Scientific Programming with Python*. Springer.

LinuxHint. (n.d.). [Programming]. Linux Hints. https://linuxhint.com

Lutz, M. (2006). *Programming Python*. O'Reilly Media, Inc.

Lutz, M. (2013). *Learning Python*. O'Reilly Media, Inc.

Mirjalili, S. (2019). *PYTHON MACHINE LEARNING - THIRD EDITION: machine learning and deep learning with python, scikit... -Learn, and tensorflow 2*. Packt Publishing Limited.

Pardalos, P. M., Piero Conca, Giuffrida, G., Nicosia, G., & Springerlink. (2016). Machine Learning, Optimization, and Big Data. *Second International Workshop, MOD 2016*. Springer International Publishing.

Platt, E. L. (2019). *Network Science with Python and NetworkX Quick Start Guide*. Packt Publishing Ltd.

Rajamani, S. K., & Iyer, R. (2022). Development Of an Android Mobile Phone Application for Finding ClosedLoop, Analytical Solutions to Dense Linear, Algebraic Equations for The Purpose of Mathematical Modelling in Healthcare and Neuroscience Research. *NeuroQuantology: An Interdisciplinary Journal of Neuroscience and Quantum Physics*, *20*, 4959–4973. doi:10.6084/m9.figshare.c.6156024.v1

Vanderplas, J. T. (2017). *Python data science handbook: Essential tools for working with data*. O'reilly, Cop.

Varoquaux, G., Buitinck, L., Louppe, G., Grisel, O., Pedregosa, F., & Mueller, A. (2015). Scikit-learn. *GetMobile: Mobile Computing and Communications*, *19*(1), 29–33. doi:10.1145/2786984.2786995

Wade, C. (2020). *Hands-on gradient boosting with XGBoost and scikit-learn: Perform accessible machine learning and extreme gradient boosting with python*. Packt Publishing.

Weisberg, S. (2013). *Applied Linear Regression*. John Wiley & Sons.

Zaccone, G. (2017). *Deep learning with TensorFlow*. Uk Packt Publishing.

Zhang, Y., & Rajapakse, J. C. (2009). *Machine Learning in Bioinformatics*. John Wiley & Sons.

Chapter 17
A Thorough Investigation of Various Goals and Responses for Mobile Software-Defined Networks

Somesh Kumar Sahu
https://orcid.org/0000-0003-2171-0338
LTI, India

Chandra J.
CHRIST University (Deemed), India

Kiran Muloor
CHRIST University (Deemed), India

Debabrata Samanta
https://orcid.org/0000-0003-4118-2480
Rochester Institute of Technology, Kosovo

ABSTRACT

Cloud computing has caused some companies to modify their IT infrastructure and maintenance procedures and may eliminate their current hardware altogether. Conventional methods of setting up a switch or router may be error-prone and unable to make full use of the capabilities of current network architectures. As many intelligent networking designs as possible must be developed for intellectualization, activation, and customization in future networks. Due to software-defined networking (SDN) technology, it's possible to control, secure, and optimize network resources, eliminating the rigid coupling between the control plane and the data plane in traditional network architectures. Here, the chapter explores the problems, difficulties, and potential solutions associated with software-defined networks (SDN), a novel concept in computer networking. Through SDN, the network gains the ability to be programmable, quick, and adaptable thanks to its separation of data and its ability to control traffic.

DOI: 10.4018/978-1-6684-8582-8.ch017

INTRODUCTION

Networking is an exciting intellectual and technical area because technology allows people and their computers to connect with each other. A network is a group of computers (called nodes) and transmission lines (called links) that allow communication over long distances (Cabaj et al., 2018). Navigating hardware switches has become more difficult as network size and demand increase. Companies running highly virtualized systems with huge networks have found manually configuring individual network soft switches to be extremely difficult and time-consuming (Guleria, 2017). Switches, routers, and other physical hardware are used in traditional networks to make connections and operate the network. However, this is where software-defined networking (SDN) comes into play.(IEEE et al., 2015). AS SDN is considered a popular alternative to conventional networks because it allows IT administrators to add additional bandwidth and physical infrastructure services at no additional cost. SDN is a network strategy that provides network administrators with the ability to configure, monitor, modify, and manage network operations using open interfaces such as the Open Flow protocol programmatically. The application, control, and data layers make up the three tiers of this design.

TRADITIONAL NETWORKING

A traditional network is constrained by both the size of the network and the requirements of various services. It is important to note that these requirements cover reliability, virtualization, flow management, engineered traffic, and implementation of policy requirements.(Al-Haddad & Velazquez, n.d.) There has been an increase in the complexity of network architecture and management over the last few years. We are in the process of connecting everything, thanks to enabling technologies like cloud computing, portability, and newly emerging ideas such as IoT (internet of things)(Al et al., 2022). Due to the complexity and reliability of these new concepts, improved data center infrastructure as well as more bandwidths are required, as they call for a simpler, more resilient network infrastructure. For managing network traffic, the most widely used type of networking is traditional networking, which uses switches, routers, as well as dedicated hardware with fixed functions to provide network management services (Haji et al., 2021) The gadgets on this list perform different functions and complement each other to provide support to the network as a whole. It is common for traditional networks to suffer from issues of scalability as a result of their inherent nature. In the majority of switching hardware and software, provisioning APIs are rarely if ever available, and most of the switching hardware and software that are available are proprietary and difficult to access. It is important to note that traditional networks and proprietary provisioning software are very much compatible with each other. Unfortunately, traditional networks cannot be modified regularly, while hardware-centric networking may not offer all the features that traditional networks can offer.(Jimson et al., 2017). In traditional networking, the functions and functionality are implemented in multiple ways, including the use of application-specific integrated circuits (ASICs), dedicated hardware, and dedicated devices that use switches, routers, and application, delivery controllers. With the increase in popularity of the internet, traditional networking solutions are not able to keep up with the computing demands of modern society, so to be able to keep up we need to increase our agility and flexibility to function effectively (Sezer et al., 2013). As far as these objectives are concerned, software-defined networking (SDN) appears to be able to achieve them.

SOFTWARE-DEFINED NETWORKING (SDN)

In essence, SDN refers to software-defined networking, a concept that allows for software-based controllers to effectively manage the traffic on a network, as well as the hardware infrastructure that supports it. Since SDN fundamentally alters the way networking is conducted, it will be especially useful in situations where traffic in a conventional network must pass through multiple switches and routers. As a result of SDN, the brain is centralized rather than having its intelligence distributed across the network. With SDN, everything will be divided between the various control planes and data planes (forwarding function), allowing command and control to be centrally controlled from a single node by distributing everything between them(Wang et al., 2019).

HOW ARE SDN AND TRADITIONAL NETWORKING DIFFERENT?

Essentially, the main difference between SDN and traditional networking is that the former is primarily based on software, whereas the latter is based on hardware in most cases. Due to its software-based nature, SDN has the advantage of scalability, flexibility, and the ability to give users more control and simpler resource management as a result of it being a software-based system (Haji et al., 2021). A traditional network, as opposed to SDN, generates connections between network devices and manages the network through the use of routers, switches, and other hardware and physical components. Figure 1 shows details about SDN and traditional networking differences.

Figure 1. Traditional vs SDN Networking

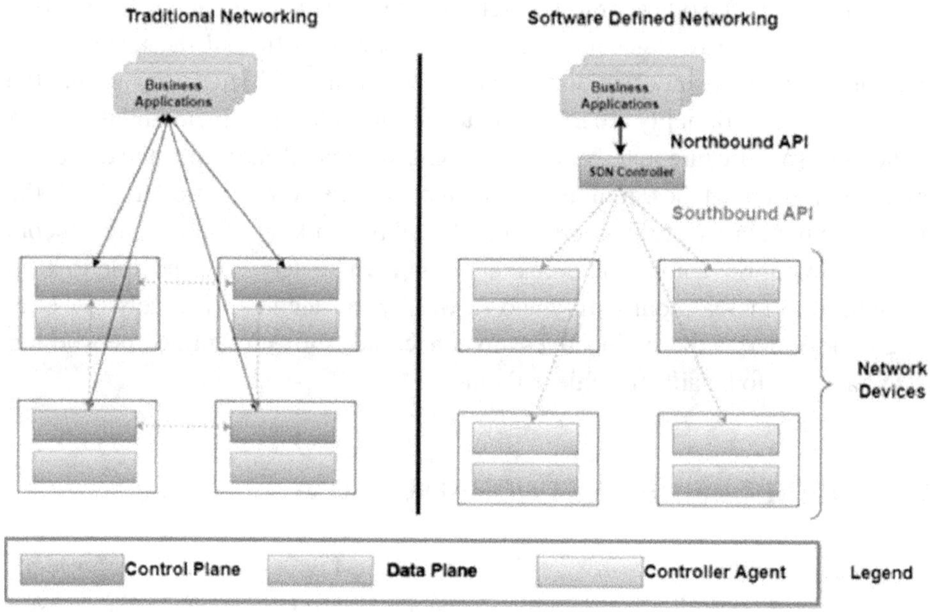

Compared to conventional networking, SDN is more capable of interacting with hardware that uses the network. SDN allows users to provision new devices using software rather than physical infrastructure (Gupta et al., 2022). As part of this system, IT administrators are also able to control network paths and network services. As a result of the use of virtualization, the main distinction between SDN and conventional networking is emphasized due to the use of virtualization. An abstract version of your network is created when your network is virtualized using SDN technology (Satheesh et al., 2020) A single location can be used for provisioning all of your resources using this virtualization software.

WHY DID WE NEED SDN?

Even though IT teams are tasked with creating and maintaining IT applications and infrastructure, their efforts should ultimately support the following key business objectives for their companies (*What Is Network Virtualization? Everything You Need to Know*, n.d.).

As one of the primary benefits of SDN, centralized management is one of its most important features. With SDN networks, data center managers have the option of centrally managing the network using a central management tool, which provides them with significant advantages. In addition, it eliminates the hurdles posed by conventional systems, all from one place, thus increasing agility for both the provisioning of physical and virtual networks, all at the same time.

In terms of security, an SDN controller gives network engineers a centralized location to manage and monitor the overall security of the network. Many benefits can be gained from the use of SDN, even though the trend toward the use of virtualization has made protecting networks from outside threats more challenging (Vissicchio & Vanbever, 2014) SDN controllers guarantee that the network will implement security guidelines and data within the network under the data security guidelines. The management system used by SDN is a single one, so it is a more secure way to implement the network.

Savings on operating and capital costs - SDN networks provide users with a massive reduction in operating and capital costs. It has been traditionally used to ensure network availability by utilizing redundant equipment, which naturally increases the cost of operating the network. In comparison to a traditional network, a software-defined network is significantly more effective and does not require the addition of additional network switches as compared to a traditional network. In addition to this, SDN has the added benefit of being able to work well with virtualization, which in turn lowers the cost of adding new hardware by combining the two.

Featuring scalability, SDN provides users with the ability to access various network components at a centralized point, thanks to the OpenFlow agent and the SDN controller, which can manage the different network components on a centralized level. There are several options available to engineers that are not available in a conventional network setup, these options allow them to change the network infrastructure at any time without having to buy and configure resources individually.

SOFTWARE-DEFINED NETWORKING ARCHITECTURE

The SDN architecture makes the network more adaptable as well as more manageable and can help the network become more resilient. The centralization of the management of SDN is made possible by the separation of the control plane from the data forwarding operation in discrete networking devices(Latif et al., 2020). A new network architecture has been developed which is completely programmable, dynamic, manageable, affordable, and adaptable, making it the perfect solution for a changing world in which demands are always changing.(Priyadarsini & Bera, 2021; Taherian-Fard et al., 2022; Usman et al., 2019). Then SDN architecture is composed of three layers (APP, Control, and Data) as shown in Figure.2 along with three APIs (Northbound, Southbound, and East-West)(Sharifian et al., 2021). Figure 2 shows details about Software-defined networking architecture.

Figure 2. SDN Architecture

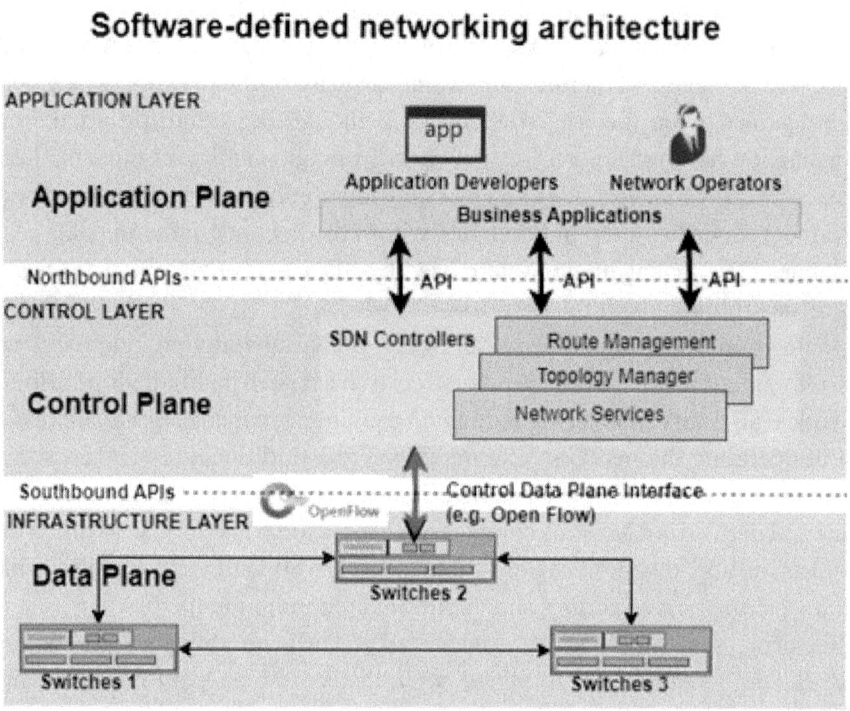

SDN network components can be represented by three layers:

Users can utilize the application layer to deliver data regarding requests specific to specific networks through the use of network "apps" that are offered to them by the application layer. For intelligent traffic analysis to be possible at line speed, data-driven models, rather than predefined protocols, are used in the data plane of the application (Joe, 2020). This feature is called Intelligence Data Plane (IDP), and in the coming decades, such a feature is expected to have a significant impact on many different networking architectures (Zhou et al., n.d.). It has recently been discovered that the newly developed programmable switches can provide a great deal of support for IDP hardware implementation (IEEE et al., 2015)

Two main categories of the prior art can be distinguished in this regard:

(i) Using the data plane to extract useful flow information, while placing the learning-based traffic analysis on the control plane.
(ii) embedding learning models in the data plane while ignoring flow-level features that are vital to achieving high learning accuracy.

Figure 3 shows details about the workflow of the Data Plane.

Figure 3. Workflow of Data Plane

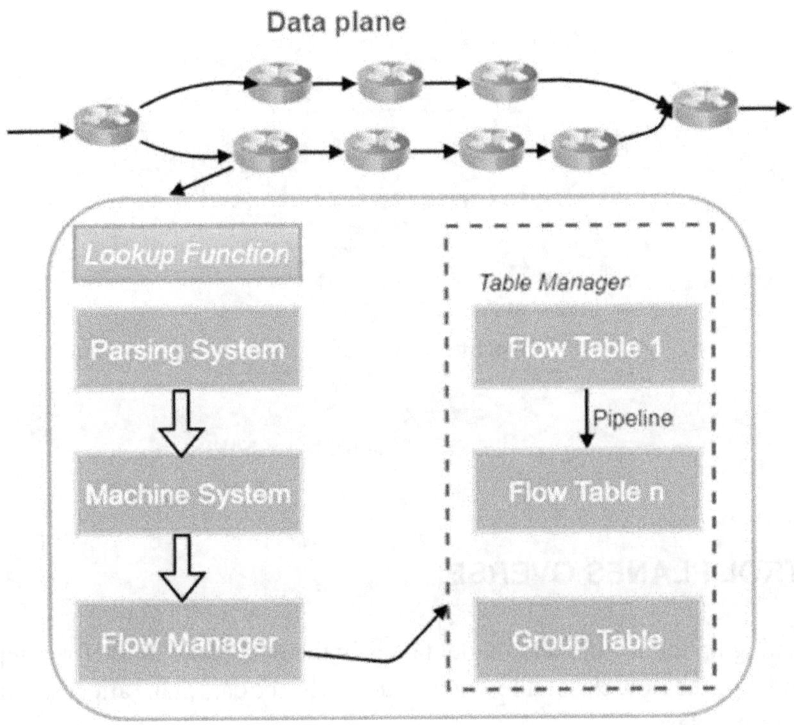

The data traffic's destination is determined by the SDN controllers, which are part of the control layer (Gupta et al., 2022). Calculating a router's forwarding state is important for routing, traffic engineering, and detecting and recovering failures. Figure 4 shows details about the workflow of the Control Plane.

IDP achieves line-speed traffic analysis compared to state-of-the-art technology, which deploys learning models on the network control plane. However, state-of-the-art technology can only send and receive traffic at about 10 Gbps (Gupta et al., 2022). A further disadvantage of putting the control plane on the critical path of the traffic analysis is that it adds additional reaction time. The majority of designs rely on programmable switches to gather vital flow data, such as the distribution of interpacket delay across packets and then conduct traffic analyses on the control plane as a result of the collected data. In the absence of a control plane and a data plane, the traffic is not analyzed at line speed since there is a latency associated with the interaction between them.

Figure 4. Workflow of Control Plane

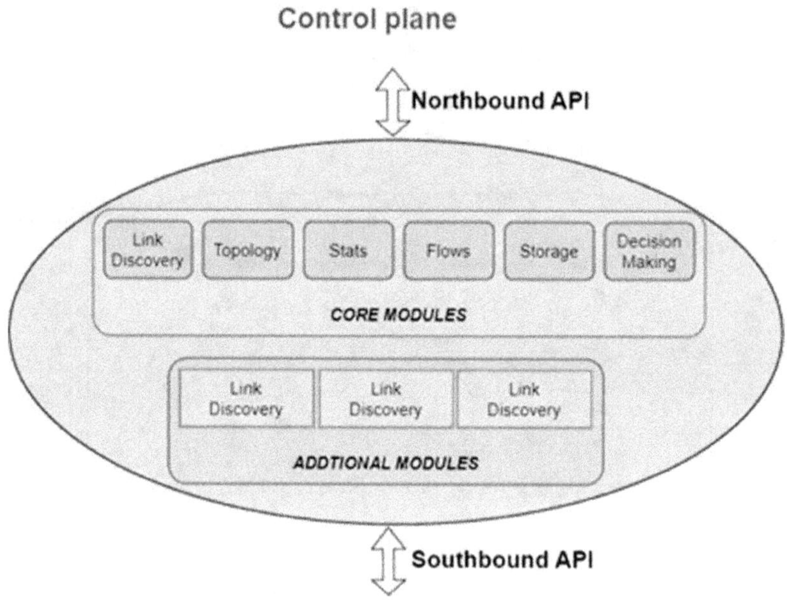

CONTROL PLANES OVERSEE

(i) Setting up the feature and model tables in the data plane in the first step.
(ii) updating the flow class table when requested by the data plane after establishing the flow class.

A packet that does not match the flow class table will instead be sent through the regular pipeline for inference by the regular model in case the A packet does not match. Although there is a latency associated with updating the flow class table, the analysis of the traffic does not suffer from that latency. Consequently, the control plane is kept out of the crucial path of packet classification, which means that line-speed traffic analysis is ensured.

A network "application" is available at the application layer that users can use when requesting data that is specific to the network. Used to efficiently operate, maintain, and fine-tune the network, the application, and the management plane on a daily basis, as well as configure and fine-tune them. Furthermore, the program is also responsible for disseminating the network reports that you run on your computer (Gupta et al., 2022). As a result, it helps in the provisioning of devices, the configuration of ACLs, and the engineering of traffic patterns. Figure 5 shows details about the workflow of the Management Plane.

Figure 5. Workflow of Management Plane

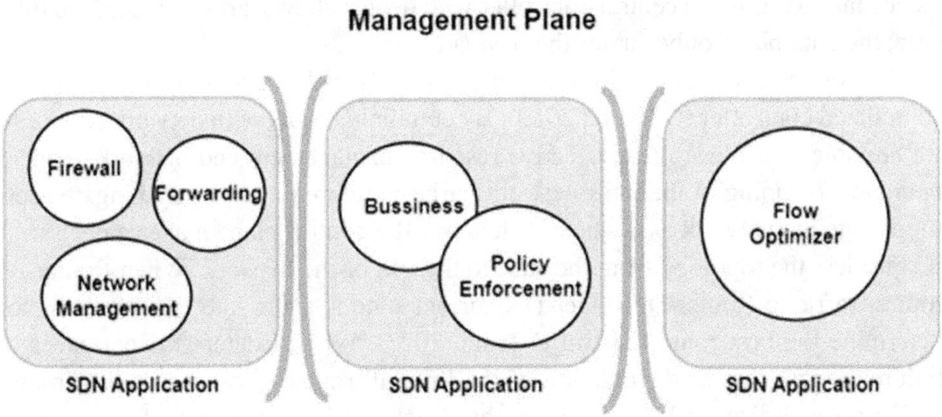

Basically, SDN provides orchestration of network management by dissecting the control plane from the data plane in order to separate them. By dividing the system into the corresponding separate planes, SDN aims to create a computer network while assisting clients in virtualizing their hardware: The control plane is frequently used for managing device designs that are remotely connected to the software-defined network because it provides the performance and fault management of NetFlow according to the type of deployed protocols (Gupta et al., 2022). Data planes can speed up the delivery of traffic to the right place at the right time. Control planes use flow control to determine the traffic path streams to be taken by traffic before it reaches the data planes, in order to direct its path (Wang et al., 2019). A software-defined network can be managed and interacted with effectively through this interface, making it easier for network administrators to manage and maintain.

SDN: SEPARATION OF CONTROLLER & DATA LAYER

This relies on four fundamental SDN Networks principles:

Flexible Networking & IP Routing: SDN enables packets and traffic to reach their destinations, and it does this by using software and dynamic algorithms in a fully flexible and agile manner. The SDN protocol makes it possible to achieve reachability in a much more efficient and time-efficient way than manual routing, which wastes a lot of time and resources.

In a conventional network, the router is the brain of the network, and it also serves as a data forwarding layer, or what is known as the data plane. Although the router still controls the data forwarding plane in this case, the data forwarding plane is still controlled by the router. The data plane is only used to send the payload to its destination; it is the central controller who decides how traffic is routed; in this case, the data plane only carries the payload.

Provides a centralized view of resources and offloads the brain functions to the centralized controller (Oo et al., 2019). By combining SDN with network services, we are able to achieve more effective resource management and network service operation. By doing so, the entire network can be monitored centrally, making it easier to manage the network as a whole. Whenever the control plane makes a decision, it considers the topology from the start to the end of the network. When Router-A routes traffic to Router-D, Router-D's current state is taken into consideration to determine the best route (Biswal & Patel, 2022). When Router-D experiences a failure, congestion, or degradation of the link, all routers are instructed to route traffic through Router-X in the event that Router-D fails. Figure 6 shows details about the more details and explanation on Separation of the controller and Data layer

Figure 6. SDN. Separation of the controller and Data layer

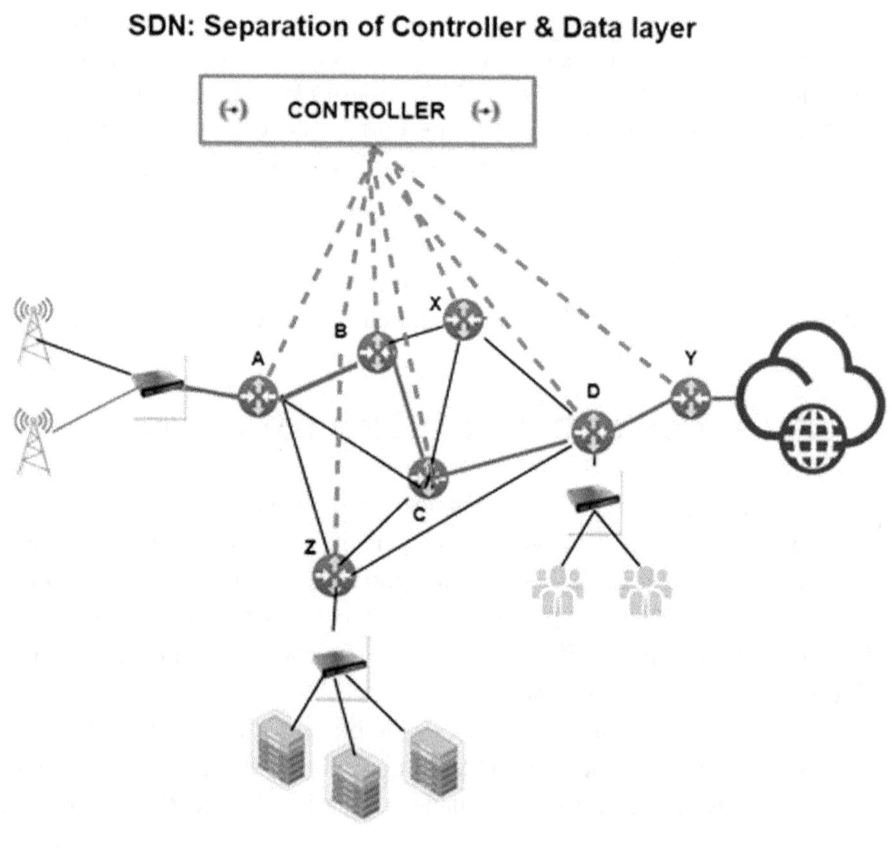

With centralized control planes, network management becomes directly programmable with agile for any need, centrally managed, and programmable, and applications and network services are shielded from infrastructure changes by centralized control planes. With the help of SDN, networks are becoming programmable, enabling operators to support a variety of applications such as dynamic bandwidth provisioning, auto-scaling, and building protection paths, to name a few. Higher-level applications can flexibly manage the controller in many ways, for example, by directly exposing the controller to the end-user on the web portal so that they can either provision or de-provision bandwidth at their discretion. One of the true strengths of SDN is abstraction; the entire logic flow of the network is so automated that network applications can send commands to the SDN controller, which will then modify the configuration of the network resources according to the command. All of this takes place in seconds.

SDN CONTROLLERS

The architecture of an SDN controller includes the following components:

In terms of the Northbound Interface, it is non-standard and takes different shapes depending on the application. There are a variety of APIs that can be used by applications to communicate with the SDN network, and these APIs can be used in languages like Java, Python, and REST (Representational State Transfer) (Aladaileh et al., 2022).

An interface that is used for Southbound traffic has been standardized and utilizes the OpenFlow protocol. SNMP, CLI, and other protocols can be used in "closed SDN" solutions, as well as additional protocols such as SNMP.

A set of modules are available for managing network elements, including topologies, flow tables, discovery modules for discovering network elements, SDN management modules for managing network elements, usage statistics tables, etc. Figure 7 shows details about the more details and explanation on SDN Controllers architecture

An SDN controller is the software that manages a set of "standardized" switches that are now available on the market and which are managed by SDNs. As a result of SDN network devices becoming less computationally complex and more commoditized, which is another word for lower in price, the devices become less computationally complex and less expensive as a result. SDN Controllers construct comprehensive views of the network to determine how flows should be routed from one point of the network to the next to decide how they should be routed (Bhardwaj & Panda, 2022). Through its Northbound interface, it uses APIs to communicate with applications (which submit requests for opening and closing flows) in order to exchange information between them. It can manage the network nodes (SDN devices) by using the OpenFlow protocol and its Southbound interface.

An SDN architecture cannot be fully implemented without the OpenFlow protocol. It is a key component of the entire topology. Stanford University developed the OpenFlow specification over the course of a few years with a group of intelligent individuals. The Open Networking Foundation, or ONF as it is also known, was established in 2011 in order to provide support for SDN and OpenFlow. To begin standardizing the OpenFlow protocol, numerous working groups have been established to carry out the process.

Figure 7. SDN Controller Architecture

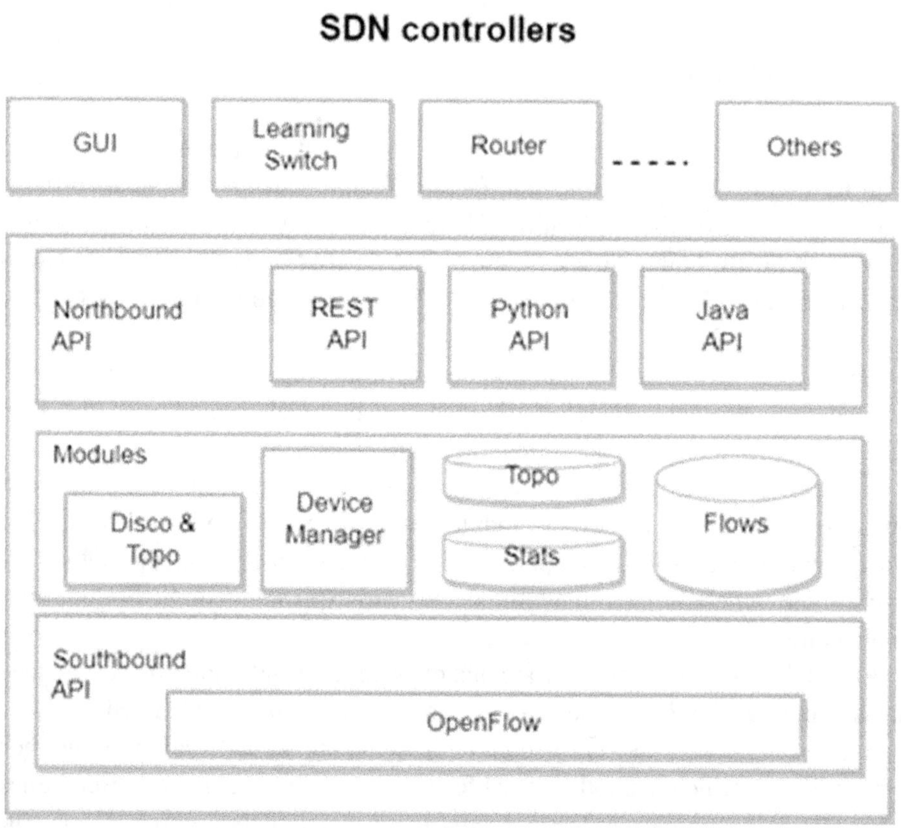

OPENFLOW (OF)

All network devices are managed and tracked with the help of OF (OpenFlow), a programmable network interface protocol. Figure 8 shows details about the more details on OpenFlow

Figure 8. OpenFlow

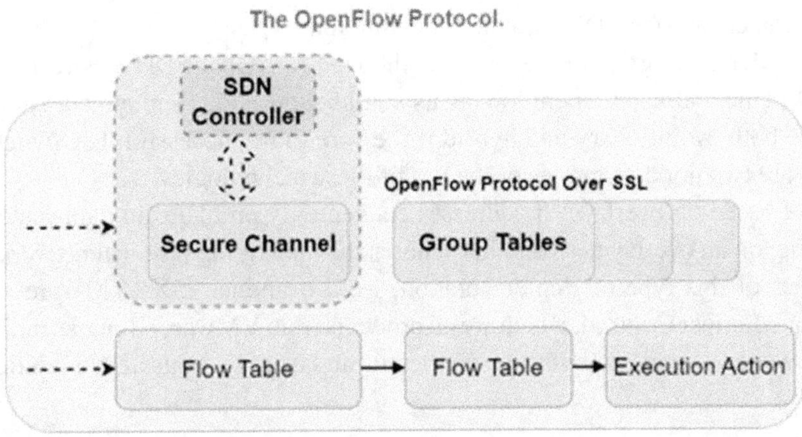

SDN standards such as OF were among the first to be developed. In the case of SDN architectures, the protocol that enables direct communication between SDN controllers and forwarding planes at the time was the first to be defined(Satheesh et al., 2020). The OF protocol can be used to program a switch so that it operates the same way as a legacy switch, preventing the need for manual reconfiguration in the event that the network changes in the future. An OF switch is made up of three main components: a secure channel, a flow table, and a collection of tables. In the group table, there is several flow entries that are grouped into a single identifier that denotes a network node based on the flow entries.(Braun & Menth, 2014; Zhai et al., 2020) This abstraction allows common output actions to be applied to flow entries, which can then be changed in an effective way as a result of a common output action. An OF switch compares incoming packets against multiple flow table entries until a match is found, and then a set of actions are performed that apply to the flow entry for which the packet was received. The OF switch performs a set of operations depending on the specific flow entry that matches with an incoming packet, once it has found a match between the packet and one of the entries in the flow table. Businesses and service providers can better control their networks by taking advantage of SDN to respond quickly to changing business requirements (Al et al., 2022)

DIFFERENT SDNS

Various implementations of the SDN concept have been developed since the first implementation was developed, depending on how the controller layer interacts with the devices.

We can categorize SDN into the following four types:

Open SDN: Using OpenFlow as the southbound API, Open SDN uses a centralized control plane and used OpenFlow as its southbound API. Deployment strategies must be both evolutionary and hybrid if they are to be successful. BigSwitch and ONOS are two good examples of the use of these technologies.

SDN based on overlays: It addresses connectivity problems in data centers by operating on an overlay network, but ignores the underlying physical networks. As examples of this type of implementation, we'll examine NSX (VMWare), NEC VTN, and Juniper Contrail, which are all products from VMware (Bouguerra, 2021). Figure 9 shows details about the more details and brief on Four SDN Families.

Figure 9. Four SDN Families

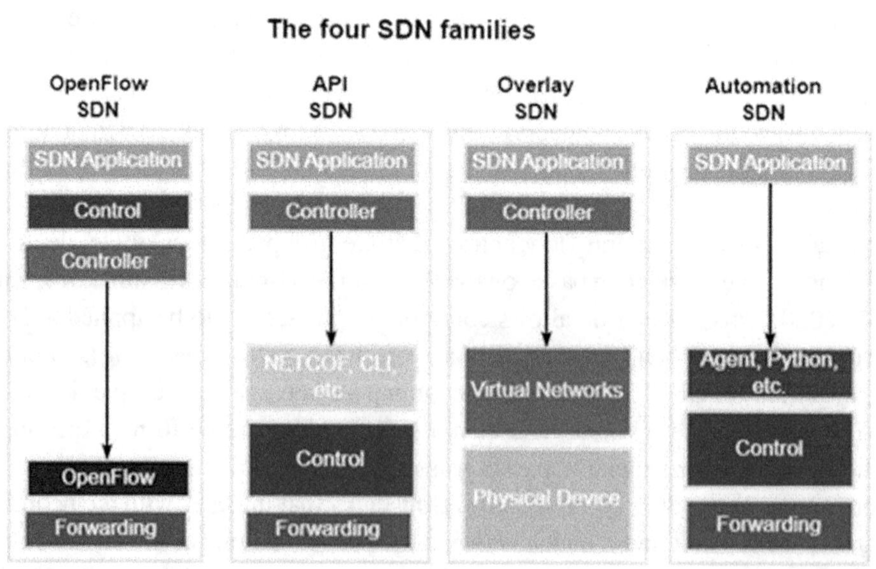

In the context of unopened SDN or API-based SDN, the main topic here is network programming. To enable the programming of network nodes, including the pre-existing hardware infrastructure, NetConf, OpFlex, and SSH for the southbound APIs are used. As an example of their application, they are providing OpenDaylight, APIC-DC, EM, and Tail-f (NSO) examples.

The main focus of this course is network programming. NetConf, OpFlex, and SSH are used for southbound APIs to program network nodes, including preexisting hardware infrastructure. They have developed applications such as OpenDaylight, APIC-DC, EM, and Tail-f (NSO).

Automation based on SDN (or hybrid SDN): Covers both legacy and SDN-capable hardware. A key objective of the project is to make it as cost-effective as possible. It makes use of generic white-label components that support a wide range of operating systems and automation tools (such as agents, Python, and ONIE, the Open Network Install Environment), all of which are white-label components (Sharifian et al., 2021).

COMMON USE CASES FOR SDN

DevOps

It is through SDN's ability to automate application updates and deployments that DevOps can be supported. There is a possibility that this strategy will require the automation of IT infrastructure components as the DevOps apps and platforms are deployed.

Networks on Campus

It can be challenging to manage campus networks, especially given the ongoing requirement to combine Ethernet and wireless networks at the same time. There are numerous benefits of using SDN controllers for campus networks, including the ability to manage and automate the network in a central place, enhanced security, and network-wide quality of service at the application level (Zhai et al., 2020)

Networks for Service Providers

As a consequence of SDN, service providers will be able to automate and simplify network provisioning for complete network and service management.

Safety/Security in the Data Center

The use of SDN enables more focused security and simplifies the process of managing firewalls. It is typical for businesses to use perimeter firewalls to protect their data centers in order to keep them secure. However, businesses can add virtual firewalls to protect virtual machines and build a distributed firewall system as a way of protecting virtual machines. This added firewall protection layer aids in preventing a breach in one virtual machine from spreading to another. Administrators can view, modify, and control network activity with SDN centralized control and automation, which lowers the risk of a breach.

IOT

Many sensors (millions of them) continuously populate and transfer data to real-world infrastructures (the Internet of Things) through complex networks the Internet of Things. It is important to note that devices that constitute the Internet of Things, including storage, processing, and communication resources, lack the computational capacity to perform computation-intensive tasks at the local level efficiently. There has been a wide variety of smart items have been introduced to the market in recent years, including automobiles, wearable technology, measuring sensors, home appliances, medical equipment, and industrial goods (Wilson et al., 2019). SDN provides a boost in the solution of complicated problems and assists in orchestrating IoT services in a more effective way. Implementing the Internet of Things is challenging due to its large data generation, complicated infrastructure, security risks, and requirements from recently developed technologies (Rafique et al., 2020).

Satellite-Terrestrial Networks (STN)

There has been a lot of attention paid to satellite networks as one of the most important means of communication in the past few years(Lin et al., 2019). Since satellite networks offer high bandwidths, worldwide coverage, and continuous services, they are widely employed in emergency aid, navigation, and location applications. The advancement of STNs requires innovative information and communication technologies. A software-defined and virtualized STN, also known as a software-defined and virtualized STN, is gaining support from the networking community. It is anticipated that the paradigm shift known as SDN will reshape future networking architectures. Through network abstraction, network programming, and the separation of data and control layers, SDN improves STNs.(Qiu et al., 2019).

WHY ARE COMPANIES SHIFTING TO SDN?

Data centers continue to change, and traditional networking is having a hard time keeping up with the changes, which is why vendors are turning to SDN. A few examples of these can be found below.

Considering the proliferation of cloud services, we need to ensure that users are able to access infrastructures, applications, and IT resources without restriction. Hence, there is an increasing need for more bandwidth, processing power, and storage a result of that.

Concepts of cloud computing offer effective and efficient methods for addressing computational problems that require large amounts of resources. Cloud data center performance may be limited by conventional methods, architectures, and processing techniques. Software-Defined Cloud Computing (SDCC) is a technology where virtualization services to all network resources in a data center are software-defined, and in which software-defined networking and cloud computing are integrated (Abbasi et al., 2019). The SDCC concept promotes the centralization of networking functions within data centers, thereby altering the current situation. Developing software-driven cloud infrastructure requires open and interoperable networking hardware, software, storage, security, and network traffic management.

In addition, IT has evolved into a consumer good, and as a result of the BYOD (bring your own device) trend, networks need to be adaptable and secure enough to protect assets and data, while simultaneously adhering to regulations and standards that govern compliance as well.

Furthermore, SDN architectures allow network control to be programmable, particularly via open protocols such as OpenFlow (Petale et al., 2020). The advantage of this is that businesses are able to implement software control at the edges of their network by using aware software. It allows access to network switches and routers instead of using the closed and proprietary firmware that is usually used in order to configure, manage, secure, and optimize network resources in order to facilitate configuration, management, and optimization of network resources.

With SDN, users are able to control devices through centralized controllers, providing a centralized view of underlying networks, easy configuration, and reduced costs. In the absence of advanced network techniques and modern networking hardware, Cloud computing, one of the ingredients of the Cyber-Physical System, and the Communication Network for CPS would not have been popularized (Snehi& Bhandari, 2021).

However, traditional networking technology is incapable of keeping pace with the growing demand because it must follow the product cycles and interfaces unique to vendor-specific environments, making it difficult to keep up with this demand. Network operators encounter a number of difficulties when it comes to altering the programming of their networks, which can result in a number of problems. Traditional methods of increasing the capacity of a network or adding and moving devices to a network have been very laborious and time-consuming in the past. The data has to be accessed manually by the user for each console and device in order to be accessed. SDN is becoming a viable alternative to traditional data centers because it gives administrators the ability to instantly configure resources and bandwidth from anywhere and increases the data center's efficiency, flexibility, and resilience. It also does away with the need to spend money on additional infrastructure.

There is a wireless spectrum deficit in the telecommunications sector as the spectrum is already crowded. Wireless radio spectrum is crucial for standard grid modernization. Therefore, the telecommunications sector has a wireless radio spectrum deficit. Utility companies are therefore leasing telecommunications resources from current telecommunications providers rather than building their own infrastructure (Rehmani et al., 2019). Telecommunications companies are changing their business practices because of SDN. In the financial services industry, SDN's ability to connect to many trading participants, maintain low latency, and support a highly secure network infrastructure is essential. Market participants generally rely on legacy networks, which may be unpredictable, difficult to manage, slow to deliver, and vulnerable to attacks. Using SDN technology, financial services companies can create more efficient and effective platforms for trading applications.

CONCLUSION

As with traditional networks, they are complex and difficult to manage because of the amount of complexity involved. One of the major problems with this design is the fact that the control plane and the data plane are vertically integrated and vendor-specific, which is one of the reasons for this design. Furthermore, another important factor is the degree to which the line products and versions of these devices are closely related to the common networking devices. SDN is a ground-breaking innovation that has broken through the architectural limitations of conventional networking. SDN enables businesses to use software and hardware from a variety of vendors to create specialized network services and infrastructure as well as enhance network connectivity for such functions as customer service, internal communications, and document sharing while allowing them to use software and hardware from a variety of vendors as well. As organizations automate their networks, they will save a great deal of time, become more flexible, and provide a faster and more efficient way to roll out new services, applications, and business models. An SDN can enable organizations to develop new applications, services, and business models faster and more easily than ever before.

REFERENCES

Abbasi, A. A., Abbasi, A., Shamshirband, S., & Theodore, A. (2019). *Software-defined Cloud Computing : A Systematic Review on Latest Trends and Developments. XX*(Dc). IEEE. doi:10.1109/ACCESS.2019.2927822

Al, M., Adhami, H., Eddin, A., & Mohamed, A. (2022). Toward integrating software defined networks with the Internet of Things : A review. *Cluster Computing*, *25*(3), 1619–1636. doi:10.100710586-021-03402-4 PMID:34512120

Al-haddad, R., & Velazquez, E. S. (n.d.). *A Survey of Quality of Service (QoS) Protocols and Software-De fi ned Networks (SDN)* (Vol. 1). Springer International Publishing. doi:10.1007/978-3-030-01177-2

Aladaileh, M. A., Anbar, M., Hintaw, A. J., Hasbullah, I. H., Bahashwan, A. A., & Al-Sarawi, S. (2022). Renyi Joint Entropy-Based Dynamic Threshold Approach to Detect DDoS Attacks against SDN Controller with Various Traffic Rates. *Applied Sciences (Switzerland)*, *12*(12), 6127. doi:10.3390/app12126127

Bhardwaj, S., & Panda, S. N. (2022). Performance Evaluation Using RYU SDN Controller in Software-Defined Networking Environment. *Wireless Personal Communications*, *122*(1), 701–723. doi:10.100711277-021-08920-3

Biswal, S. P., & Patel, S. (2022). Introduction to Software Defined Networking. *Software Defined Networks*, (pp. 1–27). doi:10.1002/9781119857921.ch1

Bouguerra, F. (2021). Data centre networking: SDN fundamentals. *Canonical*. https://canonical.com/blog/data-centre-networking-sdn-fundamentals

Braun, W., & Menth, M. (2014). *Software-Defined Networking Using OpenFlow: Protocols, Applications and Architectural Design Choices*, 302–336. MDPI. doi:10.3390/fi6020302

Cabaj, K., Gregorczyk, M., & Mazurczyk, W. (2018). Software-defined networking-based crypto ransomware detection using HTTP traffic characteristics. *Computers & Electrical Engineering*, *66*, 353–368. doi:10.1016/j.compeleceng.2017.10.012

Guleria, A. (2017). *Traffic Engineering in Software Defined Networks : A Survey. January*. Semantic Scholar.

Gupta, N., Maashi, M. S., Tanwar, S., Badotra, S., Aljebreen, M., &Bharany, S. (2022). *A Comparative Study of Software Defined Networking Controllers Using Mininet*. Research Gate.

Haji, S. H., Zeebaree, S. R. M., Ameen, S. Y., & Shukur, H. (2021). Comparison of Software Defined Networking with Traditional Networking. *Comparison of Software Defined Networking with Traditional Networking.*, (June), 1–18. doi:10.9734/ajrcos/2021/v9i230216

Ieee, F., Rothenberg, C. E., Ieee, M., Azodolmolky, S., Ieee, S. M., Uhlig, S., & Ieee, M. (2015). Software-Defined Networking : A Comprehensive Survey. *Proceedings of the IEEE, 103*(1), 14–76. doi:10.1109/JPROC.2014.2371999

Jimson, E. R., Nisar, K., & Hanafi, M. (2017). Bandwidth Management using Software Defined Network and Comparison of the Throughput Performance with Traditional Network. *JIITA*, 71–76.

Joe, H. (2020). What is Software Defined Networking (SDN)? Definition. *SDxCentral.* https://www.bmc.com/blogs/software-defined-networking/#%0Ahttps://www.sdxcentral.com/networking/sdn/definitions/what-the-definition-of-software-defined-networking-sdn/

Latif, Z., Sharif, K., Li, F., Karim, M., Biswas, S., & Wang, Y. (2020). Journal of Network and Computer Applications A comprehensive survey of interface protocols for software defined networks. *Journal of Network and Computer Applications, 156*(February), 102563. doi:10.1016/j.jnca.2020.102563

Lin, Z., Lin, M., Wang, J. B., de Cola, T., & Wang, J. (2019). Joint Beamforming and Power Allocation for Satellite-Terrestrial Integrated Networks with Non-Orthogonal Multiple Access. *IEEE Journal of Selected Topics in Signal Processing, 13*(3), 657–670. doi:10.1109/JSTSP.2019.2899731

Oo, M. M., Kamolphiwong, S., Kamolphiwong, T., &Vasupongayya, S. (2019). *Advanced Support Vector Machine- (ASVM-) Based Detection for Distributed Denial of Service (DDoS) Attack on Software Defined Networking (SDN). 2019.* Semantic Scholar.

Petale, S., Member, G. S., & Thangaraj, J. (2020). Link Failure Recovery Mechanism in Software Defined Networks. *Link Failure Recovery Mechanism in Software Defined Networks., 8716*(c), 1–8. doi:10.1109/JSAC.2020.2986668

Priyadarsini, M., & Bera, P. (2021). *Software defined networking architecture, traffic management, security, and placement : A survey. 192*(June 2020).

Qiu, C., Yao, H., Richard Yu, F., Xu, F., & Zhao, C. (2019). Deep Q-Learning Aided Networking, Caching, and Computing Resources Allocation in Software-Defined Satellite-Terrestrial Networks. *IEEE Transactions on Vehicular Technology, 68*(6), 5871–5883. doi:10.1109/TVT.2019.2907682

Rafique, W., Qi, L., Yaqoob, I., Imran, M., Rasool, R. U., & Dou, W. (2020). Complementing IoT Services through Software Defined Networking and Edge Computing: A Comprehensive Survey. *IEEE Communications Surveys and Tutorials*, *22*(3), 1761–1804. doi:10.1109/COMST.2020.2997475

Rehmani, M. H., Davy, A., Jennings, B., & Assi, C. (2019). Software Defined Networks-Based Smart Grid Communication: A Comprehensive Survey. *IEEE Communications Surveys and Tutorials*, *21*(3), 2637–2670. doi:10.1109/COMST.2019.2908266

Satheesh, N., Rathnamma, A. M., Rajeshkumar, G., Sagar, P. V., Dadheech, P., Dogiwal, S. R., Velayutham, P., & Sengan, S. (2020). Flow-based Anomaly Intrusion Detection using Machine Learning Model with Software Defined Networking for OpenFlow Network. *Microprocessors and Microsystems*, *103285*, 103285. doi:10.1016/j.micpro.2020.103285

Sezer, S., Scott-Hayward, S., Chouhan, P., Fraser, B., Lake, D., Finnegan, J., Viljoen, N., Miller, M., & Rao, N. (2013). Are we ready for SDN? Implementation challenges for software-defined networks. *IEEE Communications Magazine*, *51*(7), 36–43. doi:10.1109/MCOM.2013.6553676

Sharifian, M., Abdolvand, N., & Harandi, S. R. (2021). Context-Based Expert Finding in Online Communities Using Ant Colony Algorithm. *Journal of Information Systems and Telecommunication*, *8*(30), 130–139. doi:10.29252/jist.8.30.130

Snehi, M., & Bhandari, A. (2021). Vulnerability retrospection of security solutions for software-defined Cyber-Physical System against DDoS and IoT-DDoS attacks. *Computer Science Review*, *40*, 100371. doi:10.1016/j.cosrev.2021.100371

Taherian-fard, E., Niknam, T., Sahebi, R., Javidsharifi, M., Kavousi-fard, A., & Aghaei, J. (2022). A Software Defined Networking Architecture for DDoS-Attack in the Storage of Multimicrogrids. *IEEE Access: Practical Innovations, Open Solutions*, *10*(July), 83802–83812. doi:10.1109/ACCESS.2022.3197283

Usman, M., Ali, I., Khan, S., & Khurram, M. (2019). Journal of Network and Computer Applications A survey on software defined networking enabled smart buildings : Architecture, challenges and use cases. *Journal of Network and Computer Applications*, *137*(March), 62–77. doi:10.1016/j.jnca.2019.04.002

Vissicchio, S., & Vanbever, L. (2014).. . *Opportunities and Research Challenges of Hybrid Software Defined Networks.*, *44*(2), 70–75.

Wang, H. U. A., Rafique, W., & Anwar, Z. (2019). Cyberpulse : A Machine Learning Based Link Flooding Attack Mitigation System for Software Defined Networks. *IEEE Access: Practical Innovations, Open Solutions*, *7*, 34885–34899. doi:10.1109/ACCESS.2019.2904236

What is Network Virtualization? Everything You Need to Know. (n.d.). TechTarget. https://www.techtarget.com/searchnetworking/What-is-network-virtualization-Everything-you-need-to-know

Wilson, G., Pereyda, C., Raghunath, N., de la Cruz, G., Goel, S., Nesaei, S., Minor, B., Schmitter-Edgecombe, M., Taylor, M. E., & Cook, D. J. (2019). Robot-enabled support of daily activities in smart home environments. *Cognitive Systems Research*, *54*, 258–272. doi:10.1016/j.cogsys.2018.10.032

Zhai, P., Song, Y., Zhu, X., Cao, L., Zhang, J., & Yang, C. (2020). Distributed Denial of Service Defense in Software Defined Network Using OpenFlow. *ICCC*, 1274–1279.

Zhou, G., Liu, Z., & Fu, C. (n.d.). *An Efficient Design of Intelligent Network Data Plane*. Research Gate.

Chapter 18
Mobile Application-Based Sign Language Detector for Deaf People

Ramani Selvanambi
Vellore Institute of Technology, Vellore, India

Marimuthu Karuppiah
https://orcid.org/0000-0001-7379-2641
Presidency University, Bengaluru, India

Islabudeen M.
https://orcid.org/0000-0002-0811-450X
Presidency University, Bengaluru, India

ABSTRACT

In everyday life, we use hand gestures for several activities, like calling someone or showing directions to somebody, and don't even realise or consciously acknowledge them as a part of life. But for people with hearing disabilities, efficiently contributing to society is a much more important part of their everyday routine. They are their major medium of communication and an incredible way for human communication. Vision-based hand gesture recognition procedures have many demonstrated advantages compared with conventional gadgets. In this chapter, the authors discuss various available products to predict what a user is trying to convey through hand gestures using American Sign Language (ASL). While being able to help deaf people, this chapter also focuses on the proposed method or tool that can also be used for the practice of ASL by regular people to learn and practice sign language. This concept can also be implemented in schools easily to design ASL into their curriculum.

DOI: 10.4018/978-1-6684-8582-8.ch018

INTRODUCTION

Around a billion individuals worldwide suffer from partial or total deafness and lack the means to adequately interact with one another. The lives of deaf individuals are fraught with difficulties, starting with communication difficulties, which are the source of all other problems with school, employment, job interviews, and social stigma. The most efficient way for people with hearing disabilities to communicate is through the use of Sign language. There are numerous standards in various nations, and sign language is far from being universal. Also, regional variances are unique (Moryossef et al, 2020). Due to the variations across areas and states, there are several instances of experienced interpreters using the incorrect terminology. This may not appear to be a major concern, yet it has led to misconceptions during trials or when patients visit a clinic, for example. One in four deaf people now leave their jobs because of prejudice as a result of social problems relating to deafness. Their inadequate communication is the primary cause of deaf people's continued social exclusion.

Due to ineffective communication, education is one of the biggest obstacles the deaf individuals must overcome. 98% of deaf people don't receive sign language instruction, thus they rely on it to communicate (Shah et al., 2021). One of the greatest worries for all NGOs and other communities is that this is one of the most difficult problems people have to deal with. This chapter discusses the proposed methodology that work as just a step in assisting good communication between hearing and hearing-impaired people.

American Sign Language (ASL) is by far the most common form, consisting of signs for the 26 alphabets, 10 digits and numerous gestures for phrases. The traditional method to approach this problem using machine learning is to use image segmentation techniques to identify signs (Sangeethalakshmi et al., 2021). This approach works by taking a frame from a camera and segmenting the hand region using techniques like thresholding, cascade filtering and greyscaling. This is passed on to convolutional neural networks which use filters to reduce the size of the images and classify them into classes/alphabets. In our approach, we will use landmarking models that track the motion of the user's hands relative to their face. The main points which will be tracked are the joints in the fingers, wrist and shoulders. Their relative position in the frame shall be tracked by noting their coordinates and normalising their values. This data will then be used to perform feature extraction to form the data which will be used by our deep learning model.

The following techniques are typically used to identify sign gestures:

The signer must wear a hardware glove while the hand movements are being recorded in this glove-based method.

Method based on vision that can be further divided into static and dynamic recognition. Whereas dynamic is a real-time live capture of the motions, statics deals with the recognition of static gestures (2d-images). This entails using the camera to record movement.

Although while the glove-based method has an accuracy of over 90%, it appears a little awkward for application in everyday situations (Zhang et al., 2013).

BACKGROUND SURVEY

Sharma and Kumar (2021), have used the 3D convolution neural network to recognize the ASL in volumetric data like videos. They have used the Boston ASL LVD (Lexicon video dataset) for their model. The system was used to recognize the ASL in real-time from dynamic videos by converting each video into frames. Pre-processing takes a lot of time, which is the main drawback of this method. The video is first converted into several frames, then the frames are changed to specific sizes for processing, after that the frames are changed into grayscale and if any noise is present that is also removed at this stage and finally the frames are then combined again into a video and could be used for training the model.

(Abdulhussein & Raheem, 2020), deep learning techniques, are used to classify the ASL gestures. The first solution to recognize the static ASL is by using the Robert edge detection method and the second solution proposed within this paper uses the Convolution neural networks to recognize the static ASL gestures for alphabets. The system is limited only to recognizing 25 letters of the English alphabet and it does not provide any further extension to recognize words and phrases. Also using the Robert edge detection method used in this paper is susceptible to noise and can greatly affect our results.

Borg and Camilleri (2019) have proposed a multi-layer RNN for sign language detection. The system uses a 2-stream convolutional neural network (CNN) that automatically extracted features and takes video and motion as input. The accuracy of the system was not very high and vanishing gradients are an issue with RNNs that makes learning from lengthy data sequences difficult. The parameter updates become minor as the gradient gets smaller, meaning the model doesn't actually learn anything new.

Ye, Tian, Huenerfauth, and Liu (2018), the 3D convolution neural network is used to recognize the American Sign Language gestures. They also have used the multi-modality features with the 3DCNN to further improve the capturing of the temporal boundaries in the continuous video. This system gave only a moderate accuracy of 69% and also the use of CNN increased its computation cost and also required a large dataset for classification.

Saquib, and Rahman (2020), the ASL recognition is done through the use of data gloves containing some suitably positioned sensors. Recognition of both the static and dynamic characters of ASL is possible using the developed system. The sensors used in the data gloves were expensive. The system failed to distinguish range based query i.e. range of flex sensors were not able to distinguish certain alphabets and there was an overlap for characters 'S' and 'X'.

Chong and Lee (2018) have used SVM and deep neural networks (DNN) to recognize the ASL. A leap motion controller (LMC) is used for forming the dataset for the model training. The SVM gave 77.79% and the DNN gave 88.79% accuracy in detecting the ASL gestures. The cost of using an LMC for data collection is high. Also, the SVM algorithm used in the model was not able to provide good accuracy for the detection of the ASL.

Farooq et al. (2021) proposed the methodology using machine translation for sign language. This includes statistical, semantic and neural techniques. It also expires the use of transfer learning to speed up the process. Though this work they give an outline of the various methods which can be employed to approach this problem, it is only an overview that doesn't talk about the specifics of each technique.

Bansal et al. (2021) discussed three different approaches to analyse gestures to help aid communication. They analyse Google's Mediapipe, Azure's Kinect and AlphaPose and solutions to track more complex hand gestures and generate features for transformer models down the line. Overall the Mediapipe model achieves a 91.1% accuracy whereas the AlphaPose achieved an 84.8% accuracy. When using Azure's Kinect to visualize features, some signs which contained "above" and "in" were confused as they are similar. It also has issues with swift hand movements. AlphaPose had issues with missing hand features for certain signs, specifically some joints.

(S. Amaliya, et.al, (2021)) aims to compare OpenPose, TensorflowLite, and MediaPipe approaches to hand key point recognition. It describes the general accuracy of tracking with the TensorflowLite in general situations vs. the precision of the MediaPipe model for more intricate tracking. When displaying just the arm using photographs, the OpenPose and Tensorflow Lite models were able to extract features from almost all the samples. But the mediapipe model had difficulty in land marking a lot of images without any motion.

(Bragg, D, et.al, (2019)) uses Hidden Markov Models but for dynamic hand gestures. It uses the Adaboost algorithm for detecting the user's hand. Combining condensation and partitioned sampling, they formed a contour-based hand tracker. They captured invariant curve moments and orientation for representation of the trajectory of hand gestures. Based on the result, they concluded that their algorithm outperformed traditional hand recognition methods. The Viterbi algorithm is time and memory intensive to compute. The HMM needs a larger seed than the straightforward Markov models and must be trained on a set of seed sequences.

Du et al. (2020) mainly focuses on methods to extract the correct Region of Interest for accurate gesture detection. It weakens the impact of unimportant features and passes the adapted image along with the CNN model. They efficiently use micro-Doppler, elevation, and azimuth angle information to improve the performance and robustness of the recognition. They provide a two-dimensional scaling-based data augmentation method based on the gesture movement mechanism to lessen the complexity of micro-Doppler characteristics. Hence gets the correct ROI and then uses micro-Doppler features for correct prediction.

Al-Hammadi (2020) suggests an alternative to the traditional CNN methodology for hand gesture recognition. The author claims that the image dataset present for gesture recognition is not sufficient enough to predict such a diverse range of gestures. Hence he uses transfer learning for the automatic identification of hand gestures. To globalise the local features that the 3DCNN model has learnt, the authors create several fusion strategies and evaluate their effectiveness. For the signer-independent mode, it obtained recognition rates of 84.38%, 34.9%, and 70% on the three datasets, respectively. These accuracies are not satisfactory even for static gestures and the project fails to talk about dynamic gestures.

Abiyev et.al, (2020) created a system based on the Single Shot Detection algorithm (SSD) combined with a detection model consisting of the Inception model and an SVM classifier for classification. The Inception model is a CNN architecture that is used to extract features for future classification. This system is limited to more static data due to the overhead of the multiple classifiers involved in the process. It also focuses on the SSD model for purely speed proposed but doesn't elaborate on the variations of the SSD model which works on smaller images to speed up the process.

VISION-BASED VIDEO RECOGNITION SYSTEM

Image capture, picture pre-processing, segmentation, feature extraction, and classification are the five phases that make up eyesight-based sign language recognition (SLR). The first most phase in the sign language recognition, which is image acquisition, can really be carried out using either custom-created datasets or public datasets that are available for free. The following stage includes pre-processing to enhance the image's clarity and remove any excess noise, (Kamal et al, 2019). After pre-processing, the region of interest is segmented and retrieved from the complete image. The input image region is converted into feature vectors for recognition in the fourth stage, which is referred to as feature extraction. In order to identify the target sign, classification, the final phase of vision based SLR entails comparing the latest sign image's attributes to those that have previously been recorded in the database.

In order to collect sign images for categorization, researchers have used a variety of image capture tools. Some of these devices are the camera or webcam, Kinect, data glove, and jump motion controller. A camera or webcam is the tool that most researchers use the most because, unlike data glove based, it enables finer and more human-computer interaction that is natural without the need for extra devices. Although relatively expensive and inconvenient for the users, data collection using data gloves has been shown to be more accurate. Kinect is efficient and widely utilised. It simultaneously offers a depth video feed and colour video stream. It recovers 3D hand motion trajectories and detaches the background from the real sign image with ease. But, Kinect's flaw is its financial impact because it is so expensive.

For image acquisition, the input device for viewing the data of hands or fingers is a computer camera. A gesture is a symbol that conveys physical behaviour or emotional expression. It can be a body gesture or merely a hand gesture. Gestures can be static gestures or dynamic gestures. In static gestures, the posture of the body or the gesture of the hand represents a symbol. (Singha, J., & Das, K, (2015)) In a dynamic gesture, the body movement or the hand movement symbolizes some messages. Here we are using only hand gestures. Image classification software and/or hardware technologies transform biological vision into artificial vision systems.

Image Pre-processing: This module mainly consists of acquiring the image of hand gestures from datasets and camera captures and converting the image from the RGB format to the Hue-Saturation View (HSV) format to enable the tracking of HSV markers as shown in Fig.1. HSV markers are plotted to help the model decipher hand gestures. The image will have to be denoised and the background disregarded to make sure irrelevant information doesn't skew our results.

Model Training: This module will use the GridSearchCV/RandomSearchCV method to establish a base set of parameters to evaluate our models. We will create a Convolutional Neural Network (CNN) while varying the number of neurons and the pooling limits to achieve the most optimal result.

For the normal camera, the skin colour and certain feature points are combined to accurately detect and classify the hand area. For foreground picture separation from backdrop and background noise removal we use HSV (hue, saturation, and value) Trackbars which are given certain HSV Values which could detect the skin colour and would ignore all other colours in the video feed which are either above or below the given threshold. Generally, the HSV value of the skin colour is 315, 94, and 37. The given Region of Interest (ROI) is then cropped and converted into an image which is sent to the model to test the given input. The model detects the hand gesture and provides the required output after examining it and comparing it with all the other images recognised. This is a classification approach and is used to classify the gesture in different classes.

Figure 1. Vision-based video recognition system process

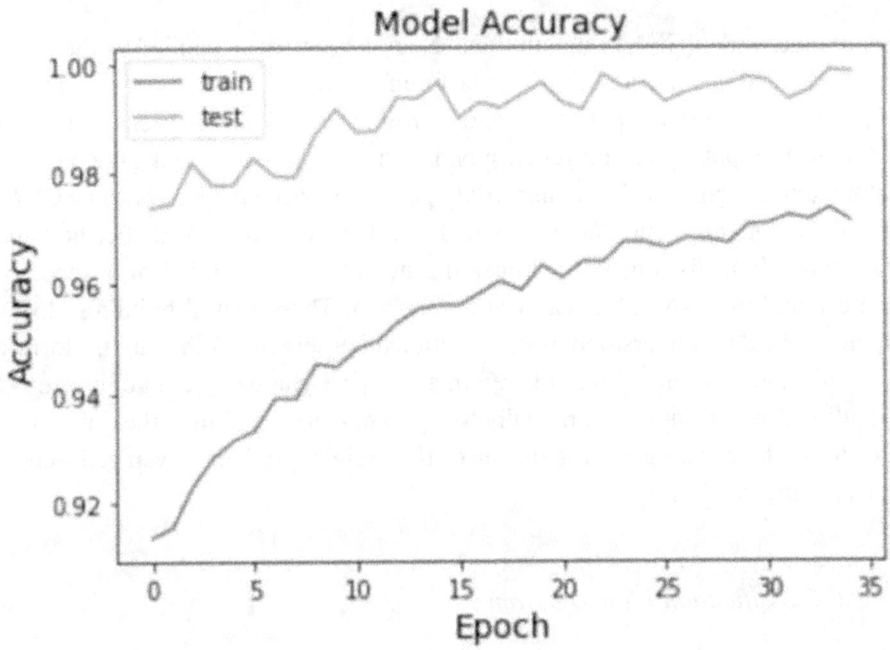

Real-Time Prediction: This module will be responsible for acquiring real-time images through a webcam or any other means of video (mp4 etc.) and processing it before passing them to our trained model for predictions. It will overlay the Label outputted by the model over the image/video to aid in the understanding of sign language.

The actual capturing of the live feed and sending the relevant content to the model to detect the output from the gesture are discussed in the Fig.2. The OpenCV library is used to initially map a Region of Interest (ROI) where the required gesture will be captured. To remove the background noise and separate the foreground image from that of the background we use HSV Track Bars which are given certain HSV Values which could detect the skin colour and would ignore all other colours in the video feed which are either above or below the given threshold. The given ROI is then cropped and converted into an image which is then sent to the model to test the given input. The model then sends the frame which then detects the hand gesture and provides the required output after examining it and comparing it with all the other images recognised.

CONVOLUTIONAL NEURAL NETWORK

The neural networks (CNN) Convolutional Neural Network as implied by the name are an AI method that are created using the brain's anatomical structure, as depicted in Fig. 3. It entails a system of neurons, or learning cells. These neurons figure out how to convert input signals into corresponding output signals, framing the premise of automated recognition. The connectivity pattern between the neurons in a CNN, a type of feedforward neural network which is artificial, is modelled after how the visual human brain is organised. Repeating neural blocks applied over space or time are found in CNNs (Nandy, A et al, (2020)). These neural building blocks for images could be understood as convolutional kernels in 2D that are uniformly made applicable to each region of given image. These can be thought of as the 1D convolutional kernels applied gradually for speech. During training, these repeated blocks have shared weights, for instance, the weight gradients averaged across various picture patches.

Figure 2. Classification Flow Diagram

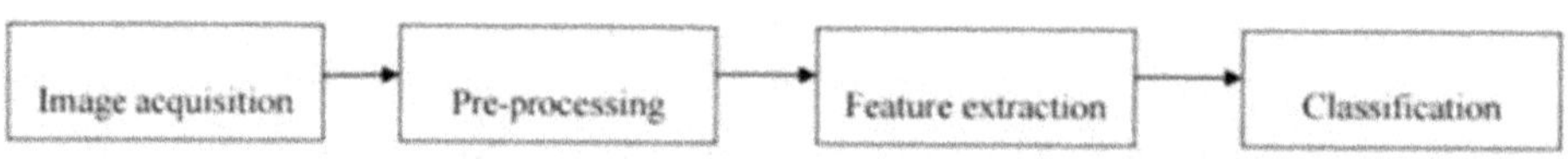

Convolution filters are the first layers that a signal enters. As shown in Fig. 3, in the convolution phase, the network tries to categorise the incoming signal by making references to what it has previously learned.

Convolution layer inputs can be "smoothened" to lessen their susceptibility to noise and fluctuations. Subsampling is a technique for achieving this smoothing, and it involves taking the means or the maximum of a sample of the signals. Examples of subsampling techniques (for the picture signals) include reducing the image's size or the contrast between the colours in the RGB channels.

Pooling layer is another element of a CNN. Spatial scale of representation can gradually be shrunk in order to minimize the various parameters and computations in network. Each element is handled independently by the pooling layer. Signal flow from one layer to the next is controlled by the activation layer, simulating how neurons fire in our brains. More neurons would be activated by output signals that have a strong connection to prior knowledge, allowing signals to propagate for identification more efficiently. The network's bottom layers are completely associated, implying that the neurons of the layers before it are associated with each neurons in later layers. This is similar to high-level thinking, where all possible routes from the input to the yield are considered. Utilizing the openly accessible ASL dataset of letter sets, which is utilized to prepare our CNN model alongside utilizing data augmentation to get the ideal outcomes. The Neural Network was prepared for 30 epochs and effectively accomplished a precision of 97% on the test data set.

Figure 3. CNN Model Summary

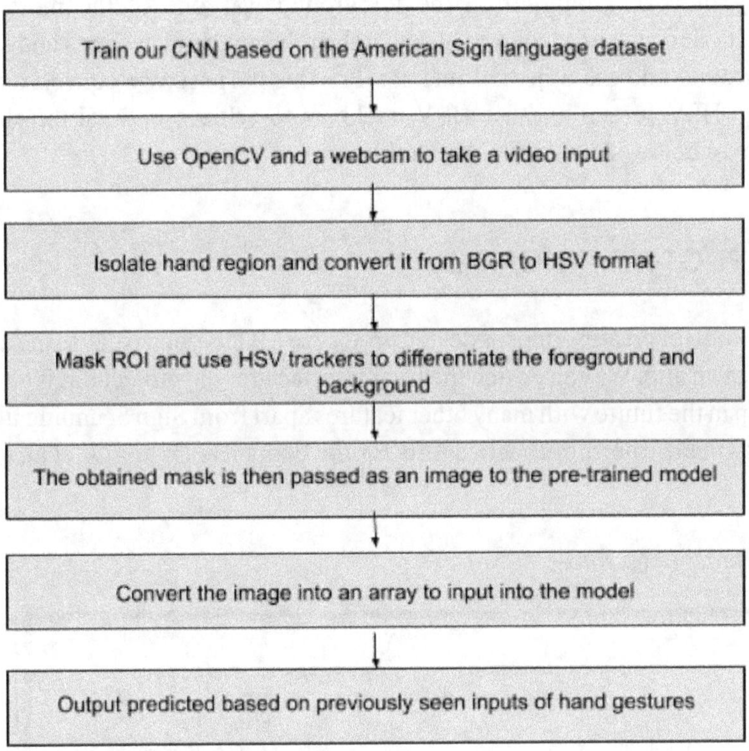

OPEN COMPUTER VISION LIBRARY

A computer camera is the input device for viewing the information of hands or fingers in vision-based approaches. Vision-Based techniques need only one camera, thus realising an unaided, natural exchange of information between people and computers. (Tripathi, K., & Nandi, N. B. G, (2015)) These frameworks, which describe artificial vision frameworks that are executed in software and/ or hardware, will often supplement natural vision. This is a challenging problem in order to achieve real-time performance, these frameworks must be background-invarient, lighting-intensive, individual and camera-independent. In addition, these frameworks must have been streamlined to fulfil prerequisites, involving precision and power. (Elmezain, M et al, (2008)) Based on a hidden Markov model, an automated system that recognises Arabic digits (0–9) in both isolated and continuous gestures was proposed.

THRESHOLDING OPERATIONS

Separating the background from the foreground was one of the most crucial components since we needed to only detect the relevant hand gestures and separate the background and the object of interest. For this purpose, we convert any RGB image into HSV format and use HSV trackbars to achieve optimal thresholding. The figure. 4 below shows HSV thresholding.

FIGMA PROTOTYPING TOOL

Figma is a tool for creating apps as design prototypes. It lets you visualise the ideas for a website or an app. We can design the User Interface for the project if it is converted into an app in the future with many other features apart from Sign Scanning like ASL courses for the learners, or Practice tests for the beginners (Wang, X et al, (2012)).

Figure 4. Making Gestures

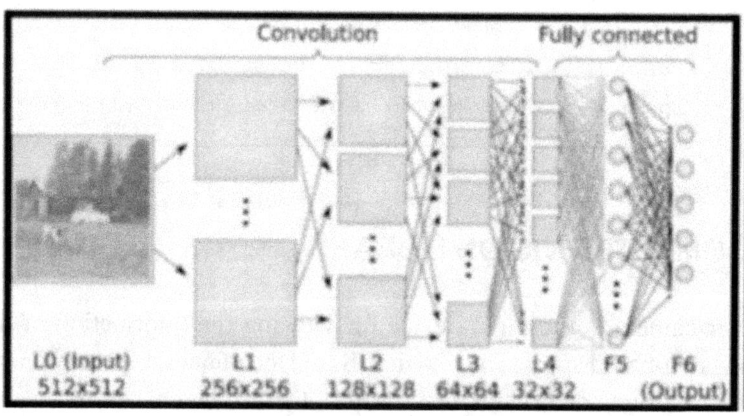

FUNCTIONAL REQUIREMENTS

Hand Detection: Using landmarking models once can track the motion of the user's hands relative to their face.

Filtered Object Detection: The main points which will be tracked are the joints in the fingers, wrist and shoulders. Their relative position in the frame shall be tracked by noting their coordinates and normalizing their values (Nikam, A. S., & Ambekar, A. G (2016)).

Hand Calibration: To perform automatic adjustments while capturing hand gestures which would help to eliminate errors and provide us with accurate results.

Gesture Recognition: Machine learning algorithms like LSTM and CNN can be to recognize dynamic gestures.

Accuracy: The system should be able to recognize both static and dynamic gestures correctly with high accuracy.

NON-FUNCTIONAL REQUIREMENTS

Efficiency in Computation: Software shall minimize the use of CPU as well as system memory and is able to run on low-end devices.

Extensibility: The software shall support future developments and the addition of more functionality to the application.

Portability: The software could run on any operating system as well as on android devices which makes it highly portable.

Performance: The software uses Google's MediaPipe library which helps it to minimize the calculation and provide results in much less time.

Reliability: The software works robustly and detects gestures with utmost accuracy even in relatively poor lighting conditions.

Usability: The software is simple and easy to use by all users with minimum instructions. The interface should be self-understood.

OPERATIONAL REQUIREMENTS

Economics: This application requires less investment for development and can run on any low-end device. Hence it's beneficial for people who want to learn ASL or communicate with deaf people.

Environmental: This application provides a smart environment where a person can effortlessly communicate with deaf people even if he doesn't understand ASL gestures.

Social: This application allows deaf people to have a normal social life like others where they can communicate with anyone they want to without feeling left alone by society.

Ethical: This application helps to bring deaf people close to the general public.

Health and Safety: This application could be used in hospitals and clinics to help doctors understand the problem from which the deaf person is suffering so as to provide better treatment.

Sustainability: This application is easy to use and can be easily maintained for a longer period of time. New functionalities can be easily added to it in future.

DESIGN APPROACH

The main objectives in this design approach is to create a model to accurately track a user's hand gestures and relative location to their body. Accurately train a deep learning model to recognize common ASL gestures and characters. Develop a mobile application that could be utilised to perform these translations in actual-time.

Most papers focus on improving accuracy by using more intricate architecture or transformer models which are inherently slower. This poses a problem for real-time processing as the latency is far too high. In the papers which do focus on faster mobile implementations, we see the comparison of various pose estimation packages such as MediaPipe, Azure Kinect and OpenPose. It is concluded that MediaPipe is the most optimal package when the accuracy of intricate movements is of priority (Ronchetti, F, (2016)).

MediaPipe Library: MediaPipe is a library created by Google which enables the tracking of a user's hand, palm and body using pre-trained models on millions of frames. It offers various models such as the Hands model which tracks 21 key points on each hand of the user, the Face model which can track a user's facial expression and finally the Holistic model which is a combination of them.

MediaPipe is apt for the following reasons:

- Allows accurate tracking of intricate movements of a user's fingers and hands.
- Possible to expand to work on both Python and JavaScript, thus enabling easier and more efficient implementation in Web Applications.
- Uses multithreading to accelerate performance on mobile devices hence making real-time tracking possible.

American Sign Language(ASL): It is communication form consisting of gestures and signs for the English alphabet which enables people who are hearing impaired and vocally challenged to converse and interact with others (Cooper, H. M, (2012)). Being one of the most popular versions of language of sign, it is general convention used for communication. It means that our application will be based on ASL to have the greatest reach.

Figure 5. ASL Alphabets

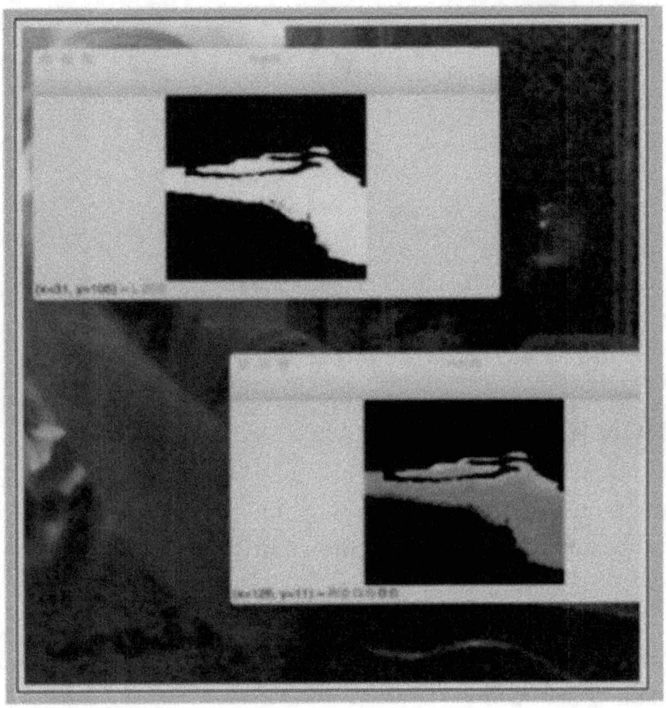

Holistic Model: The holistic model is a machine learning model trained by Google to recognize a user's face, shoulders and hands. This is tracked using 543 landmarks consisting of per hand 21 hand landmarks, 33 pose landmarks, and 468 face landmark. It is a probabilistic model which means that it uses a confidence threshold to map the landmarks. This threshold for both plotting and tracking is defined at a default 50% but has been set to 70% for our application.

Figure 6. Landmarks on Each Hand

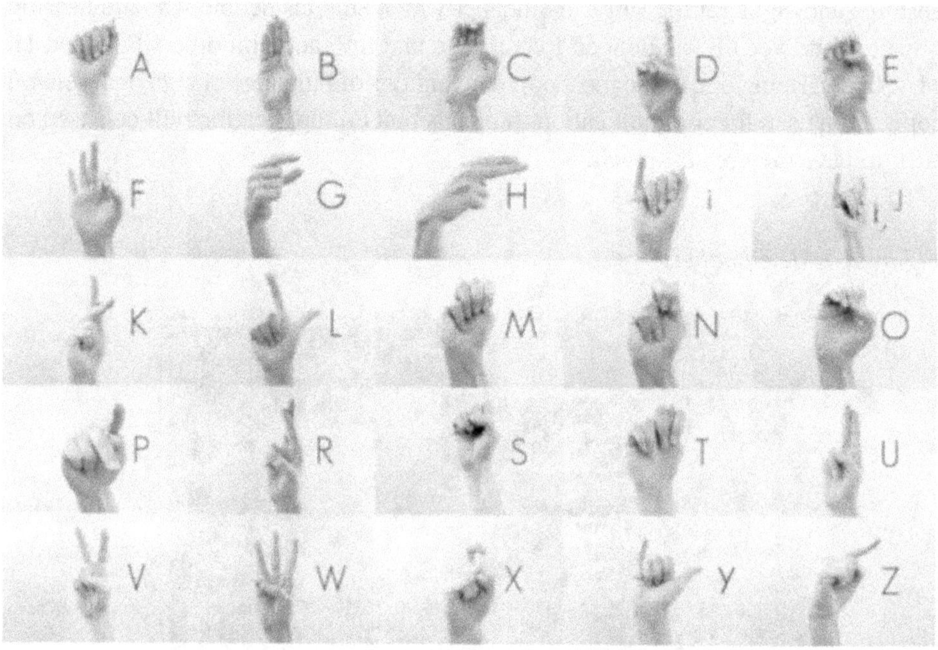

APPLICATION RESULTS

We have created three different iterations of our model to compare their performances and choose the best performing one as shown in Table 1 and Table 2. The models we made are:

1. Only Coordinates: This initial model uses only the normalized coordinates (X, Y) from the video feed.
2. Two Angles: This model considers an extra two angles, namely
    a. (Thumb tip - Wrist base - Pinky tip) - This differentiates between 'A', 'S' and the 'Y' symbols
    b. (Index finger tip - Index finger base - Middle finger tip) - This helps differentiate between the 'U' and 'V' symbols.
3. Distance + Angles: Ths model adds on the distance between landmarks along with the angles to help differentiate between symbols at different angles.
    a. (Thumb tip - Index finger tip) - This distance helps differentiate between 'O' and 'C'
    b. (Thumb tip - Pinky tip) - Helps differentiate between 'A', 'I' and 'J'

Figure 7. Demonstration of Hand gestures for words

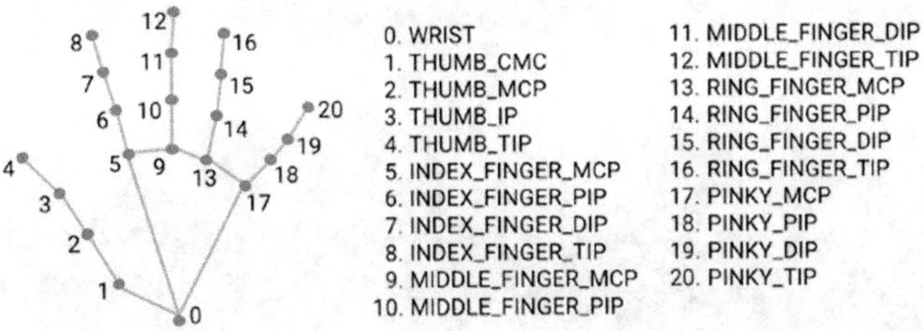

Our app provides real time recognition and translation of ASL gestures as shown in Fig.6 and Fig.7. Users can also use our application to learn about various ASL gestures and characters. The application frontend is built using XML and JAVA. For the backend we have used Google's firebase Realtime database to store all the data needed by our application. The login and signup page uses the firebase email and password authentication method to ensure safe and secure user login. The various components of our application are as follows:-

- Splash Screen: - This is the first screen that a user sees when he opens up the app.
- On boarding Screen: - This screen gives the user an overview of the app and also tells them about the benefits of using the application.
- Login/ Signup Screen: - This screen helps a new user to create an account that would be used to login. The existing user can simply use their login credentials to sign in to the application. The users can also use forgot password option to reset their password in case they forgot their password.
- Choose Scan/Learning Screen: - Once the user has successfully logged in then he is provided with 2 options. He can either use the scan option for the real-time ASL translation of gestures or can use the learn ASL option if he wants to learn about various ASL phrases and characters.
- Choose Characters/Phrases Screen: - If the user chooses the learn ASL option then he is provided with 2 options from which he can choose to either learn ASL characters like alphabets or learn ASL gestures and phrases.
- Learn Characters/Phrases Screen: - This screen helps the users to learn various ASL gestures and phrases as shown in Fig.8. It provides the image or short video of the gesture along with description of the gesture to help the user to easily learn various ASL gestures.

Figure 8. Demonstration of Hand gestures for Alphabets

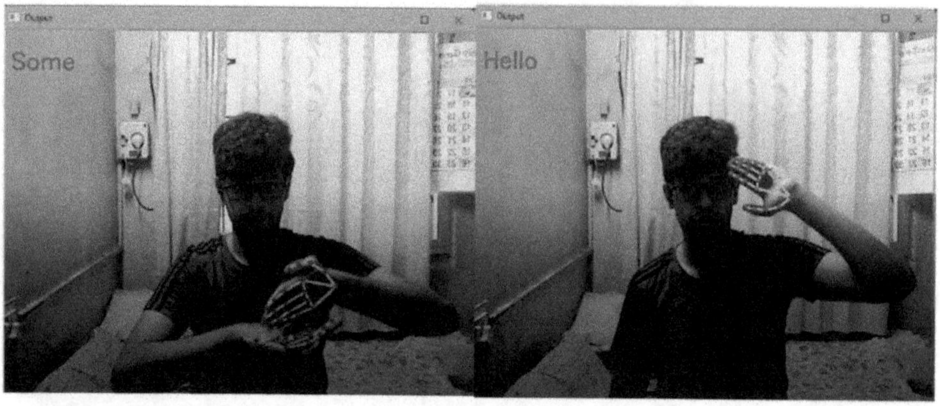

Figure 9. Learn Characters/Phrases Screen

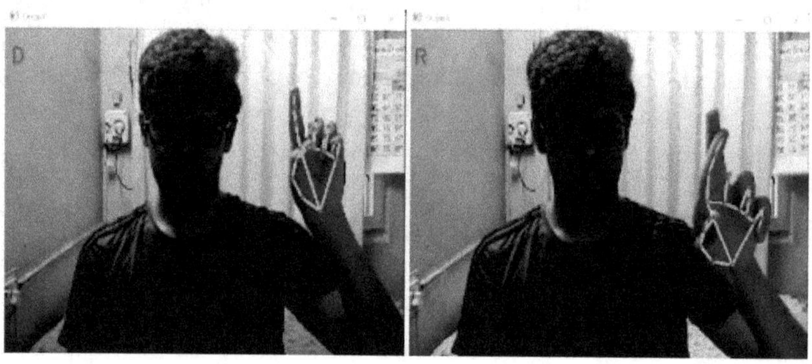

Tables 1 and 2 display the outcomes for the various characters and words using the methods such as only coordinates, distance and angles, Random forest, XGBoost.

The trained model is trained and tested on the ASL dataset containing 34627 images of gestures with varying backgrounds, colours and orientations. Of this, 80-20 is the split between training and testing/validation in the dataset. Metrics used by us to judge the model we train are the Accuracy, loss and the F-score calculated as shown in Table 3. We want to maximise the accuracy and F-score while minimising the loss.

We achieved a final test accuracy of 96.82% and a training accuracy of 97.12% over 35 epochs. This was achieved after augmenting the dataset using rotation, skew and zoom to better fit real-world examples. We achieved a final test loss of 0.0091 and a training loss of 0.0859 over 35 epochs as shown in Fig 10. This was achieved after augmenting the dataset using rotation, skew and zoom to better fit real-world examples.

Table 1. Alphabet model comparison

Model	Metric	A	B	C	D	E	F	G	H	I	J	K	L
Only Co-ordinates	Acc	0.959	0.987	0.836	1.000	0.977	0.994	0.86	0.942	0.714	0.981	1.000	0.803
	F1	0.979	0.993	0.882	0.67	0.893	0.997	0.918	0.89	0.772	0.867	0.984	0.888
	Recall	1.0	1.000	0.932	0.503	0.823	1.0	0.985	0.843	0.841	0.776	0.969	0.994
Two Angles	Acc	0.961	0.949	0.942	1.000	0.994	0.986	0.977	0.992	0.706	0.887	0.988	1.000
	F1	0.81	0.974	0.967	1,000	0.997	0.954	0.981	0.973	0.821	0.937	0.988	0.997
	Recall	0.7	1,000	0.993	1,000	1,000	0.923	0.985	0.955	0.981	0.993	0.988	0.994
Distance + Angles	Acc	0.952	0.993	1,000	1,000	0.994	0.994	0.942	0.992	0.969	0.963	0.994	1.000
	F1	0.976	0.997	0.983	1,000	0.997	0.997	0.963	0.945	0.984	0.967	0.991	0.997
	Recall	1.000	1.000	0.966	1.000	1,000	1,000	0.985	0.903	1.000	0.970	0.988	0.994
Random Forest	Acc	0.909	0.993	0.993	1.000	1.000	0.994	1.000	1.000	0.994	1.000	1.000	1,000
	F1	0.952	0.997	0.986	1.000	0.997	0.997	0.992	0.992	0.994	0.981	0.994	0.997
	Recall	1.000	1.000	0.980	1.000	0.994	1.000	0.985	0.985	0.994	0.963	0.988	0.994
XGBoost	Acc	0.989	0.971	0.932	0.983	0.900	0.851	0.937	0.976	0.967	0.941	0.979	0.993
	F1	0.774	0.797	0.879	0.863	0.846	0.883	0.918	0.934	0.955	0.881	0.915	0.943
	Recall	0.636	0.676	0.831	0.769	0.797	0.917	0.901	0.903	0.943	0.828	0.859	0.897

Table 2. Gesture Model Comparison

Model	Metric	Eat	Friend	Hello	Help Me	Home	How	Love You	My
Only Co-ordinates	Acc	0.982	1.000	1.000	1.000	1.000	0.982	0.992	0.981
	F1	0.988	0.973	1.000	0.993	0.993	0.988	0.996	0.991
	Recall	0.994	0.948	1.000	0.986	0.986	0.994	1.000	1.000
Distance + Angles	Acc	0.876	0.961	0.909	0.950	0.992	0.946	0.963	0.938
	F1	0.906	0.946	0.952	0.964	0.935	0.897	0.778	0.965
	Recall	0.939	0.932	1.000	0.978	0.884	0.884	0.653	0.993
Random Forest	Acc	0.938	1.000	1.000	0.993	1.000	1.000	1.000	0.994
	F1	0.968	0.981	1.000	0.993	0.997	0.997	1.000	0.997
	Recall	1,000	0.963	1.000	0.993	0.993	0.994	1.000	1.000
XGBoost	Acc	0.981	0.948	0.993	0.913	0.986	0.988	0.952	0.897
	F1	0.963	0.948	0.997	0.935	0.986	0.982	0.964	0.946
	Recall	0.945	0.948	1,000	0.958	0.986	0.976	0.975	1.000

Figure 10. Accuracy Graph for Different Epochs

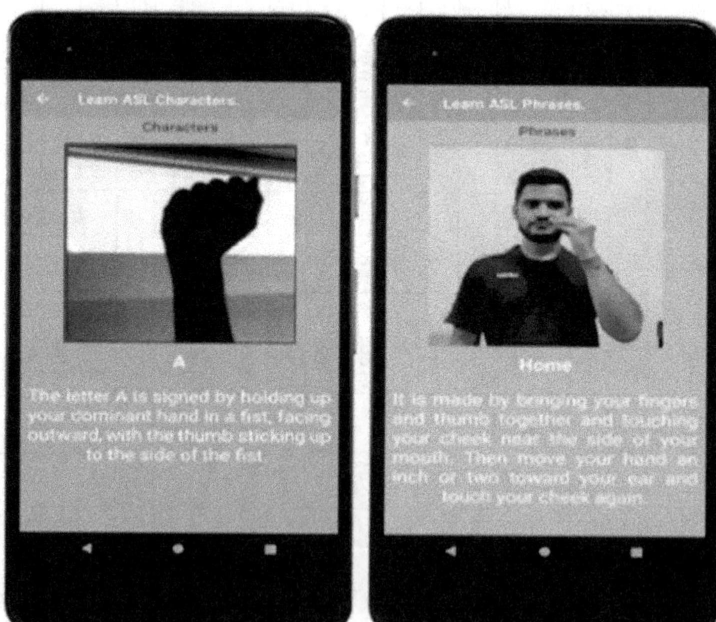

Table 3. Statistical Measures

Epoch	Train Accuracy	Train Loss	Test Accuracy	Test Loss
10	94.47%	0.1646	92.15%	0.0355
20	96.31%	0.1125	94.62%	0.0165
30	97.04%	0.0887	96.72%	0.0096
40	97.12%	0.0859	96.82%	0.0091

CONCLUSION

In this chapter we have discussed the Vision based approach to detect dynamic gestures of the users using inherently different approaches. We also look into the various architectures that we can adopt to implement the deep learning module of our project with the main method being Neural Networks. We also experiment with Random Forest classification and Boosting techniques to test their viability. We used a novel approach of leveraging Google's Mediapipe library to translate ASL phrases in real-time, enabling us to dynamically detect hand movements along with static gestures. The application explained in this chapter not only improves the time and space efficiency of the ASL translation problem but also helps other users learn ASL in an easy and interactive way. While this app is solely based on a web application and an android application, it can be extended to Apple iOS operating system. The accuracy can be improved using exhaustive tests based on different lighting conditions and skin tones. The app can be scaled to include sign languages from different parts of the world like Chinese Sign Language, British sign language and Indian Sign Language etc.

REFERENCES

Abdulhussein, A. A., & Raheem, F. A. (2020). Hand gesture recognition of static letters American Sign Language (ASL) using deep learning. *Engineering and Technology Journal*, *38*(6), 926–937. doi:10.30684/etj.v38i6A.533

Abiyev, R. H., Arslan, M., & Idoko, J. B. (2020). Sign language translation using deep convolutional neural networks. [TIIS]. *Transactions on Internet and Information Systems (Seoul)*, *14*(2), 631–653.

Al-Hammadi, M., Muhammad, G., Abdul, W., Alsulaiman, M., Bencherif, M. A., & Mekhtiche, M. A. (2020). Hand Gesture Recognition for Sign Language Using 3DCNN. *IEEE Access: Practical Innovations, Open Solutions*, *8*, 79491–79509.

Amaliya, S., Handayani, A. N., Akbar, M. I., Herwanto, H. W., Fukuda, O., & Kurniawan, W. C. (2021). Study on Hand Keypoint Framework for Sign Language Recognition. *7th International Conference on Electrical, Electronics and Information Engineering (ICEEIE),* (pp. 446~451). IEEE.

Bansal, D., Ravi, P., So, M., Agrawal, P., Chadha, I., Murugappan, G., & Duke, C. (2021). Copycat: Using sign language recognition to help deaf children acquire language skills. In *Extended Abstracts of the 2021 CHI Conference on Human Factors in Computing Systems*, (pp. 1~10). CHI.

Borg, M., & Camilleri, K. P. (2019). Sign language detection "in the wild" with recurrent neural networks. *IEEE International Conference on Acoustics, Speech and Signal Processing (ICASSP).* IEEE.

Bragg, D., Koller, O., Bellard, M., Berke, L., Boudreault, P., Braffort, A., & Ringel Morris, M. (2019). Sign language recognition, generation, and translation: An interdisciplinary perspective. In *Proceedings of the 21st International ACM SIGACCESS Conference on Computers and Accessibility*, (pp. 16~31). ACM.

Chong, T.-W., & Lee, B. G. (2018). American Sign Language Recognition Using Leap Motion Controller with ML Approach. *Sensors (Basel)*, *18*, 3554.

Cooper, H. (2012). EngJon Ong, Nicolas Pugeault, and Richard Bowden. Sign language recognition using subunits. *Journal of Machine Learning Research*, *13*, 2205–2231.

Du, C., Zhang, L., Sun, X., Wang, J., & Sheng, J. (2020). Enhanced Multi-Channel Feature Synthesis for Hand Gesture Recognition Based on CNN with a Channel and Spatial Attention Mechanism. *IEEE Access: Practical Innovations, Open Solutions*, *8*, 144610–144620.

Elmezain, M., Al-Hamadi, A., Appenrodt, J., & Michaelis, B. (2008). A hidden markov model-based continuous gesture recognition system for hand motion trajectory. In *2008 19th international conference on pattern recognition*, .(pp. 1~4). IEEE.

Farooq, U., Rahim, M. S. M., Sabir, N., Hussain, A., & Abid, A. (2021). Advances in machine translation for sign language: Approaches, limitations, and challenges. *Neural Computing & Applications*, *33*(21), 14357–14399.

Kamal, S. M., Chen, Y., Li, S., Shi, X., & Zheng, J. (2019). Technical approaches to Chinese sign language processing: A review. *IEEE Access: Practical Innovations, Open Solutions*, *7*, 96926–96935.

Moryossef, A., Tsochantaridis, I., Aharoni, R., Ebling, S., & Narayanan, S. (2020). Real-time sign language detection using human pose estimation. In Computer Vision–ECCV 2020 Workshops: UK.

Nandy, A., Prasad, J. S., Mondal, S., Chakraborty, P., & Nandi, G. C. (2010). Recognition of isolated indian sign language gesture in real time. In *Information Processing and Management: International Conference on Recent Trends in Business Administration and Information Processing, BAIP 2010*, (pp. 102~107).

Nikam, A. S., & Ambekar, A. G. (2016). *Sign language recognition using image based hand gesture recognition techniques. In 2016 online international conference on green engineering and technologies*. IC-GET.

Ronchetti, F., Quiroga, F., Estrebou, C. A., & Lanzarini, L. C. (2016). Handshape recognition for argentinian sign language using probsom. *Journal of Computer Science and Technology*, 16.

Sangeethalakshmi, K., Shanthi, K. G., Raj, A. M., Muthuselvan, S., Taha, P. M., & Shoaib, S. M. (2021). Hand gesture vocalizer for deaf and dumb people. *Materials Today: Proceedings*.

Saquib, N., & Rahman, A. (2020). Application of machine learning techniques for real-time sign language detection using wearable sensors. In *Proceedings of the 11th ACM Multimedia Systems Conference*, (pp. 178~189).

Shah, F., Shah, M. S., Akram, W., Manzoor, A., Mahmoud, R. O., & Abdelminaam, D. S. (2021). Sign language recognition using multiple kernel learning: A case study of Pakistan sign language. *IEEE Access: Practical Innovations, Open Solutions*, *9*, 67548–67558.

Sharma, S., & Kumar, K. (2021). ASL-3DCNN: American Sign Language recognition technique using 3-D convolutional neural networks. *Multimedia Tools and Applications, 80*(17), 26319~26331.

Singha, J., & Das, K. (2015). Automatic Indian Sign Language recognition for continuous video sequence. *ADBU Journal of Engineering Technology, 2*(1).

Tripathi, K., & Nandi, N. B. G. (2015). Continuous Indian sign language gesture recognition and sentence formation. *Procedia Computer Science*, *54*, 523–531.

Wang, X., Xia, M., Cai, H., Gao, Y., & Cattani, C. (2012). Hidden-markov-models-based dynamic hand gesture recognition. *Mathematical Problems in Engineering*.

Ye, Y., Tian, Y., Huenerfauth, M., & Liu, J. (2018). Recognizing American Sign Language gestures from within continuous videos. In *Proceedings of the IEEE Conference on Computer Vision and Pattern Recognition Workshops*, (pp. 2064-2073). IEEE. 10.1109/CVPRW.2018.00280

Zhang, C., Yang, X., & Tian, Y. (2013). Histogram of 3d facets: A characteristic descriptor for hand gesture recognition. In *2013 10th IEEE international conference and workshops on automatic face and gesture recognition (FG)*, (pp. 1~8). IEEE.

Chapter 19
Affecting Computing in Multimodal Mobility

Priyal Dangi
CHRIST University (Deemed), India

Muskan Garg
Mayo Clinic, USA

ABSTRACT

Computational models that simulate human emotions have witnessed a substantial development in recent years for widening the spectrum of applications. Emotional computation is becoming crucial in human-to-computer interactions with exponential growth of artificial intelligence. Normally referred to as emotion recognition, it is widely believed that the prospective detection of a person's emotional state of mind should be computed from their facial expressions. Face-movement combinations may express many different emotion types, for instance, hate, anger, panic, joy, grief, surprise, shock, to name a few. The goal and emphasis of this manuscript is the deployment of different algorithms and computation models for emotions. Considerable advancements in this domain of emotion recognition can be made through AI model development that discusses the challenges of the system and Facial Action Coding as an integral part of the models.

INTRODUCTION

Artificial Intelligence (AI) for emotion computation is an interesting study with a prime focus on developing algorithms and models to enable recognition, interpretation, and response to human emotions. The challenge for AI is to detect and understand

DOI: 10.4018/978-1-6684-8582-8.ch019

these cues and use them to infer the emotional state of a person. The relationship between cognition, emotion and behaviour patterns has undergone a shift as a result of contemporary research in the psychology, information theory, and sciences of emotion. Emotional computation is a key component of affective computing, which is the study of how computers can detect, interpret, and respond to human emotions. The AI research community encourages a frequent adaptive role that emotions play in human behaviour to determine the adaptability and capability of a system. Emotional computing encompasses human emotions and sentiments for emotion recognition & sentiment classification, respectively. Many experts presume emotional computing as a fundamental domain of affective computing, built on different modalities of data such as text, image, video, and audio. The physical information for emotional computation is illustrated by vision data (YouTube videos) and physical signals (ECG, EEG, Sensors), thereby making it challenging to portray a person's inner feelings with their active concealment from facial expressions, aural tones, body motions, etc. Emotional computation employs two major sub-tasks as:

1. *Recognition of emotions*: Emotion recognition is the process of identifying and interpreting the emotional state of a person based on their facial expressions, body language, and other nonverbal cues. This process involves using machine learning algorithms to analyses and classify the various emotional cues that a person exhibits.
2. *Analysis of sentiments*: Sentiment analysis is the process of using natural language processing (NLP) and machine learning techniques to identify and extract subjective information from textual data, and to classify that information into positive, negative, or neutral sentiment. The goal of sentiment analysis is to automatically determine the emotional tone behind a piece of text, such as a review, a tweet, or a news article.

Two models for the better understanding of one's emotion and feelings which are

1. *Discrete/Basic emotion model*: The discrete/basic emotion model is a theory of human emotion that proposes that there are a small number of basic emotions that are innate and universal across cultures that can be identified through specific facial expressions, physiological responses, and subjective experiences.
2. *Dimensional emotion model*: The dimensional emotion model is an alternative theory of human emotion that proposes that emotions can be described along several continuous dimensions, rather than as discrete and categorical states that can be characterized by their intensity or arousal level, their valence (positive or negative), and their motivational or action tendencies.

On considering the databases with multiple inputs, feature extraction from physical or EEG signals, and techniques for multimodal recognition, the thoughts influence appropriate behaviour through different algorithms to deploy the responsible AI models.

FACIAL EXPRESSION- WITH THE EMOTION CATEGORY AND ITS DESCRIPTION

Face movements exhibit different emotion categories. For instance, anger represents the facial prototype with glazed eyes, widened eyes, and pursed lips. Different emotion categories and gestures with their description in Table 1.

Table 1. Different emotion category with their description

Emotion Category/State	Description (with different factors & units)
Amusement	Rodin smile, lips apart, retracted head
Anger	Eyes were wide, lips were pursed, and brows were wrinkled.
Boredom	Lips apart, jaw clenched, tongue bulging, eyes strained, nose furrowed
Confusion	Head bent, eyelids drawn shut, and brows wrinkled
Disgust	Lips apart, jaw tightened, tongue protruding, eyes squinted, nose wrinkled
Embarrassment	Eyes closed, a restrained smile, a lowered head, hand touches face
Fear	Lips parted and stretched, upper eyelid lifted, lower eyelid strained, brows arched and drawn together.
Happy	Different factors and units of smile
Shame	Lowered head and eyes clenched
Surprise	Lips parted, upper eyelid lifted, brows raised, and jaw down

EMOTION MODELS

1. Discrete Model

AI models that classify emotions into discrete classes rather than as continuous or dimensional values are known as discrete emotion models. A discrete model in emotional computation is a model that encodes emotions as a collection of make a distinction or tags. This implies that feelings are thought of as unique, independent objects that may be categorised according to specific standards. The basic emotion model, which asserts that there are only a small number of fundamental emotions that are shared by individuals from all cultures, including happiness, anger, surprise, sorrow, disgust, and fear, is an illustration of a discrete emotion model.

(b) The Plutchik's wheel of emotions is another illustration; it likewise classifies eight basic emotions, as well as the related complex emotions, and the opposites complex emotions. These kinds of models are frequently used in applications like sentiment analysis, where the objective is to determine the overall emotional tone of text or speech, or in computer vision, where the objective is to determine emotions from facial gestures.

Figure 1. The Plutchik's wheel

2. Dimensional Model

AI models that describe emotions as continuous or dimensional values rather than as discrete categories are known as dimensional emotion models. The dimensional model is a framework for interpreting emotions in emotional computation that sees them as being made up of two core aspects: valence and arousal. To express emotions, these models frequently employ a number of dimensions or elements, such as Valence/polarity (positively or negatively), arousal/excitation (excitement or calm), and domination (control or submission), which are known as PAD Model. Figure 2 shows the description of PAD Dimensional Model

Figure 2. The Dimensional Model (PAD)

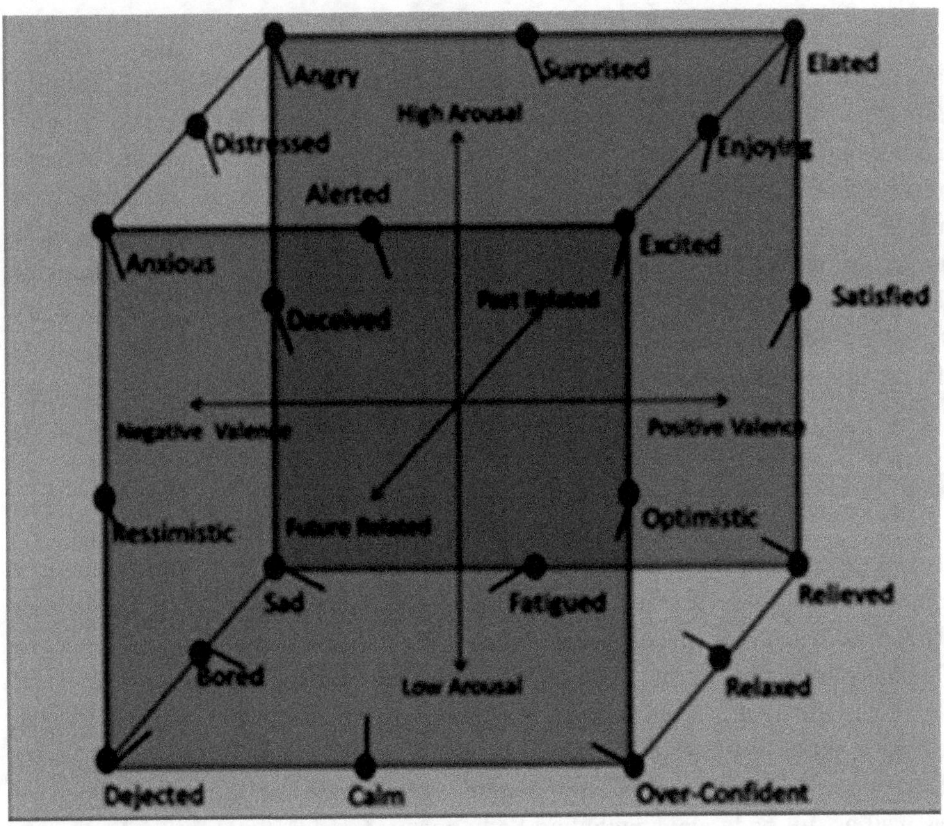

When attempting to identify emotions in text or speech, these models are frequently utilized in natural language processing (NLPs) and voice recognition. Although they are more complex than discrete emotion models and therefore are ideally able to reflect the complexities of human emotion, their implementation also demands more data and processing capabilities. A potential resource of Affective Norms for English Words (ANEW) gives valence/polarity, arousal/ excitation, and dominance values to a wide number of English words on the basis of the judgments from human participants, is an illustration of a dimensional emotion model. The Self-Assessment Manikin (SAM) non-verbal pictorial assessment approach that directly measures the valence and arousal on a 9-point scale, is another illustration in the picture.

FACIAL EMOTION RECOGNITION

The study of facial expressions of emotion in non-verbal interaction facilitates and increases the effectiveness of spoken interaction. It is important to equip the systems with identical knowledge to that of humans in order for it to be capable of identifying facial expression, making it essential to improve efficient automatic facial recognition system, image processing, etc. One could offer the maximum level of user satisfaction and criticism to enhance present technologies by evaluating these emotions, sentiments and feelings. there are different techniques such as machine learning/ deep learning can be utilized for this. to develop a number of FER systems that have been verified for conveying and identification of the information from facial representations.

BODY GESTURE RECOGNITION

Body gesture detection enables applications that allow for natural human-computer interaction. Body gesture recognition is the method of identifying and interpreting human body movement and gestures, frequently in actual, using automated systems. This technology use sensors or cameras to record and examine body gestures and motions. It then uses AI and machine learning algorithms to decipher the significance of these movements. It attempts to reveal a person's concealed emotional state using upper-body visual cues like hand movements, head positions, and eye gaze, as well as full-body signs including body postures and skeleton gestures. Due to low recognition rate, remote sensor positioning, real-time processing performance concerns, intrusive interactive tracking technologies, and the cost of motion capture for gesture representation, this approach to interaction is still challenging to implement. After obtaining the input image, classified gesture recognition system into primarily three processes. Those actions are: Features evaluation, extraction, and method of extraction. The BGR systems can be classified into machine learning and deep learning. Recent years have seen a lot of interest focused on this area of research based on the feature extraction and then the emotion recognition are carried out end-to-end. The input of a given data is mapped into the emotional feature set and the output of the emotions is obtained with responsible AI models.

PHYSIOLOGICAL GESTURE RECOGNITION

A simple approach of capturing human body language, speech, text, and facial gestures, incorporate the fluctuations in physiological parameters, corresponding to shifts in human feelings and emotions, making it possible for people to identify, understand, and imitate various human emotional responses. The utilisation of recognizing emotions and sentiments through physiological signals and indicators are widely adopted in fields like social welfare, health care, and defensive driving and training. One process includes analysing physical cues from people, such as their position, voice, movements, and facial gestures, which can easily be collected. Consider a scenario where people experiencing bad emotions, still smile in a formal public situation. The internal signals are used by the physiological signals such as EEG, ECG, EMP, SKT (skin temperature), SKC (skin conductance), RESP etc.

1. EEG: Electroencephalography, or EEG, can be used to identify and analyse the brain's electrical activity that matches to particular motions or movements of the body. The EEG signal's transmissions can be used to distinguish between various emotional states, including joy, sadness, fear, and worry.
2. ECG: the ECG signal can be used to recognise various emotional states, including joy, sadness, and fear. Changes in heart rate variability, for instance, have been connected to emotional responses like stress and worry.
3. SKT: Emotional shifts can also be linked to variations in skin temperature. For instance, the "fight - or - flight" response might be triggered in the body when a person feels an emotional reaction like anxiety, fury, or stress.
4. SKC: Skin conductance can be employed as a measure of emotion and stress. When an individual feels an emotive response such as anxiety, tension, or pleasure, it can elicit a physical biological response, which can require adjustments in the functioning of the sweat glands located on the skin
5. RESP: Respiratory signal which is a physiological signal frequently used to monitor a person's breathing habits.

It is attempted to establish a uniform and fixed link between emotional changes and physiological signs on well-established physiological signal databases.

PUBLICLY AVAILABLE DATASETS FOR COMPUTING

Publicly accessible dataset is an essential component to facilitate future research and development in emotional computation. We investigate the existing and well-established datasets for textual, speech/audio, visual, physiological, and multimodal data. The existing literature suggests following datasets for emotional computing (Garg, 2022):

1. **EmoReact**[1]: A dataset of videos and physiological signals recorded from participants watching emotionally evocative film clips. The dataset includes ratings of valence, arousal, and dominance, as well as physiological signals such as heart rate, electrodermal activity, and respiration. EmoReact is a multimodal dataset that includes both subjective and objective measures of emotion, and has been used for research in affective computing and psychology.
2. **Affectiva-MIT Facial Expression Dataset (AM-FED)**[2]: A dataset of facial expressions recorded from participants watching emotionally evocative video clips. The dataset includes recordings of facial expressions, along with ratings of valence, arousal, and dominance. The dataset was collected by researchers at MIT and Affectiva, and has been used for research in affective computing and computer vision. This dataset consists of 242 facial videos (168,359 frames) recorded in real world conditions.
3. **The RECOLA Dataset**[3]: A dataset of audio visual recordings of participants in naturalistic social interactions. The dataset includes recordings of facial expressions, speech, and physiological signals, as well as ratings of valence, arousal, and dominance. The RECOLA dataset has been used for research in multimodal emotion recognition and affective computing. The database consists of 9.5 hours of audio, visual, and physiological (electrocardiogram, and electrodermal activity) recordings of online dyadic interactions between 46 French speaking participants.
4. **The SEMAINE Database**[4]: A dataset of audio visual recordings of participants in naturalistic social interactions. The dataset includes recordings of facial expressions, speech, and physiological signals, as well as annotations of emotions and communicative signals. The SEMAINE database has been used for research in multimodal emotion recognition and affective computing, as well as for the development of affective conversational agents.

5. **YFCC100M**[5]: YFCC100M (Yahoo Flickr Creative Commons 100 Million) is a large-scale multimedia dataset that contains 100 million images and videos that were collected from the photo-sharing website Flickr. The dataset was released by Yahoo in 2014 for research and educational purposes.
6. **Moodo**:A dataset of the two key innovations are a multimodal method that integrates user context and combines emotional and colour perception to music. In non-musical and musical contexts, there are significant differences in the emotion-color associations and their valence-arousal evaluations.
7. **Iemocap**: A dataset of Multimodal Emotion Recognition which consists of 151 videos of recorded dialogues, with 2 speakers per session across the dataset. Nine emotions (angry, aroused, fear, sad, astonished, frustrated, pleased, disappointed, and neutral) as well as valence, arousal, and dominance are marked for each segment.
8. **The humaine database**: A dataset of tagged multimodal database with natural speech is Humaine database. The database's goal is to give the public access to a resource where they may look at tools that could be useful for describing emotion as well as see and hear the shapes that emotion takes in routine activity and interaction which has 48 clips.
9. **TASS 2020**: A dataset for Sentiment analysis purposes. It is divided into a number of subsets made for the different tasks suggested in the various editions over the years. It outlines the many strategies used by the participating teams, the important learnings from their systems, and the outcomes attained for all the suggested solutions.
10. **AffectNet**: A large-scale dataset of images with annotations for valence, arousal, and eight basic emotions. The dataset contains over one million images, and includes images from a variety of sources, such as social media, search engines, and photo sharing websites. AffectNet is one of the largest and most diverse datasets for multimodal emotion recognition (Mollahosseini, 2017). There are more than a million facial photos on AffectNet.

PHASES OF EMOTION RECOGNITION

The following phases are commonly incorporated into the process of emotion computation using the given algo*rith*ms:

1. **Data collection:** The available information of text, pictures/ images, videos, audio, and other sensor-specific data of *persons with different emotions* is composed as a collection of data for predicting emotions. As discussed earlier, many datasets are publicly available which can be further used for cross-

sectional studies of emotion and sentiment detection. The dataset is deployed for the next phase of feature extraction.

2. **Feature extraction and selection:** To depict the emotion in a given data, the features extraction and selection mechanism supports the quantification of unstructured data, especially through bio-inspired and linguistically computational approaches of computational intelligence, leveraged on AI. Feature extraction mechanism is handled by identifying handcrafted features and/ or the embedding-based data representation for semantically enhanced information from a given dataset. The selected features are further deployed to the next phase of training the model.
3. **Training of the Model:** The algorithm is trained over selective features using a labelled training dataset where each algorithm has its own way of training the model. The explanation for training of different algorithms for emotional computation is given in the upcoming sections for better understanding on different types of training.
4. **Evaluation of the Model:** The model's performance is assessed and evaluated through precision, recall, F1-measure, and accuracy on a validation dataset. It can be used to detect emotions in language, but how successfully relies on the standard and volume of training data, as well as the particular algorithms and methods employed.
5. **Deploying the Model:** The algorithm can be used in a real implementation to detect and compute the emotions in pictures/images after it has been trained and evaluated or validated. The model can be used when it has been trained and assessed. This can entail text input and emotional recognition predictions for users. Figure 3 represents the steps incorporated for the process of emotion computation.

Figure 3. Steps for the emotional computation

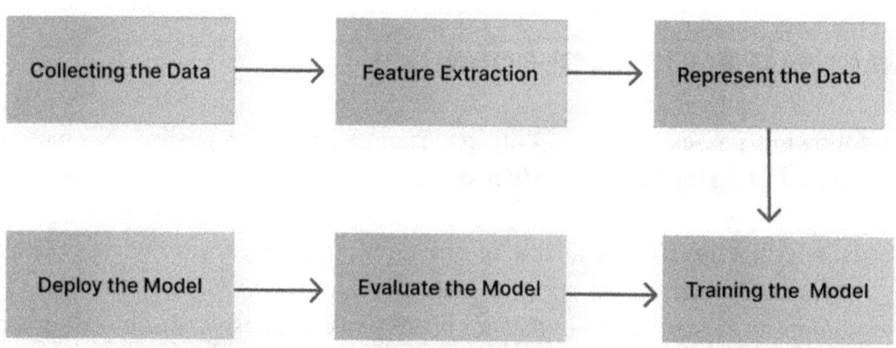

PROPOSED ALGORITHMS FOR EMOTIONAL COMPUTATION

1. Support Vector Machine

Support Vector Machines (SVM) is a popular machine learning technique used for emotion computation in affective computing. It is a supervised learning algorithm that can be trained on labeled data to classify new data based on the learned patterns.

In the context of emotion computation, SVM can be used to learn patterns in the data that are associated with different emotional states. For example, it can learn to identify facial expressions, voice intonations, or other features that are associated with particular emotions. To use SVM for emotion computation, the first step is to prepare a training dataset of labeled data, where each sample is associated with a label that represents the emotional state that it conveys.

The dataset can include any type of data that is relevant for recognizing emotions, such as audio recordings, video clips, or physiological signals. Next, an SVM model is trained on the labeled dataset to learn the patterns that are most relevant for emotion recognition. During training, the model learns to identify a boundary in the feature space that separates the different emotional states.

The boundary is chosen to maximize the margin between the different classes, i.e., the distance between the boundary and the closest points from each class. Once the SVM model has been trained, it can be used to classify new data into one of several emotional states. The model takes the input data, applies the learned boundary to extract relevant features, and then uses the boundary to make a prediction.

2. KNN (K-Nearest Neighbour)

K-nearest neighbour (KNN) has produced outstanding results in physiological data relying on identification of emotion. However, certain problems persist. The simple majority just among the closest neighbours is too simplistic to tackle difficult class distributions (like data distribution); factors that contribute equally to resemblance will diminish accuracy of classification; samples at class boundaries are now more likely to be misidentified when k is high.

Initially, the weighting method, a new approach that utilizes the distance of long haul neighbours, is recommended to weight the categorization. It tends to increase going to the polls precision by allowing closer neighbours to contribute significantly to the classification and by using farthest neighbours to lessen the classes of non-target classes weights.

The KNN experiment demonstrates greater accuracy. What should the value of nearest neighbours (k) be that produces the maximum accuracy across my mind while I create the KNN classification model. This is a crucial question since the selection of k affects the accuracy of classification.

3. Random Forest

Random Forest is a machine learning technique that can also be used for emotion computation in affective computing. It is an ensemble learning method that combines multiple decision trees to make predictions.

Random Forest (RF) is a group of binary decision trees built using a series of training sample. It's the substitution of a random selection from the training sample N. The creation of each of the tree's nodes additionally applies a random descriptor selection from the entire sample p. Consequently, it steers clear of large features (p > N) and minimizes the correlation between the classifiers inside that forest while retaining the strength.

The usage of the out of error estimate is one of the most crucial aspects of RF. A sample set known which was not used in the training of the present tree. Consequently, it is exploited in error estimates. The accuracy of the tree classification is improved by this internal assessment task of the generalization error.

Because a single classifier is prone to overfitting and has poor generalization capabilities, the ensemble learning method is developed to increase accuracy of classification. The basic novice under the concept of ensemble methods should be "excellent but distinct," meaning that each person should have a relatively high recognition or computation rate that is distinct from the other.

The process of creating numerous decisions made trees using the conventional approach might result in decision forests which are neither significantly different from one another or lower recognition rates for the obtained one's decision forests, both of which will have an impact on the outcome.

4. CNN (Convolutional Neural Network)

Convolutional Neural Networks (CNN) are a popular machine learning technique used for emotion computation in affective computing. CNNs are a type of neural network that is designed to analyze visual imagery and are often used in image and video analysis applications.

In order to work with visuals without impacting accuracy of the system, we can minimise processing costs by using the convolution algorithm. These are extensively employed to detect emotions in visuals. CNNs are programmed to categorize facial expressions and gestures to predetermined emotions for this purpose (like happy, angry, etc.). It is important to note and to know that the network's performance has the significant alteration by the design of CNN. These designs have been optimized for the classification of images and applications which are then performed greatly.

Input picture data is often entered into the model, which then executes pools and convolution operations for the extraction of the features. The features are subsequently categorized and classified using fully linked layers. In minimizing a loss function, to calculates the disparity between the model's predictions which are then are adjusted throughout training.

In terms of presenting characteristics of extraction and classification of convolution data, CNN varies significantly from the classical neural networks. Both forward & backward propagating ways are the two processes that make up CNN's core model

DATA PREPARATION PERFORMANCE MEASURES FOR EMOTIONAL ANALYSIS USING VARIOUS EMOTION MODELS AND ALGORITHMS

Despite the lack of universal consensus on how to represent affective states, most emotional analyses are trained and assessed using one of the two categories of emotion models: discrete models or dimensional models. Typically, discrete or dimensional identifiers are employed interchangeably to fit the input data when creating the emotional databases. The most common measures used for categorical emotion and sentiment classification are accuracy, recall, precision, and recall. Among the various evaluation measures or metrices for the emotional classification, the F1-Score appears to be the most effective and the best option.

The internal character of emotions, which renders them unavailable to exterior systems, presents a significant challenge in the development of these emotional machines. As a result, ongoing annotation depending on dimensional models is desirable, and several annotation systems have been created to facilitate this. Several (different-modal i.e, discrete and dimensional model) datasets with continuous annotation have now been developed for solving this difficulty.

APPLICATIONS AND THE USE CASES WITH UTILIZATION OF THE EMOTIONAL COMPUTATION AI

The market for AI and computer vision has witnessed an increase in demand for AI emotion computation. Despite, not yet widely used, incurring the visual emotional tasks simultaneously is anticipated to have a significant impact on real-life applications.

1. Opinion Mining:

It seeks to identify people's attitudes, feelings, sentiment, and opinions from the inputs. The focus of traditional sentiment analysis is mostly on text.

2. Analysis of medical opinion:
   The patient's health, symptoms, and treatments are the focus of medical opinion. Multiple fields, including that of the treatment of mental illnesses, remote medical services, and interpersonal behaviour, can benefit from its evaluation and extraction.
3. Detection and prevention of fraud:
   Just to take in the case of Insurance Co. to avoid the problems and determine whether a client is dishonest while filing a claim, insurance companies can use voice analysis and detection.
4. Driving efficiency:
   The performance of driving by autonomous vehicles can also be evaluated using the emotional computation. The system can monitor the passengers' human emotions and determine whether they appear worried or happy with the driving experience due to mics and cameras installed in the car.
5. Workplace layout:
   To enhance the physical worksite and environment design and pleasure, companies can monitor their people at work and assess the sentiment in corporate social networks and discussion.

Humans have been able to read emotions and recognise facial gestures and movements for a very long time. However, it is novel to employ these motions for enhancing services and generating business advantages. Although using technology to recognise and detect faces and measure emotions is not simple due the complexity induced with subjective nature of the task, the development of face detection and emotion recognition and sentiment analysis systems is feasible.

COMMON REPRESENTATION OF EMOTION RECOGNITION WITH A DIAGRAM

In emotion recognition, the input data is usually represented as a feature vector that captures the relevant information from the raw data. The feature vector can include various types of information depending on the modality of the input data, such as visual features for facial expressions, audio features for speech or tone of voice, or physiological features for signals from biosensors. Figure 4 shows the basic and general depiction of emotion recognition in the flow with a diagram.

Figure 4. Basic depiction emotional recognition

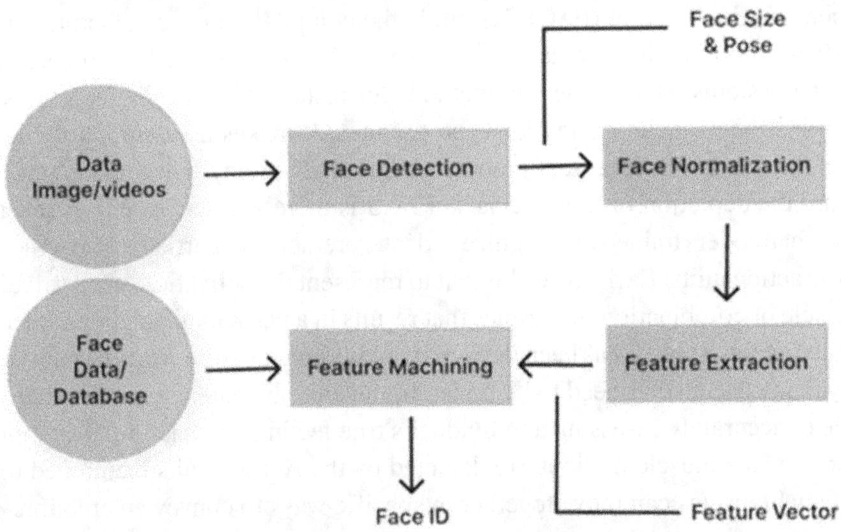

Monitoring the Facial Gestures

An essential behavioural indicator for the analysis of, cognitive functions, emotions and social contact is facial expression. a method for assessing face movement and expression objectively in the context of individual acts. To handle this, the AI research community uses digital image data.

Contraction of facial muscle are to produce movements that are apparent to the unaided eye, altering the space among facial features and causing the skin to fold and wrinkle over an internal skeletal system. The principles of making a facial movement vary depending on the individual, including anatomical specifics. To measure, assess and detect the facial movements in a scientific analysis, there are three standard methods. Electromyography (EMG), the most accurate and sensitive method of measuring facial movements and expressions, finds electrical activity resulting from significant muscle activity. In this method, using electrodes placed, a technique has been devised to identify facial muscle activity as precisely and clearly as feasible.

FACIAL ACTION CODING SYSTEM

The Facial Action Coding (FAC) System is the compilation of facial expression-related facial motions, thereby identifying the participant's expressed emotions using the coding systems. One of the few methods for instantly identifying emotions is the examination of facial expressions and gestures. It makes no assumptions as to whether those motions represent emotions/sentiments or any other related events which are the depiction of mind because FACS is merely descriptive. For several times, human coders trained to recognize and interpret sensory particular movements known as action units. Each AU is thought to represent the activation of a particular face muscle or combination of muscles that results in a particular facial motion and expression. After a signal has been delivered, additional measurements like surveys and psychometric testing need to be taken. In the end, this latency creates another obstacle to accurately assessing a individual's true feelings in reaction to an input. The various face muscle motions are depicted by the Action Units mentioned top. These facial muscles can move together in specific ways to convey an emotion.

Difficulties With Emotion Recognition

According to estimates, the application and advancement of technology for facial recognition will continue to grow. Despite the fact that many firms successfully use the technologies, the development of AI models have major challenges. With the emotional computational approach, there are several challenges in emotion recognition and detection by facial analysis that must be overcome in order to properly realise the full capabilities of the entire algorithm or technique used.

With the emotional computational approach, there are several challenges in emotion recognition and detection by facial analysis that must be overcome in order to properly realise the full capabilities of the entire algorithm or technique used. The accuracy and precision of predicting and monitoring has been a difficulty in the initial phases of practically every type of AI-based recognition (voice recognition, face recognition, etc). This lack of precision diminishes over time as technology improves, more data is gathered, new findings are made, and new procedures are adopted, leaving noise problems in its wake. The major obstacles are:

- There isn't always a corresponding emotion below that of a facial expression and features.
- Because framed expressions, as employed by the majority of study participants, are not genuine and real, algorithms and techniques developed using them may not be applicable to the given expressions which are real.

- Variations in the light, facial misalignment or position, an occlusion of any kind, etc.
- This is a result of how light interacts with the face, the spectator, and the variances in how faces seem and look. Lighting, scale, posture and image processing factors are included.

Challenges and issues linked to face recognition and sentiment analysis research are also regarded as scientific data gathering and objective evaluation of face recognition systems. A range of recognition standards have been published recently in this regard to assist with the development of the systems and algorithms for face recognition beneath various imaging situations including both 2D and 3D.

CONCLUSION

In summation, the field of emotional computation is witnessing the exponential growth and huge potential to alter how people engage with technologies. To properly control the development, advancement and application of this technology is essential to take into account its moral and societal consequences. This manuscript primarily examines and investigates the use of the emotion models (discrete and dimensional model) and discusses the deployment of baseline methods on publicly available datasets for emotional computation. The operational recommendations include data collection, feature extraction, training models, validating model and its deployment. The current study is anticipated to advance affective computing's use and assistance in its better comprehension. This work on emotional computation is a potential guide to evaluate the advancements in face detection and recognition work and comprehend the difficulties faced by face recognition in practical situations. In this paper, we've examined the recent developments in facial recognition while also talking about a number of difficulties and crucial elements that could have a big impact on how well the technology works with the emotional computation for development of responsible AI models.

Ethical Considerations: The advancement of technology which can detect, comprehend, and adapt to human emotions as affective computing. While creating and utilizing affective computing, there are ethical concerns to be thought about, just like with any modern innovation. In general, ethical concerns are very important in the creation and application of emotional computational power. To guarantee that innovation is utilized in a manner that is fair, ethical, and respects the rights and dignity of every person, developers and users must interact. Where some of the most important ethical issues with affective computing can have privacy, bias, transparency, security, informed consent, human autonomy.

REFERENCES

Bhardwaj, S.,(2021). A Study on Human Emotion Detection.. *2021 International Conference on Advance Computing and Innovative Technologies in Engineering (ICACITE)*, (pp. 909–13). IEEE Xplore

CSDL. (n.d.).| *IEEE Computer Society*. IEEE. https://www.computer.org/csdl/journal/ta.

Egger, M., Ley, M., & Hanke, S. (2019). Emotion Recognition from Physiological Signal Analysis: A Review. [ScienceDirect]. *Electronic Notes in Theoretical Computer Science, 343*(May), 35–55. doi:10.1016/j.entcs.2019.04.009

Fang, X., & Zhan, J. (2015, June). Sentiment Analysis Using Product Review Data. [BioMed Central]. *Journal of Big Data, 2*(1), 5. doi:10.118640537-015-0015-2

Garg, M., Wazarkar, S., Singh, M., & Bojar, O. (2022, June). Multimodality for NLP-Centered Applications: Resources, Advances and Frontiers. *In Proceedings of the Thirteenth Language Resources and Evaluation Conference* (pp. 6837-6847). IEEE.

Giri, S. (2022).Emotion Detection with Facial Feature Recognition Using CNN & OpenCV. *2022 2nd International Conference on Advance Computing and Innovative Technologies in Engineering (ICACITE),* pp. 230–32. IEEE Xplore 10.1109/ICACITE53722.2022.9823786

Horvat, M., ... "An Overview of Common Emotion Models in Computer Systems." 2022 45th Jubilee International Convention on Information, Communication and Electronic Technology (MIPRO), 2022, pp. 1008–13. IEEE Xplore 10.23919/MIPRO55190.2022.9803498

Krishna, K. V. (2022). Speech Emotion Recognition Using Machine Learning.. *2022 6th International Conference on Computing Methodologies and Communication (ICCMC),* (pp. 1014–18*. IEEE Xplore 10.1109/ICCMC53470.2022.9753976

Liu, C., & Kirubakaran, S.Daniel, A. J. (2022, April). Deep Learning Approach for Emotion Recognition Analysis in Text Streams. [IJTHI]. *International Journal of Technology and Human Interaction, 18*(2), 1–21. www.igi-global.com. doi:10.4018/IJTHI.313927

Ma, H., & Yarosh, S. (2021). A Review of Affective Computing Research Based on Function-Component-Representation Framework. [IEEE Xplore]. *IEEE Transactions on Affective Computing*, 1–1. doi:10.1109/TAFFC.2021.3104512

MIT. (n.d.). *Publications List Pre 2020.* MIT Media Lab, https://www.media.mit.edu/groups/affective-computing/publications-list/.

Mollahosseini, A., Hasani, B., & Mahoor, M. H. (2017). Affectnet: A database for facial expression, valence, and arousal computing in the wild. *IEEE Transactions on Affective Computing*, *10*(1), 18–31. doi:10.1109/TAFFC.2017.2740923

Pathar, R., & Human Emotion Recognition Using Convolutional Neural Network in Real Time." 2019 1st International *Conference on Innovations in Information and Communication Technology (ICIICT), 2019*, (pp. 1–7). IEEE Xplore 10.1109/ICIICT1.2019.8741491

Robinson, P., & el Kaliouby, R. (2009, December). Computation of Emotions in Man and Machines. [PubMed Central]. *Philosophical Transactions of the Royal Society of London. Series B, Biological Sciences*, *364*(1535), 3441–3447. doi:10.1098/rstb.2009.0198 PMID:19884138

Rudovic, O., Zhang, M., Schuller, B., & Picard, R. (2019). Multi-modal Active Learning From Human Data: A Deep Reinforcement Learning Approach. In *2019 International Conference on Multimodal Interaction (ICMI '19)*, ACM, 10.1145/3340555.3353742

Sarker, I. H. (2021, March). Machine Learning: Algorithms, Real-World Applications and Research Directions. [Springer]. *SN Computer Science*, *2*(3), 160. doi:10.100742979-021-00592-x PMID:33778771

Tao, J., & Tan, T. (2005). Affective Computing: A Review. In J. Tao & ... (Eds.), *Affective Computing and Intelligent Interaction* (pp. 981–995). Springer. doi:10.1007/11573548_125

Tawsif, K., & ... . (2022, August). A Systematic Review on Emotion Recognition System Using Physiological Signals: Data Acquisition and Methodology. [ijournalse.org]. *Emerging Science Journal*, *6*(5), 1167–1198. doi:10.28991/ESJ-2022-06-05-017

Wankhade, M., Rao, A. C. S., & Kulkarni, C. (2022, October). A Survey on Sentiment Analysis Methods, Applications, and Challenges. [Springer]. *Artificial Intelligence Review*, *55*(7), 5731–5780. doi:10.100710462-022-10144-1

What Is Multimodal Emotion Recognition (2021). IGI Global. https://www.igi-global.com/dictionary/multimodal-emotion-recognition/19631.

Yan, W. (2021). *"Research on Emotions Based on Brain-Computer Technology: A Bibliometric Analysis and Research Agenda." Frontiers in Psychology* (Vol. 12). Frontiers.

Yoon, W.-J., & Park, K.-S. (2007). A Study of Emotion Recognition and Its Applications.. Modeling Decisions for Artificial Intelligence,, pp. 455–62. Springer doi:10.1007/978-3-540-73729-2_43

ENDNOTES

http://multicomp.cs.cmu.edu/resources/emoreact-dataset/

2 https://www.affectiva.com/facial-expression-dataset/

3 https://diuf.unifr.ch/main/diva/recola/index.html

4 https://semaine-db.eu/

5 http://projects.dfki.uni-kl.de/yfcc100m/

6 http://mood.musiclab.si/en/dataset

7 https://sail.usc.edu/iemocap/

8 http://universal.elra.info/product_info.php?cPath=25&products_id=2063

9 http://tass.sepln.org/tass_data/download.php

Chapter 20
Exploring Best Practices in Mobile App Design Patterns and Tools:
A User-Centered Approach

Rajesh Kanna Rajendran
CHRIST University (Deemed), India

Mohana Priya T.
CHRIST University (Deemed), India

ABSTRACT

Design patterns are reusable solutions to common design problems that provide a consistent user experience across different apps. This article explores the best practices in mobile app design patterns and tools with a focus on the user-centered approach to design. Design patterns such as navigation bars, tab bars, list views, and card views are discussed, along with design tools such as Sketch, Figma, Adobe XD, and InVision. The problem is to ensure that mobile app design is centered around the needs and preferences of the user, rather than the designer or the technology, and that the right design patterns and tools are used to create interfaces that are familiar and easy to use. The chapter emphasizes the importance of conducting user research to understand the needs and preferences of the target audience and using design patterns and tools to create interfaces that are familiar and easy to use. Mobile apps have become an integral part of our lives, and designing a successful mobile app is a challenging task that requires a thorough understanding of user needs and preferences.

DOI: 10.4018/978-1-6684-8582-8.ch020

INTRODUCTION

Mobile app design has come a long way since the first iPhone was released in 2007. With the rise of mobile technology, app design has become an essential part of our daily lives. According to Statista, there were 218 billion app downloads in 2020 alone, and this number is expected to grow in the coming years. As the number of mobile apps continues to increase, it is crucial for designers to explore best practices in mobile app design patterns and tools to ensure user satisfaction and retention. In this chapter, we will explore the importance of a user-centered approach in mobile app design, as well as the benefits of incorporating best practices and tools in the design process (Feng et al. 2012).

Mobile app development design patterns refer to best practices and reusable solutions for common problems that arise during the development of mobile applications. Here are some common design patterns used in mobile app development:

Model-View-Controller (MVC): This pattern separates the app into three components - model (data and business logic), view (user interface), and controller (handles user input and interacts with the model).

Singleton: This pattern ensures that only one instance of a class exists in the app at a time. It is often used for creating shared resources like databases or network managers.

Observer: This pattern allows objects to be notified when a particular event occurs. For example, a view controller can be notified when a data model change.

Factory: This pattern provides a way to create objects without specifying their exact class. It is useful for creating objects that belong to different classes but share common properties.

Adapter: This pattern allows incompatible classes to work together by creating a bridge between them. For example, an adapter can be used to display data from a database in a table view.

Delegate: This pattern allows one object to delegate tasks to another object. For example, a table view delegate can be used to handle user interaction with the table view.

Strategy: This pattern allows an algorithm to be selected at runtime based on user input or other factors. It is often used for handling different user input scenarios.

Facade: This pattern provides a simplified interface to a complex subsystem. For example, a facade can be used to provide a simple API for accessing complex network services.

Command: This pattern encapsulates a request as an object, allowing it to be stored and executed at a later time. It is useful for implementing undo/redo functionality.

LAYERS OF MOBILE APP DEVELOPMENT ARCHITECTURE

Presentation Layer

The aim of this layer is to look into how to present the application to end users. When designing this layer, the mobile app developers must identify the correct client type for intended infrastructure. Additionally, the client's deployment restrictions must also be kept in mind. Another necessity is selecting the correct data format and using robust data validation mechanisms for protecting the apps from invalid entry (Lee et al. 2012).

Business Layer

This layer looks into elements on the business front. In layman words, it looks into the way businesses are presented to the end users. This consists of business components, workflow, and the entities under two sub-layer hood: Domain model and Service. The service layer looks into the definition of common application function set that are available to the end users. While the domain model layer looks into the knowledge and expertise linked to specific problem areas (Microsoft, 2013).

Data Layer

The data access layer must meet the application requirements and should help in offering efficient and secure data transactions. Mobile app developers should also consider the maintenance side of the data while ensuring that the data layer can be modified easily with the changing business requirements. This layer consists of the data specific components such as access components, utilities, helpers, and the service agents. The three elements of mobile architecture patterns are placed under two subheads: Persistence Layer and Network Layer. The former offers simplified data access which is stored in the mobile app backend architecture, the latter is needed for making networking calls. Figure 1 shows Mobile App Development Architecture (Yumin et al. 2008).

THE USER-CENTERED APPROACH

Mobile app design should always prioritize the user experience. A user-centered approach involves designing with the user's needs, wants, and preferences in mind. This approach involves research and analysis of the target audience to identify their behaviors, habits, and pain points. Once the designer has a clear understanding

Figure 1. Mobile App Development Architecture
(Source: Mobile App Architecture - The Basis of App Ecosystem (appinventiv.com))

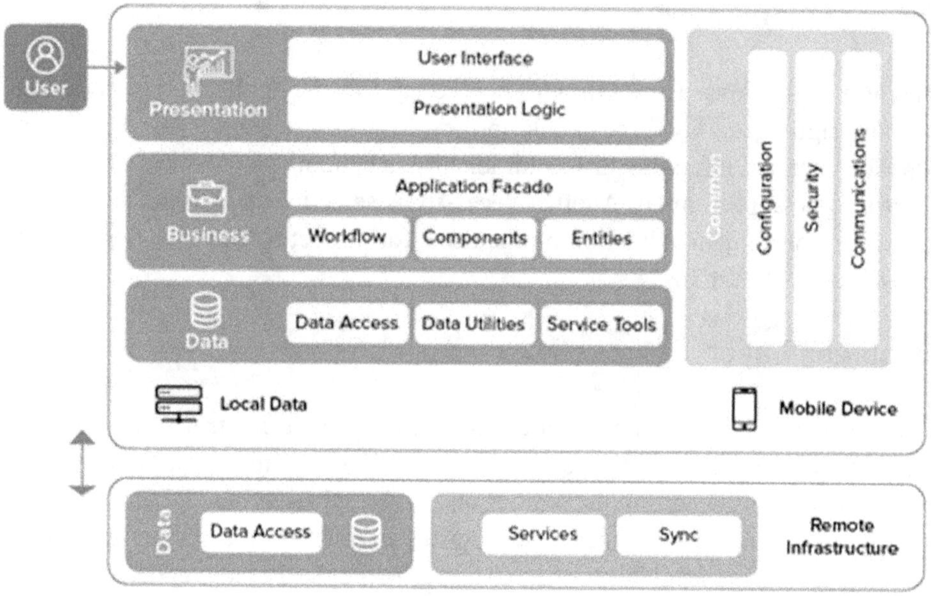

of the user, they can begin to create user personas and user journeys to guide the design process. The user-centered approach is a design methodology that focuses on understanding the needs, goals, and behaviors of users to create products and services that meet their needs. This approach involves placing the user at the center of the design process, with the goal of creating products and services that are intuitive, easy to use, and effective (Lanchun et al. 2017).

The user-centered approach involves several key principles, including:

Understanding User Needs: This involves gathering data and insights about users, including their goals, tasks, and behaviors. User research techniques such as surveys, interviews, and usability testing can be used to gain this understanding.

Involving Users in The Design Process: Users should be involved in the design process from the beginning, with their feedback and input informing the design decisions. This can be done through user testing, focus groups, and other participatory design techniques.

Designing For Usability: Designers should create products and services that are easy to use and navigate, with a focus on minimizing cognitive load and ensuring that users can accomplish their tasks quickly and efficiently.

Iterative Design: The user-centered approach involves an iterative design process, with designs being refined and improved based on user feedback and testing (Zhaoxin et al. 2016).

Continuous Improvement: Products and services should be continually improved based on user feedback and usage data, with the goal of creating a better user experience over time.

By following the user-centered approach, designers can create products and services that are more effective, more usable, and more satisfying for users. This approach can lead to increased user engagement, loyalty, and satisfaction, as well as improved business outcomes for organizations that adopt it.

BEST PRACTICES IN MOBILE APP DESIGN PATTERNS

Mobile app design patterns are standardized ways of solving common design problems. Best practices in mobile app design patterns are established guidelines that have been proven to enhance the user experience. In this chapter, we will explore some of the most common mobile app design patterns, including navigation, forms, and notifications, and how to implement them in a user-centered way (Jing et al. 2017).

The following are some best practices in mobile app design patterns that can help developers create high-quality and user-friendly mobile apps:

Consistency: It is important to maintain consistency in the app design, including the use of colors, typography, and layout. Consistency makes the app easier to use and navigate, and it helps to establish a strong visual brand.

Simplicity: The app design should be simple and intuitive, with a clear and easy-to-understand user interface. The app should be designed with the user in mind, with a focus on minimizing cognitive load and ensuring that users can accomplish their tasks quickly and efficiently.

Navigation: The app navigation should be simple and straightforward, with clear and easy-to-use menus, buttons, and other controls. Users should be able to navigate through the app with ease, without getting lost or confused.

Performance: The app should be designed for optimal performance, with fast load times and minimal lag. This can be achieved through the use of optimized code, caching, and other performance optimization techniques.

Feedback: The app should provide clear and meaningful feedback to users, including feedback on user actions, errors, and other interactions. This feedback helps to keep users engaged and informed, and it can help to prevent user frustration and confusion.

Accessibility: The app should be designed to be accessible to users with disabilities, including support for screen readers, high-contrast mode, and other accessibility features.

Security: The app should be designed with security in mind, with a focus on protecting user data and preventing unauthorized access. This can be achieved through the use of encryption, secure login protocols, and other security features.

TOOLS FOR MOBILE APP DESIGN

Mobile app design requires a range of tools to create, test, and optimize designs. In this chapter, we will explore some of the most popular tools for mobile app design, including wireframing and prototyping tools, design systems, and usability testing tools. There are several tools available for mobile app design, and choosing the right one depends on your design needs and preferences. Here are some popular tools for mobile app design:

Sketch: Sketch is a vector-based design tool that is popular among UI/UX designers. It has a large library of plugins and add-ons that make it easy to design mobile app interfaces.

Adobe XD: Adobe XD is another vector-based design tool that is popular among mobile app designers. It allows for quick prototyping and wireframing and has a wide range of features that make it easy to design complex interfaces.

Figma: Figma is a web-based design tool that allows for collaboration among team members. It has a powerful design engine that makes it easy to create complex interfaces and also supports real-time collaboration.

InVision Studio: InVision Studio is a powerful design tool that allows for quick prototyping and wireframing. It also supports collaboration and has a range of features that make it easy to create complex designs.

Marvel: Marvel is a web-based design tool that is popular among mobile app designers. It allows for quick prototyping and wireframing and has a range of features that make it easy to create complex interfaces.

Procreate: Procreate is an iPad app that is popular among digital artists and designers. It allows for drawing, sketching, and painting and has a range of features that make it easy to create complex designs.

Canva: Canva is a web-based design tool that is popular among non-designers. It has a range of templates and design elements that make it easy to create simple mobile app interfaces.

Affinity Designer: Affinity Designer is a vector-based design tool that is popular among graphic designers. It allows for creating complex illustrations, icons, and logos for mobile apps.

Axure RP: Axure RP is a tool for creating wireframes, prototypes, and specifications for mobile app design. It allows for interactive prototyping and has a range of features that make it easy to create complex user interfaces.

NAVIGATION APPROACH

While majorly accounted for the app architecture designing front, the element would call for an expertise in both backend and frontend. On the basis of your understanding of who the customers are and what are their app requirements, you should analyze which one of these would be good for your app:

- Stacked navigation bar
- Single view
- Scroll views
- Tab controller
- Search based
- Model controller
- Gesture driven

THE DESIGN PROCESS

A successful mobile app design process involves a series of steps, from research and ideation to testing and implementation. In this chapter, we will provide an overview of the design process and how to incorporate user-centered design principles and best practices in each stage. The design process is a systematic approach to problem-solving that is used to develop effective solutions for a wide range of design challenges. It typically consists of several key stages, which include:

Research: This involves gathering information and insights about the problem or challenge you are trying to solve, as well as the users and stakeholders involved.

Analysis: This involves synthesizing the research to identify key insights, user needs, and pain points that can inform the design solution.

Ideation: This involves generating a wide range of potential design solutions through brainstorming, sketching, and other ideation techniques.

Prototyping: This involves creating physical or digital prototypes of the most promising design concepts to test and refine the solution.

Testing: This involves gathering feedback from users and stakeholders on the prototype to identify areas for improvement and refine the design solution.

Implementation: This involves refining the design solution based on feedback and finalizing the design for implementation.

Evaluation: This involves measuring the effectiveness of the design solution in addressing the problem or challenge and identifying opportunities for further improvement.

Throughout the design process, it's important to maintain a user-centered approach and focus on designing solutions that meet the needs and goals of the users and stakeholders involved. Collaboration and communication with team members and stakeholders are also key to a successful design process.

MOBILE APP DESIGN CASE STUDIES

In this phase, we will examine some successful mobile app design case studies, including the design of the Airbnb app and the redesign of the Instagram app. We will explore how these apps incorporate user-centered design principles and best practices to create a seamless user experience.

Here are some mobile app case studies that showcase the design process in action:

Headspace: Headspace is a popular meditation and mindfulness app that helps users reduce stress and anxiety. The design process for Headspace involved researching user needs and preferences, analyzing market trends, ideating potential app features and functionalities, creating and testing prototypes with users, and refining the design solution based on feedback. The result is a user-friendly app with a simple and intuitive design that provides a seamless meditation experience.

Duolingo: Duolingo is a language learning app that helps users learn a new language through gamified lessons and quizzes. The design process for Duolingo involved researching user needs and preferences, analyzing market trends, ideating potential app features and functionalities, creating and testing prototypes with users, and refining the design solution based on feedback. The result is an engaging and effective language learning app that has been downloaded by millions of users around the world.

Airbnb: Airbnb is a popular travel and accommodation booking app that allows users to book unique and affordable accommodations around the world. The design process for Airbnb involved researching user needs and preferences, analyzing market trends, ideating potential app features and functionalities, creating and testing prototypes with users, and refining the design solution based on feedback. The result is a user-friendly and visually appealing app that provides a seamless booking experience.

Shazam: Shazam is a music recognition app that allows users to identify songs by simply holding up their phone to the music. The design process for Shazam involved researching user needs and preferences, analyzing market trends, ideating potential app features and functionalities, creating and testing prototypes with users, and refining the design solution based on feedback. The result is a highly accurate and efficient music recognition app with a simple and intuitive design.

In all of these case studies, the design process was a systematic approach to problem-solving that involved gathering information, ideating potential solutions, prototyping and testing those solutions, and refining the design based on feedback. The result is a user-friendly and effective app that meets the needs and goals of the users and stakeholders involved.

CONCLUSION

Mobile app design is a critical aspect of creating successful mobile applications. The mobile app development process involves a systematic approach to problem-solving that is used to develop effective solutions for a wide range of design challenges. This process typically includes researching user needs and preferences, analyzing market trends and competitors, ideating potential app features and functionalities, creating and testing prototypes with users, and refining the design solution based on feedback. Throughout the mobile app development process, it's important to maintain a user-centered approach and focus on designing solutions that meet the needs and goals of the users and stakeholders involved. Collaboration and communication with team members and stakeholders are also key to a successful mobile app development process. Ultimately, the goal of the mobile app development process is to develop an app that is user-friendly, visually appealing, and provides a seamless user experience. By following a structured and iterative design process, mobile app developers can create effective solutions that meet the needs of their target audience and achieve their business goals.

REFERENCES

Feng, F. (2012). Research on Android architecture and application development [J] [theoretical Edition]. *Information and Computation*, (10), 55–56.

Google. (2013, August 12). Android Developers. http://developer.android.com

Zhao, J. (2017). Research on Android system architecture and application development. *Automation and instrumentation*, (01), 86-87 + 90.

Lee, H., & Chuvyrov, E. (2012). Beginning Windows Phone App Development. *Apress*. https://appinventiv.com/blog/mobile-app-architecture-explained/

Microsoft. (2013, June 16). *Visual Studio Resources*. Microsoft. https://msdn.microsoft.com/en-us/vstudio/cc136611

The Mobile App Architecture Guide. (n.d.). Net Solutions. https://www.netsolutions.com/insights/mobile-app-architecture-guide/

Yao, Y. & Liu, W. (2008). *Research on Android architecture and application development.* Computer system application.

Zhang, L. (2017). Research on application development based on Android platform architecture. *Computer programming skills and maintenance*, (06), 52-53 + 85.

Zhang, Z. (2016). Research on Android architecture and application development. *Inner Mongolia Science and technology and economy,* (17), 72-73 + 78.

Compilation of References

Abbas, N., Gravell, A. M., & Wills, G. B. (2008). Historical roots of agile methods: Where did "agile thinking" come from? *In International conference on agile processes and extreme programming in software engineering,* (pp. 94-103). Springer.

Abbasi, A. A., Abbasi, A., Shamshirband, S., & Theodore, A. (2019). *Software-defined Cloud Computing : A Systematic Review on Latest Trends and Developments. XX*(Dc). IEEE. doi:10.1109/ACCESS.2019.2927822

Abdulhussein, A. A., & Raheem, F. A. (2020). Hand gesture recognition of static letters American Sign Language (ASL) using deep learning. *Engineering and Technology Journal, 38*(6), 926–937. doi:10.30684/etj.v38i6A.533

Abhilasha, S. A. (2013). Test effort estimation in regression testing, in: Innovation and Technology in Education (MITE). *IEEE International Conference in MOOC,* (pp. 343–348). IEEE.

Abiyev, R. H., Arslan, M., & Idoko, J. B. (2020). Sign language translation using deep convolutional neural networks. [TIIS]. *Transactions on Internet and Information Systems (Seoul), 14*(2), 631–653.

Abraham, C., & Michie, S. (2008). A taxonomy of behavior change techniques used in interventions. *Health Psychology, 27*(3), 379–387. doi:10.1037/0278-6133.27.3.379 PMID:18624603

Acerbis, R., Bongio, A., Butti, S., & Brambilla, M. (2015). Model-driven development of cross-platform mobile applications with WebRatio and IFML. *Proc. MOBILESoft,* (pp. 170-171).

Aditama, P. W., & Setiawan, I. N. A. F. (2020, February). Indigenous bali on augmented reality as a creative solution in industrial revolution 4.0. *Journal of Physics: Conference Series, 1471*(1), 012008. doi:10.1088/1742-6596/1471/1/012008

Ahirwar, K. (2019). *Generative Adversarial Networks Projects, 18,* 45. Packt Publishing Ltd.

Ahmed, M. A., Butt, W. H., Ahsan, I., Anwar, M. W., Latif, M., & Azam, F. (2017). A novel natural language processing (NLP) approach to automatically generate conceptual class model from initial software requirements. Proc. 8th Int. Conf. Inf. Sci. Appl. (ICISA). IEEE.

Ahsan, I., Butt, W. H., Ahmed, M. A., & Anwar, M. W. (2017). A comprehensive investigation of natural language processing techniques and tools to generate automated test cases. *Proc. 2nd Int. Conf. Internet Things Data Cloud Comput. (ICC).* IEEE.

Akhil, P., Akshara, R., Athira, R., Kamalesh Kumar, S. P., Thamotharan, M., & Shobha Christila, S. (2022). Smart Blind Walking Stick with Integrated Sensor. *The 2nd International Conference on Innovative Research in Renewable Energy Technologies (IRRET 2022)*. https://doi.org/10.3390/materproc2022010012

Aladaileh, M. A., Anbar, M., Hintaw, A. J., Hasbullah, I. H., Bahashwan, A. A., & Al-Sarawi, S. (2022). Renyi Joint Entropy-Based Dynamic Threshold Approach to Detect DDoS Attacks against SDN Controller with Various Traffic Rates. *Applied Sciences (Switzerland)*, *12*(12), 6127. doi:10.3390/app12126127

Alakwaa, W., Nassef, M., & Badr, A. (2017). Lung cancer detection and classification with 3D convolutional neural network (3D-CNN). *International Journal of Advanced Computer Science and Applications*, *8*(8). doi:10.14569/IJACSA.2017.080853

Alali, A., Kagdi, H., & Maletic, J. (2008). What's a Typical Commit? A Characterization of Open Source Software Repositories. *Proc. the 16th IEEE Int'l Conf. Program Comprehension (ICPC'08)*, (pp. 182-191). IEEE. 10.1109/ICPC.2008.24

Alberti-Fidanza, A.; Fidanza, F.; Chiuchiu, M.; Verducci, G.; Fruttini, D. (1999). Dietary studies on two rural Italian population groups of the Seven Countries Study, Trend of food and nutrient intake from 1960 to 1991. *Eur. J. Clin. Nutr.*, *53*, 854–860.

Alderfer, C.P. (1972). *Existence, Relatedness, and Growth: Human Needs in Organizational Settings*. APA.

Al-Fahoum. (2013). A Smart Infrared Microcontroller-Based Blind Guidance System. *Active and Passive Electronic Components*, *2013*, 1–7. doi:10.1155/2013/726480

Al-haddad, R., & Velazquez, E. S. (n.d.). *A Survey of Quality of Service (QoS) Protocols and Software-De fi ned Networks (SDN)* (Vol. 1). Springer International Publishing. doi:10.1007/978-3-030-01177-2

Al-Hammadi, M., Muhammad, G., Abdul, W., Alsulaiman, M., Bencherif, M. A., & Mekhtiche, M. A. (2020). Hand Gesture Recognition for Sign Language Using 3DCNN. *IEEE Access: Practical Innovations, Open Solutions*, *8*, 79491–79509.

Al, M., Adhami, H., Eddin, A., & Mohamed, A. (2022). Toward integrating software defined networks with the Internet of Things : A review. *Cluster Computing*, *25*(3), 1619–1636. doi:10.100710586-021-03402-4 PMID:34512120

Alnasser, A., Kyle, J., Aloumi, N., Al-Khalifa, A., & Marais, D. (2019). The Twazon Arabic Weight Loss App: App-Based Intervention for Saudi Women with Obesity. *JMIR mHealth and uHealth*, *7*(5), e10923.

Alshawmar, M. (2021). A Review of the Applications of Affordance Theory in Mhealth App Research. *Proceedings of the 54th Hawaii International Conference on System Sciences*, (pp. 3595). Scholar Space. 10.24251/HICSS.2021.437

Althar, R. R., & Samanta, D. (2022). BERT-Based Secure and Smart Management System for Processing Software Development Requirements from Security Perspective. In *Machine Intelligence and Data Science Applications: Proceedings of MIDAS 2021*, , (pp. 427–440). Springer Nature. 10.1007/978-981-19-2347-0_34

Amaliya, S., Handayani, A. N., Akbar, M. I., Herwanto, H. W., Fukuda, O., & Kurniawan, W. C. (2021). Study on Hand Keypoint Framework for Sign Language Recognition. *7th International Conference on Electrical, Electronics and Information Engineering (ICEEIE),* (pp. 446~451). IEEE.

Ambriola, V., & Gervasi, V. (2006). On the systematic analysis of natural language requirements with circe. *Automated Software Engineering*, *13*(1), 107–167.

Amreen, S., Mockus, A., Zaretzki, R., Bogart, C., & Zhang, Y. (2020). ALFAA: Active learning fingerprint based anti-aliasing for correcting developer identity errors in version control systems. *Empirical Software Engineering*, *25*(2), 1–32. doi:10.100710664-019-09786-7

Ansari, A., Shagufta, M. B., Fatima, A. S., & Tehreem, S. (2017). Constructing test cases using natural language processing. *Proc. IEEE 3rd Int. Conf. Adv. Elect. Electron. Inf. Commun. Bio-Inform. (AEEICB),* (pp. 95-99). IEEE.

Antonelli, H., Silva, E., & Fortes, R. (2015). A model-driven development for creating accessible Web menus. Proc. DSAI Conf. ACM.

Antoniou, C., Dimitriou, L., & Pereira, F. B. (2019). *Mobility patterns, big data, and transport analytics: tools and applications for modeling*. Elsevier.

Anwar, M. W., Rashid, M., Azam, F., Kashif, M., & Butt, W. H. (2019, December). A model-driven framework for design and verification of embedded systems through SystemVerilog. *Design Automation for Embedded Systems*, *23*(4), 179.

An, X., & Zhao, J. G. (2018). Design of self-help guide system for intelligent tourism. *Industrial & Science Tribune*, *17*(13), 77–78.

Apeltsin, L. (2021). *Data Science Bookcamp*. Simon and Schuster.

Arafat, O., & Riehle, D. (2009). The Commit Size Distribution of Open Source Software", Proc. the 42nd Hawaii Int'l Conf. Syst. Sci. (HICSS'09), pp. 1-8, 2009.

Arafat, O., & Riehle, D. "The Commit Size Distribution of Open Source Software. *Proceedings of the 42nd Hawaii International Conference on Systems Science (HICSS'09)*, (pp. 1-8). IEEE.

Armato, S. G. II, Li, F., Giger, M. L., MacMahon, H., Sone, S., & Doi, K. (2002). Lung cancer: Performance of automated lung nodule detection applied to cancers missed in a CT screening program. *Radiology*, *225*(3), 685–692. doi:10.1148/radiol.2253011376 PMID:12461246

Arnold, U., Essig, M., & Kemper, H.-G. (2001). ìTechnologische Entwicklungen im mobilen Internet und ihre R¸ckwirkungen auf die Unternehmensstrategie,î. In A. T. Nicolai, T. Petersmann (eds.) Strategien im M-Commerce. Stuttgart.

Arseniev, D. G. (2023). Software Engineering Principles Apply to Artificial Intelligence Systems. In *Cyber-Physical Systems and Control II* (pp. 151–158). Springer International Publishing. doi:10.1007/978-3-031-20875-1_14

Arzenšek, B., & Heričko, M. (2014). Criteria for selecting mobile application testing tools. *CEUR Workshop Proceedings*, 1–8.

Ashiq, FAsif, MAhmad, M. BZafar, SMasood, KMahmood, TMahmood, M.TLee, I. H. (2022). CNN-Based Object Recognition and Tracking System to Assist Visually Impaired People. *IEEE Access: Practical Innovations, Open Solutions*, *10*, 14819–14834. https://ieeexplore.ieee.org/document/9698080. doi:10.1109/ACCESS.2022.3148036

Ashishdeep, A., Bhatia, J., & Varma, K. (2016). Software process models for mobile application development: A review. *Computer Science and Electronic Journal*, *7*(1), 150–153.

Atlassian. (n.d.). *What Is Version Control: Atlassian Git Tutorial.* Atlassian. https://www.atlassian.com/git/tutorials/what-is-version-control

Aviv, I. T. Z. H. A. K., Barger, A., Kofman, A., & Weisfeld, R. (2023). Reference Architecture for Blockchain-Native Distributed Information System. *IEEE Access: Practical Innovations, Open Solutions*, *11*, 4838–4851. doi:10.1109/ACCESS.2023.3235838

Bagassi, S., De Crescenzio, F., & Piastra, S. (2020). Augmented reality technology selection based on integrated QFD-AHP model. *International Journal on Interactive Design and Manufacturing*, *14*(1), 285–294. doi:10.100712008-019-00583-6

Bancora, M., Brambilla, M., Ripamonti, D., & Vaccarella, A. (2015). Personal social and event organization through Web and mobile apps: The fluxedo case. *Proc. ICWE Eng. Web Big Data Era*, (pp. 609-612). IEEE.

Bansal, D., Ravi, P., So, M., Agrawal, P., Chadha, I., Murugappan, G., & Duke, C. (2021). Copycat: Using sign language recognition to help deaf children acquire language skills. In *Extended Abstracts of the 2021 CHI Conference on Human Factors in Computing Systems*, (pp. 1~10). CHI.

Barbareschi, M. (2022, December). Scrum for Safety: An Agile Methodology for Safety-Critical Software Systems. *Software Quality Journal*, *30*(4), 1067–1088.

Bari, A., Chaouchi, M., & Jung, T. (2017). *Predictive analytics for dummies*. John Wiley,and Sons.

BarreraJ. M. (2023). An Extension of Istar for Machine Learning Requirements by Following the PRISE Methodology. SSRN 4358075. doi:10.2139/ssrn.4358075

Bass, L., Clements, P., & Kazman, R. (2003). *Software architecture in practice*. Addison-Wesley Professional.

BeheraT. (2022). How Blockchain Solves the Supply Chain Problems Using RFID Techniques. SSRN: https://ssrn.com/abstract=4287240 or doi:10.2139/ssrn.4287240

Behera, T., & Panda, B. S. (2023). *Master Data Management using Machine Learning Techniques: MDM Bot*. TechRxiv., doi:10.36227/techrxiv.21818040.v1

Behera, T., & Tripathi, K. (2022). *Root Cause Analysis Bot using Machine Learning Techniques*. TechRxiv., doi:10.36227/techrxiv.21588159.v2

Bernaschina, C. (2018). *ALMOsT.js: An agile model to model and model to text transformation framework*. Proc. Web Eng. ICWE.

Beysolow, T. (2018). *Applied natural language processing with Python: Implementing Machine Learning and Deep Learning Algorithms for Natural Language Processing*. Ca Apress. doi:10.1007/978-1-4842-3733-5

Bhanipati, J., Singh, D., Biswal, A. K., & Rout, S. K. (2021). Minimization of Collision Through Retransmission and Optimal Power Allocation in Wireless Sensor Networks (WSNs). In *Advances in Intelligent Computing and Communication* (pp. 653–665). Springer. doi:10.1007/978-981-16-0695-3_61

Bhardwaj, S.,(2021). A Study on Human Emotion Detection.. *2021 International Conference on Advance Computing and Innovative Technologies in Engineering (ICACITE)*, (pp. 909–13). IEEE Xplore

Bhardwaj, S., & Panda, S. N. (2022). Performance Evaluation Using RYU SDN Controller in Software-Defined Networking Environment. *Wireless Personal Communications*, *122*(1), 701–723. doi:10.100711277-021-08920-3

Bird, C., Gourley, A., Devanbu, P., Gertz, M., & Swaminathan, A. (2006). Mining email social networks. *Proceedings of the 2006 International Workshop on Mining Software Repositories*, (pp. 137-143). ACM. 10.1145/1137983.1138016

Biswal, S. P., & Patel, S. (2022). Introduction to Software Defined Networking. *Software Defined Networks*, (pp. 1–27). doi:10.1002/9781119857921.ch1

Biswal, A. K., Singh, D., & Pattanayak, B. K. (2021). IoT-Based Voice-Controlled Energy-Efficient Intelligent Traffic and Street Light Monitoring System. In *Green Technology for Smart City and Society* (pp. 43–54). Springer. doi:10.1007/978-981-15-8218-9_4

Biswal, A. K., Singh, D., Pattanayak, B. K., Samanta, D., Chaudhry, S. A., & Irshad, A. (2021). Adaptive fault-tolerant system and optimal power allocation for smart vehicles in smart cities using controller area network. *Security and Communication Networks*, *2021*, 2021. doi:10.1155/2021/2147958

Biswal, A. K., Singh, D., Pattanayak, B. K., Samanta, D., & Yang, M. H. (2021). IoT-Based Smart Alert System for Drowsy Driver Detection. *Wireless Communications and Mobile Computing*, *2021*, 2021. doi:10.1155/2021/6627217

Blanco-Silva, F. J. (2015). *Mastering SciPy*. Packt Publishing Ltd.

Bogomolova, K., van der Ham, I. J., Dankbaar, M. E., van den Broek, W. W., Hovius, S. E., van der Hage, J. A., & Hierck, B. P. (2020). The effect of stereoscopic augmented reality visualization on learning anatomy and the modifying effect of visual-spatial abilities: A double-center randomized controlled trial. *Anatomical Sciences Education, 13*(5), 558–567. doi:10.1002/ase.1941 PMID:31887792

Bok, V., & Langr, J. (2019). *GANs in Action, 34*, 40. Simon, and Schuster.

Borg, M., & Camilleri, K. P. (2019). Sign language detection "in the wild" with recurrent neural networks. *IEEE International Conference on Acoustics, Speech and Signal Processing (ICASSP)*. IEEE.

Bornholdt, S., & Schuster, H. G. (2006). *Handbook of Graphs and Networks*. John Wiley & Sons.

Bouguerra, F. (2021). Data centre networking: SDN fundamentals. *Canonical*. https://canonical.com/blog/data-centre-networking-sdn-fundamentals

Bragg, D., Koller, O., Bellard, M., Berke, L., Boudreault, P., Braffort, A., & Ringel Morris, M. (2019). Sign language recognition, generation, and translation: An interdisciplinary perspective. In *Proceedings of the 21st International ACM SIGACCESS Conference on Computers and Accessibility*, (pp. 16~31). ACM.

Brambilla, M., Koka, T., Mauri, A., & Umuhoza, E. (2017). Integrating modeling languages and Web logs for enhanced user behavior analytics. Proc. 26th Int. Conf. World Wide Web Companion, (pp. 171-175). IEEE.

Brambilla, M., Mauri, A., & Umuhoza, E. (2014). Extending the interaction flow modeling language (IFML) for model driven development of mobile applications front end. Proc. MobiWIS, (pp. 176-191). IEEE.

Brambilla, M., Mauri, A., Franzago, M., & Muccini, H. (2016). A model-based method for seamless Web and mobile experience. *Proc. 1st Int. Workshop Mobile Develop.*, (pp. 33-40). IEEE.

Brambilla, M., & Fraternali, P. (2014, September). Large-scale model-driven engineering of Web user interaction: The WebML and WebRatio experience. *Science of Computer Programming, 89*, 71–87.

Braun, W., & Menth, M. (2014). *Software-Defined Networking Using OpenFlow: Protocols, Applications and Architectural Design Choices*, 302–336. MDPI. doi:10.3390/fi6020302

Brenner, D. R., Boffetta, P., Duell, E. J., Bickeboeller, H., Rosenberger, A., McCormack, V., & Hung, R. J. (2012). Previous lung diseases and lung cancer risk: A pooled analysis from the International Lung Cancer Consortium. *American Journal of Epidemiology, 176*(7), 573–585. doi:10.1093/aje/kws151 PMID:22986146

Bressert, E. (2013). *SciPy and NumPy*. O'reilly.

Brewer, L. C., Kaihoi, B., Zarling, K. K., Squires, R. W., Thomas, R., & Kopecky, S. (2015). The use of virtual world-based cardiac rehabilitation to encourage healthy lifestyle choices among cardiac patients: Intervention development and pilot study protocol. *JMIR Research Protocols*, *4*(2), e39. doi:10.2196/resprot.4285 PMID:25857331

Bruce, P., & Bruce, A. (2017). *Practical Statistics for Data Scientists*. O'Reilly Media, Inc.

Burch, Z. A., & Mohammed, S. (2019). Exploring Faculty Perceptions about Classroom Technology Integration and Acceptance: A literature Review. The Richard W. Riley College of Education and Leadership Publications.

Burta, A., Szabo, R., & Gontean, A. (2020 July 27-28). *Object Recognition Development for Android Mobile Devices with Text-to-Speech Function Created for Visually Impaired People*. 2020 Fourth World Conference on Smart Trends in Systems, Security and Sustainability (WorldS4), London, UK. https://ieeexplore.ieee.org/document/9210381

Buschmann, F., Meunier, R., Rohnert, H., Sommerlad, P., & Stal, M. (1996). A System of Patterns. *Pattern-Oriented Software Architecture*, (*volume 1*).

Butor, M., & Howard, R. (2004). *Mobile*. Amazon, Dalkey Archive Press. https://aws.amazon.com/mobile/mobile-application-development/

Buysse, D. J., Reynolds, C. F. III, Monk, T. H., Berman, S. R., & Kupfer, D. J. (1989). The Pittsburgh Sleep Quality Index: A new instrument for psychiatric practice and research. *Psychiatry Research*, *28*(2), 193–213. doi:10.1016/0165-1781(89)90047-4 PMID:2748771

Cabaj, K., Gregorczyk, M., & Mazurczyk, W. (2018). Software-defined networking-based crypto ransomware detection using HTTP traffic characteristics. *Computers & Electrical Engineering*, *66*, 353–368. doi:10.1016/j.compeleceng.2017.10.012

Canuel, R., & Chrichton, C. (2015). Leveraging apps for research and learning: A survey of Canadian academic libraries. *Library Hi Tech*, *33*(1), 2–14. doi:10.1108/LHT-12-2014-0115

Canuel, R., MacKenzie, E., Senior, A., & Torabi, N. (2016). Apps for academic success: Developing digital literacy and awareness to increase usage. *Education for Information*, *33*(3), 1–18. doi:10.3233/EFI-160085

Cao, G., Yang, J., Zhou, Q., & Chen, W. (2012). *Software Testing Strategy for Mobile Phone, Advances and Applications in Mobile Computing*. InTech.

Cao, L., Mohan, K., Xu, P., & Ramesh, B. (2009). A framework for adapting agile development methodologies. *European Journal of Information Systems*, *18*(4), 332–343.

Carlos, A. (2012). Research on the go: Mobile tools for conducting research. *The Reference Librarian*, *53*(4), 433–440. doi:10.1080/02763877.2012.707462

Carlucho, I. (2017). Incremental Q-Learning Strategy for Adaptive PID Control of Mobile Robots. [ScienceDirect, https://doi.org/10.1016/j.eswa.2017.03.002.]. *Expert Systems with Applications*, *80*(Sept), 183–199.

Carter, A., Liddle, J., Hall, W., & Chenery, H. (2015). Mobile phones in research and treatment: Ethical guidelines and future directions. *JMIR mHealth and uHealth, 3*(4), 1. doi:10.2196/mhealth.4538 PMID:26474545

Carter, B. A., Iles, G. N., Raju, R., Afful, A. M., Maj, R., Dao, T., Terkildsen, M., Lobzin, V., Bouya, Z., Parkinson, M., Le May, S., Choy, S., Hordyniec, P., Hordyniec, B., Currie, J., Skov, T., & Peake, I. D. (2022). RMIT University's practical space weather prediction laboratory. *Journal of Space Weather and Space Climate, 12*, 28. doi:10.1051wsc/2022025

Chae, M., & Kim, J. (2001). Information Quality for Mobile Internet Services: A Theoretical Model with Empirical Validation. In V. Storey, S. Sarkar, and J. I. DeGross (eds.), *Proceedings of the Twenty-Second International Conference on Information Systems*, (pp. 43-54). IEEE.

Charland, A., & Leroux, B. (2011). mobile application Development : Web vs. native. *Communications of the ACM.*

Charlton, M. (2022). *All Stable Diffusion Artwork, 22*, 34. Independently Published.

Chaudary, B., Pohjolainen, S., & Aziz, S. (2023). Teleguidance-based remote navigation assistance for visually impaired and blind people—Usability and user experience. *Virtual Reality (Waltham Cross), 27*, 141–158. doi:10.100710055-021-00536-z

Chen, B., Seilhamer, R., Bennet, L., & Bauer, S. (2015). *Students' mobile learning practices in higher education: A multi-year study.* Educause Review. https://er.educause.edu/articles/2015/6/students-mobile-learning-practices-in-higher-education-a-multiyear-study

Chen, B., Sivo, S., Seilhamer, R., Sugar, A., & Mao, J. (2013). User acceptance of mobile technology: A campus-wide implementation of blackboard's MobileTM learn application. *Journal of Educational Computing Research, 49*(3), 327–343. doi:10.2190/EC.49.3.c

Chen, D. Y. (2018). *Pandas for everyone: Python data analysis.* Addison-Wesley.

Chen, G. (2011). Mobile research: Benefits, applications and outlooks. Internationalization. *Design and Global Development, 1*, 11–16.

Chhikara, B. S., & Parang, K. (2022). Global Cancer Statistics 2022: The trends projection analysis. *Chemical Biology Letters, 10*(1), 451–451.

Chi, A., Yik, C., & Yuen, K. (2017). *Matplotlib 2.x by example: multi-dimensional charts, graphs, and plots.* Packt Publishing Ltd.

Chinnery, G. (2006). Going to the MALL: Mobile assisted language learning. *Language Learning & Technology, 10*, 9–16.

Chon, A., Balachandar, N., & Lu, P. (2017). Deep convolutional neural networks for lung cancer detection. Standford University, 1-9.

Chong, T.-W., & Lee, B. G. (2018). American Sign Language Recognition Using Leap Motion Controller with ML Approach. *Sensors (Basel), 18*, 3554.

Claypool, M., & Claypool, K. (2006). Latency and Player Actions in Online Games. *Communications of the ACM*, *49*(11), 40–45. doi:10.1145/1167838.1167860

Clickard, C. (2019). *The internet of things*. Norwood House Press.

CodePlex Archive. (2019). CodePlex Archive. *Codeplex.* https://sharpnlp.codeplex.com/

Cohen, R., Tao Wang, & Springerlink. (2014). *Android Application Development for the Intel Platform.* Apress.

Cohen, J. (1960). A coefficient of agreement for nominal scales. *Educational and Psychological Measurement*, *20*(1), 37–46. doi:10.1177/001316446002000104

Colley, M., Rädler, M., Glimmann, J., & Rukzio, E. (2022). Effects of scene detection, scene prediction, and maneuver planning visualizations on trust, situation awareness, and cognitive load in highly automated vehicles. *Proceedings of the ACM on Interactive, Mobile, Wearable and Ubiquitous Technologies*, *6*(2), 1–21. doi:10.1145/3534609

Collins, R. (2018). *Scikit-Learn in Details*. Independently Published.

Compton, H., & Burke, D. (2018). The use of mobile learning in higher education: A systematic review. *Computers & Education*, *123*, 53–64. doi:10.1016/j.compedu.2018.04.007

Cooper, H. (2012). EngJon Ong, Nicolas Pugeault, and Richard Bowden. Sign language recognition using subunits. *Journal of Machine Learning Research*, *13*, 2205–2231.

Cross, R. (2015). *Principal component analysis*. Clanrye Intl.

Cruz, M. (2022). *A Model-Based Approach for Specifying Changes in Replications of Empirical Studies in Computer Science*. Springer. doi:10.1007/s00607-022-01133-x

CSDL. (n.d.).| *IEEE Computer Society*. IEEE. https://www.computer.org/csdl/journal/ta.

da Costa, S. S. L., Neto, V. V. G., & de Oliveira, J. L. (2014). A user interface stereotype to build Web portals. *Proc. 9th Latin Amer. Web Congr*. IEEE.

Das Gupta, R., Samanta, D., & Joseph, N. P. (2022). Rendering View of Kitchen Design Using Autodesk 3Ds Max. In *Micro-Electronics and Telecommunication Engineering: Proceedings of 5th ICMETE 2021.* Springer Nature. 10.1007/978-981-16-8721-1_33

Davari, S. (2022, March). [DC] Context-Aware Inference and Adaptation in Augmented Reality. In *2022 IEEE Conference on Virtual Reality and 3D User Interfaces Abstracts and Workshops (VRW)* (pp. 938-939). IEEE.

Davidson, J., Paulus, T., & Jackson, K. (2016). Speculating on the future of digital tools for qualitative research. *Qualitative Inquiry*, *22*(7), 606–610. doi:10.1177/1077800415622505

DeepSource. (n.d.). *The Modern Static Analysis Platform.* DeepSource, https://deepsource.io/glossary/incremental-development/

Delgado, A., Calegari, D., & Arrigoni, A. (2016, December). Towards a generic BPMS user portal definition for the execution of business processes. *Electronic Notes in Theoretical Computer Science*, *329*, 39–59.

Dennison, L., Morrison, L., Conway, G., & Yardly, L. (2013). Opportunities and challenges for smartphone applications in supporting health behavior change: Qualitative study. *Journal of Medical Internet Research*, *15*(4), e86. doi:10.2196/jmir.2583

Dey, T. (2011). A Comparative Analysis on Modeling and Implementing with MVC Architecture. *International Journal of Computers and Applications*, *1*(1), 44–49.

Dick, M., Wellnitz, O., & Wolf, L. (2005). Analysis of factors affecting players' performance and perception in multiplayer games. ACM SIGCOMM workshop on Network and System Support for Games. NetGames.

Dinov, I. D. (2018). *Springerlink, Online Service. Data Science and Predictive Analytics: Biomedical and Health Applications using R*. Springer International Publishing. doi:10.1007/978-3-319-72347-1

Du, C., Zhang, L., Sun, X., Wang, J., & Sheng, J. (2020). Enhanced Multi-Channel Feature Synthesis for Hand Gesture Recognition Based on CNN with a Channel and Spatial Attention Mechanism. *IEEE Access: Practical Innovations, Open Solutions*, *8*, 144610–144620.

Dutta, S., Das, A. K., Ghosh, S., & Samanta, D. (2022). Data Analytics for Social Microblogging Platforms. Elsevier.

EDUCAUSE. (2019). Horizon Report Preview: 2019 Higher Education Edition. *Educause*. https://library.educause.edu/-/media/files/library/2019/2/2019horizonreportpreview.pdf

EDUCAUSE. (2019). Horizon Report Preview: 2019 Higher Education Edition. EDUCAUSE. https://library.educause.edu/-/media/files/library/2019/2/2019horizonreportpreview.pdf

Egger, M., Ley, M., & Hanke, S. (2019). Emotion Recognition from Physiological Signal Analysis: A Review. [ScienceDirect]. *Electronic Notes in Theoretical Computer Science*, *343*(May), 35–55. doi:10.1016/j.entcs.2019.04.009

Elmannai, W., & Elleithy, K. (2017). Sensor-Based Assistive Devices for Visually-Impaired People: Current Status, Challenges, and Future Directions. *Sensors (Basel)*, *17*(3), 565. doi:10.339017030565

Elmezain, M., Al-Hamadi, A., Appenrodt, J., & Michaelis, B. (2008). A hidden markov model-based continuous gesture recognition system for hand motion trajectory. In *2008 19th international conference on pattern recognition*, .(pp. 1~4). IEEE.

Elomaa, T. & Helsinki, E. (2002). *Machine learning: proceedings*. Springer, C.

Enterprise Architecture As Strategy. (1999). *Creating a Foundation for Business Execution, Fahim Sachedina, Everything There Is To Know About Automated Testing Fowler, Martin. Refactoring: Improving the Design of Existing Code*. Addison-Wesley.

Escalona, M. J., & Koch, N. (2003, February). Requirements engineering for Web applications—A comparative study. *Journal of Web Engineering*, *2*, 193–212.

Eysenbach, G. (2011). Consort-EHEALTH Group. CONSORT-EHEALTH: Improving and standardizing evaluation reports of Web-based and mobile health interventions. *Journal of Medical Internet Research*, *13*(4), e126. doi:10.2196/jmir.1923 PMID:22209829

Fanfeng, Z. (2010 November 4-6). *Application Research of Voice Control in Reading Assistive Device for Visually Impaired Persons.* 2010 International Conference on Multimedia Information Networking and Security, Nanjing, China. https://ieeexplore.ieee.org/document/5670908

Fang, X., & Zhan, J. (2015, June). Sentiment Analysis Using Product Review Data. [BioMed Central]. *Journal of Big Data*, *2*(1), 5. doi:10.118640537-015-0015-2

Farooq, U., Rahim, M. S. M., Sabir, N., Hussain, A., & Abid, A. (2021). Advances in machine translation for sign language: Approaches, limitations, and challenges. *Neural Computing & Applications*, *33*(21), 14357–14399.

Feng, F. (2012). Research on Android architecture and application development [J] [theoretical Edition]. *Information and Computation*, (10), 55–56.

Feng, W.-C., Chang, F., Feng, W.-C., & Walpole, J. (2005, June). A traffic characterization of popular on-line games. *IEEE/ACM Transactions on Networking*, *13*(3), 588–500.

Flora, H. K., Wang, X., & Chande, S. V. (2014). An investigation into mobile application development processes: Challenges and best practices. *International Journal of Modern Education and Computer Science*, *6*(6), 1.

Fontichiaro, K. (2018). *Big Data*, *10*, 20. Cherry Lake Publishing.

Forastiere, M., de Pietro, G., & Sannino, G. (2016). An mHealth application for a personalized monitoring of one's own wellness: Design and development. In *Innovation in Medicine and Healthcare 2016* (pp. 269–278). Springer. doi:10.1007/978-3-319-39687-3_26

Fortino, G., Galzarano, S., Gravina, R., & Li, W. (2015). A framework for collaborative computing and multi-sensor data fusion in body sensor networks. *Information Fusion*, *22*, 50–70. doi:10.1016/j.inffus.2014.03.005

Fortino, G., Giannantonio, R., Gravina, R., Kuryloski, P., & Jafari, R. (2013). Enabling effective programming and ñexible management of efðcient body sensor network applications. *IEEE Transactions on Human-Machine Systems*, *43*(1), 115–133. doi:10.1109/TSMCC.2012.2215852

Fosnacht, K., Sarraf, S., Howe, E., & Peck, L. K. (2017). How important are high response rates for college surveys? *The Review of Higher Education*, *40*(2), 245–265. doi:10.1353/rhe.2017.0003

Frajtak, K., Bures, M., & Jelınek, I. (2015). Transformation of IFML schemas to automated tests. Proc. Conf. Res. Adapt. Convergent Syst. (RACS). IEEE.

Francino, Y. (2011). *"What Is Iterative Development? - Definition from Whatis.com." SearchSoftwareQuality.* TechTarget. https://www.techtarget.com/searchsoftwarequality/definition/iterative-development

Francis, P. (2017). *Mobile App Development Process - an in-Depth Overview.* The BHW Group. https://thebhwgroup.com/blog/mobile-app-development-process

Friedenthal, S. (2015). *A practical guide to SysML: The systems modeling language.* Morgan Kaufman.

Fukuda, T., Yokoi, K., Yabuki, N., & Motamedi, A. (2019). An indoor thermal environment design system for renovation using augmented reality. *Journal of Computational Design and Engineering*, *6*(2), 179–188. doi:10.1016/j.jcde.2018.05.007

Galanek, J. D., & Gierdowski, D. C. (2019). *ECAR Study of Faculty and Information Technology, 2019. Research report.* ECAR. https://www.educause.edu/-/media/files/library/2019/12/facultystudy2019.pdf?la=en&hash=F0F8AF83BD870239A7E5CD285FBDBE157AA51135

Galanek, J. D., Gierdowski, D. C., & Brooks, D. C. (2018). *ECAR study of undergraduate students and information technology, 2018. Research report.* ECAR. https://library.educause.edu/-/media/files/library/2018/10/studentitstudy2018.pdf?la=en&hash=C590C1F6C62B77792711BFAC1F642254A5618590

Garcia, B., Welford, J., & Smith, B. (2016). Using a smartphone app in qualitative research: The good, the bad and the ugly. *Qualitative Research*, *16*(5), 508–525. doi:10.1177/1468794115593335

Garg, M., Wazarkar, S., Singh, M., & Bojar, O. (2022, June). Multimodality for NLP-Centered Applications: Resources, Advances and Frontiers. *In Proceedings of the Thirteenth Language Resources and Evaluation Conference* (pp. 6837-6847). IEEE.

Garreta, R. & Moncecchi, G. (2013). *Learning scikit-learn: Machine learning in Python: Experience the benefits of machine learning techniques by applying them to real-world problems using Python and the open source scikit-learn library.* Packt Publishing Ltd.

Ghasemzadeh, H., Panuccio, P., Trovato, S., Fortino, G., & Jafari, R. (2014). Power-aware activity monitoring using distributed wearable sensors. *IEEE Transactions on Human-Machine Systems*, *44*(4), 537–544. doi:10.1109/THMS.2014.2320277

Gierdowski, D. C., Brooks, D. C., & Galanek, J. D. (2020) *EDUCAUSE 2020: Student Technology Report: Supporting the Whole Student.* CO: ECAR. https://www.educause.edu/ecar/research-publications/student-technology-report-supporting-the-whole-student/2020/technology-use-and-environmental-preferences

Gierdowski, D. C., Brooks, D. C., & Galanek, J. D. (2020). *EDUCAUSE 2020 Student Technology Report: Supporting the Whole Student, Research report.* CO: ECAR. https://www.educause.edu/ecar/research-publications/student-technology-report-supporting-the-whole-student/2020/technology-use-and-environmental-preferences

Gikas, J., & Grant, M. M. (2013). Mobile computing devices in higher education: Student perspectives on learning with cellphones, smartphones & social media. *The Internet and Higher Education, 19*, 18–26. doi:10.1016/j.iheduc.2013.06.002

Gilhousen, K. S., Jacobs, I. M., Padovani, R., Viterbi, A. J., Weaver, L. A., & Wheatley, C. E. (1991, May). On the capacity of a cellular CDMA system. *IEEE Transactions on Vehicular Technology, 40*(2), 303–312. doi:10.1109/25.289411

Giri, S. (2022).Emotion Detection with Facial Feature Recognition Using CNN & OpenCV. *2022 2nd International Conference on Advance Computing and Innovative Technologies in Engineering (ICACITE),* pp. 230–32. IEEE Xplore 10.1109/ICACITE53722.2022.9823786

GitHub. (n.d.). [Code Sharing]. GitHub. http://github.com

Goeminne, M., & Mens, T. (2013). A comparison of identity merge algorithms for software repositories. *Science of Computer Programming, 78*(8), 971–986. doi:10.1016/j.scico.2011.11.004

Goerzen, J. (2004). *Foundations of Python Network Programming*. Apress. doi:10.1007/978-1-4302-0752-8

Goldberg, Y. (2017). *Neural network methods in natural language processing*. Morgan & Claypool Publishers. doi:10.1007/978-3-031-02165-7

Gonzalez Vargas, J. C., Fabregat, R., Carrillo-Ramos, A., & Jové, T. (2020). Survey: Using augmented reality to improve learning motivation in cultural heritage studies. *Applied Sciences (Basel, Switzerland), 10*(3), 897. doi:10.3390/app10030897

Goodfellow, I., Bengio, Y., & Courville, A. (2016). Deep learning, (p. 44-90). The MIT Press. https://www.deeplearningbook.org/

Google. (2013, August 12). Android Developers. http://developer.android.com

Gote, C., Scholtes, I., & Schweitzer, F. (2019). git2net - Mining time-stamped co-editing networks from large git repositories. *2019 IEEE/ACM 16th International Conference on Mining Software Repositories (MSR),* (pp. 433-444). ACM, IEEE. 10.1109/MSR.2019.00070

Gotti, Z., & Mbarki, S. (2016). GUI structure and behavior from Java source code analysis. *Proc. IEEE Int. Colloq. Inf. Sci. Technol. (CiSt),* (pp. 251-256). IEEE.

Gravina, R., Alinia, P., Ghasemzadeh, H., & Fortino, G. (2017). Multi-sensor fusion in body sensor networks:State-of-the-art and research challenges. *Information Fusion, 35*, 68–80. doi:10.1016/j.inffus.2016.09.005

Grolemund, G. (2015). *Data Science withR*. O'Reilly& Associates Inc.

Gu, C. (2022, October). What Do Users Care about? Research on User Behavior of Mobile Interactive Video Advertising. [ScienceDirect, https://doi.org/10.1016/j.heliyon.2022.e10910.]. *Heliyon, 8*(10), e10910.

Guha, A., Obaidat, M. S., Samanta, D., & Islam, S. K. H. (2022). Clustering-based Optimal Resource Allocation Strategy in Title Insurance Underwriting. In *2022 International Conference on Computer, Information and Telecommunication Systems* (CITS), (pp. 1–5). IEEE. 10.1109/CITS55221.2022.9832993

Guleria, A. (2017). *Traffic Engineering in Software Defined Networks : A Survey. January.* Semantic Scholar.

Gupta, N., Maashi, M. S., Tanwar, S., Badotra, S., Aljebreen, M., &Bharany, S. (2022). *A Comparative Study of Software Defined Networking Controllers Using Mininet*. Research Gate.

Gupta, P. (2021). *Practical data science with Jupyter: Explore data cleaning, pre-processing, data wrangling, feature engineering and machine learning using Python and Jupyter*. Bpb Publications.

Gurney, K. (2018). *An Introduction to Neural Networks., 152*, 180. CRC Press.

H., D, & Pazzani, M. J. (1991*). Concept formation: knowledge and experience in unsupervised learning,* (pp. 4-12). Morgan Kaufmann.

Habibullah, K. M., Gay, G., & Horkoff, J. (2023). Non-functional requirements for machine learning: Understanding current use and challenges among practitioners. *Requirements Engineering*, 1–34. doi:10.100700766-022-00395-3

Hackeling, G. (2017). *Mastering machine learning with scikit-learn: Learning to implement and evaluate machine learning solutions with scikit-learn.* Packt Publishing Ltd.

Hagberg, A. A., Schult, D. A., & Swart, P. J. (2008). Exploring Network Structure, Dynamics, and Function using NetworkX. In G. Varoquaux, T. Vaught, & J. Millman (Eds.), *Proceedings of the 7th Python in Science Conference* (pp. 11–15).

Hahn, J. (2014). Undergraduate research support with optical character recognition apps. *RSR. Reference Services Review, 42*(2), 336–350. doi:10.1108/RSR-09-2013-0045

Haji, S. H., Zeebaree, S. R. M., Ameen, S. Y., & Shukur, H. (2021). Comparison of Software Defined Networking with Traditional Networking. *Comparison of Software Defined Networking with Traditional Networking.*, (June), 1–18. doi:10.9734/ajrcos/2021/v9i230216

Halpern, M. (2016). Mobile CPU's rise to power: quantifying the impact of generational mobile CPU design trends on performance, energy, and user satisfaction. *Proceedings of the 2016 IEEE international symposium on high performance computer architecture (HPCA).* IEEE.

Hamdani, M. (2018). A systematic literature review on interaction flow modeling language (IFML). *Proc. Int. Conf. Manage. Eng. Softw. Eng. Service Sci.*, (pp. 134-138). IEEE.

Hanbridge, A. S., Tin, T., & Sanderson, N. (2018). Information literacy skills on the go: Mobile learning innovation. *Journal of Information Literacy, 12*(1), 118–136. doi:10.11645/12.1.2322

Hands, D. S. (2004, December). A Basic Multimedia Quality Model. *IEEE Transactions on Multimedia, 6*(6), 806–816. doi:10.1109/TMM.2004.837233

Hao, J., & Ho, T. K. (2019). Machine Learning Made Easy: A Review of Scikit-learn Package in Python Programming Language. *Journal of Educational and Behavioral Statistics*, *107699861983224*(3), 348–361. doi:10.3102/1076998619832248

Harris, C. B., & Harris, I. G. (2015). Generating formal hardware verification properties from natural language documentation. *Proc. 9th Int. Conf. Semantic Comput. (IEEE ICSC),* (pp. 49-56). IEEE.

Hattori, L., & Lanza, M. (2008). On the nature of commits. Proc. the 4th Int'l ERCIM Wksp. Softw. Evol. and Evolvability (EVOL'08), (pp. 63-71). IEEE. doi:10.1109/ASEW.2008.4686322

Hatuwal, B. K., & Thapa, H. C. (2020). Lung cancer detection using convolutional neural network on histopathological images. *International Journal of Computer Trends and Technology*, *68*(10), 21–24. doi:10.14445/22312803/IJCTT-V68I10P104

Haw, M. A., MacDonald, M. E., & Colom, S. V. (2023). Big-data Efficient and Automated Science Transfer (BEAST): an open-source software architecture for arc jet data management, modeling, and automation. AIAA SCITECH 2023 Forum. ARC. 10.2514/6.2023-2712

He, C., & Mussbacher, G. (2016). Model-driven engineering and elicitation techniques: A systematic literature review. Proc. REW. IEEE.

Hennig, N. (2014). *Apps for librarians: Using the best mobile technology to educate, create and engage.* Libraries Unlimited. https://ebookcentral-proquest-com.ezproxy.waikato.ac.nz/lib/waikato/detail.action?docID=1813321

Higgins, J. P. (2016). Smartphone applications for patients' health and ðtness. *The American Journal of Medicine*, *129*(1), 11–19. doi:10.1016/j.amjmed.2015.05.038 PMID:26091764

Hindle, A., Germán, D., & Holt, R. (2008). What do large commits tell us?: a taxonomical study of large commits. *Proc. the 5th Int'l Working Conf. Mining Softw. Repos. (MSR'08),* (pp. 99-108). ACM. 10.1145/1370750.1370773

Hinton, G. (n.d.). Forward-Forward Algorithm: Some Preliminary Investigations.

Hinze, A., Vanderschantz, N., Timpany, C., Cunningham, S. J., Saravani, S. J., & Wilkinson, C. (2017a). *Use of mobile apps for teaching and research. (Working Paper Series 01/2017a).* Department of Computer Science, The University of Waikato.

Hofmann, P., & Riehle, D. (2009). Estimating Commit Sizes Efficiently. *Proc. the 5th IFIP WG 2.13 Int'l Conf. Open Source Systems (OSS'09),* (pp. 105-115). Springer. 10.1007/978-3-642-02032-2_11

Hohpe, G. (2020). *The Software Architect Elevator.* O'Reilly Media, Inc.

Honcharenko, W., Kruys, J. P., Lee, D. Y., & Shah, N. J. (1997, January). Broadband wireless access. *IEEE Communications Magazine*, *35*(1), 20–26. doi:10.1109/35.568192

Hooda, S. (Ed.). (2023). *Agile Software Development: Trends, Challenges and Applications*. Kinza Yasar, Software testing. https://www.techtarget.com/whatis/definition/software-testing

Horvat, M., ... "An Overview of Common Emotion Models in Computer Systems." 2022 45th Jubilee International Convention on Information, Communication and Electronic Technology (MIPRO), 2022, pp. 1008–13. IEEE Xplore 10.23919/MIPRO55190.2022.9803498

Hosseinzadeh, M. (2022, December). Clustering for Smart Cities in the Internet of Things: A Review. *Cluster Computing*, *25*(6), 4097–4127.

Hou, Q., Ma, Y., Chen, J., & Xu, Y. (2014). An Empirical Study on Inter-Commit Times in SVN. *Int. Conf. on Software Eng. and Knowledge Eng*, (pp. 132-137). Springer.

Huang, A., Pan, M., Zhang, T., & Li, X. (2018). Static extraction of IFML models for Android apps. *Proc. 21st ACM/IEEE Int. Conf. Model Driven Eng. Lang. Syst. Companion (MODELS)*, (pp. 53-54). ACM/IEEE.

Idris, I. (2011). *NumPy 1.5: Beginner's guide*. Packt Publishing.

Ieee, F., Rothenberg, C. E., Ieee, M., Azodolmolky, S., Ieee, S. M., Uhlig, S., & Ieee, M. (2015). Software-Defined Networking : A Comprehensive Survey. *Proceedings of the IEEE*, *103*(1), 14–76. doi:10.1109/JPROC.2014.2371999

Ilieva, M. G., & Ormandjieva, O. (2005). Automatic transition of natural language software requirements specification into formal presentation. *Proc. Int. Conf. Appl. Natural Lang. Inf. Syst. (NLDB)*, (vol. 3513, pp. 392-397). IEEE.

Islam, M. I., Raj, M. M. H., Nath, S., Rahman, M. F., Hossen, S., & Imam, M. H. (2018 April 6-8). *An Indoor Navigation System for Visually Impaired People Using a Path Finding Algorithm and a Wearable Cap*. 2018 3rd International Conference for Convergence in Technology (I2CT), Pune, India. https://ieeexplore.ieee.org/document/8529757

Jailia, M., Kumar, A., Agarwal, M., & Sinha, I. (2016). Behavior of MVC (Model View Controller) based web application developed in PHP and. NET framework. *2016 International Conference on ICT in Business Industry & Government (ICTBIG)*, 1–5.

Jenkins, C. D. (2015). Building Better Health: A Handbook of Behavioral Change. Pan American Health Organization. doi:10.2196/resprot.4008

Jeong, Y.-J., Lee, J.-H., & Shin, G.-S. (2008). Development process of mobile application sw based on agile methodology. In *Advanced communication technology, 2008. icact 2008. 10th international conference on* (*Vol. 1*, pp. 362-366). ICACT.

Jim'enez-Ruiz, E., Grau, B. C., Horrocks, I., & . Llavori, R. B. (2009). *Contentcvs: A cvs-based collaborative ontology engineering tool.* SWAT4LS.

Jimson, E. R., Nisar, K., & Hanafi, M. (2017). Bandwidth Management using Software Defined Network and Comparison of the Throughput Performance with Traditional Network. *JIITA*, 71–76.

Jobin, A., Ienca, M., & Vayena, E. (2019). The global landscape of AI ethics guidelines. *Nature Machine Intelligence, 1*(9), 389–399. doi:10.103842256-019-0088-2

Joe, H. (2020). What is Software Defined Networking (SDN)? Definition. *SDxCentral*. https://www.bmc.com/blogs/software-defined-networking/#%0Ahttps://www.sdxcentral.com/networking/sdn/definitions/what-the-definition-of-software-defined-networking-sdn/

Joshi, N., Lakier, M., Vogel, D., & Zhao, J. (2021, December). A Design Framework for Contextual and Embedded Information Visualizations in Spatial Augmented Reality. In Graphics Interface 2022.

Kaleel, S. B., & Harishankar, S. (2013). *Applying agile methodology in mobile software engineering: Android application development and its challenges*. Computer Science Technical Reports.

Kalelkar, M., Churi, P., & Kalelkar, D. (2014). Implementation of model-view-controller architecture pattern for Business Intelligence Architecture. *International Journal of Computers and Applications*, *102*(12), 16–21.

Kalimuthu, S. (2021). Sentiment Analysis on Social Media for Emotional Prediction During COVID-19 Pandemic Using Efficient Machine Learning Approach. *Computational Intelligence and Healthcare Informatics*, 215.

Kalimuthu, S., Nait-Abdesselam, F., & Jaishankar, B. (2021). Multimedia Data Protection Using Hybridized Crystal Payload Algorithm With Chicken Swarm Optimization. In Multidisciplinary Approach to Modern Digital Steganography (pp. 235-257). IGI Global. doi:10.4018/978-1-7998-7160-6.ch011

Kalpana, S., Rajagopalan, S., Ranjith, R., & Gomathi, R. (2020 July 3-4). *Voice Recognition Based Multi Robot for Blind People Using Lidar Sensor.* 2020 International Conference on System, Computation, Automation and Networking (ICSCAN), Pondicherry, India. https://ieeexplore.ieee.org/abstract/document/9262365

Kamal, S. M., Chen, Y., Li, S., Shi, X., & Zheng, J. (2019). Technical approaches to Chinese sign language processing: A review. *IEEE Access: Practical Innovations, Open Solutions*, *7*, 96926–96935.

Kane, F. (2017). *Hands-On Data Science and Python Machine Learning*. Birmingham Packt Publishing.

Kaur, A., & Kaur, K. (2022). Systematic Literature Review of Mobile Application Development and Testing Effort Estimation. *Journal of King Saud University - Computer and Information Sciences, 34*(2), pp. 1–15. ScienceDirect. doi:10.1016/j.jksuci.2018.11.002

Kaźmierczak, R., Szczepańska, A., Kowalczyk, C., Grunwald, G., & Janowski, A. (2021). Using AR technology in tourism based on the example of maritime educational trips—A conceptual model. *Sustainability*, *13*(13), 7172. doi:10.3390u13137172

Keller, J. M., Liu, D., & Fogel, D. B. (2016). *Fundamentals of Computational Intelligence*, *34*, 78. John Wiley & Sons.

Kemper, H. G. & Wolf, E. (2002). Iterative Process Models for Mobile Application Systems: A Framework. *ICIS 2002 Proceedings, 37*. PSU.

Khan, S., Nazir, S., & Khan, H. U. (2021). Analysis of Navigation Assistants for Blind and Visually Impaired People: A Systematic Review. *IEEE Access: Practical Innovations, Open Solutions*, *9*, 26712–26734. doi:10.1109/ACCESS.2021.3052415

Kienitz, P. (2021). *The Pros and Cons of Iterative Software Development.* One Beyond. https://www.one-beyond.com/pros-cons-iterative-software-development/

Kikuchi, N., Fukuda, T., & Yabuki, N. (2022). Future landscape visualization using a city digital twin: Integration of augmented reality and drones with implementation of 3D model-based occlusion handling. *Journal of Computational Design and Engineering*, *9*(2), 837–856. doi:10.1093/jcde/qwac032

Kim, J. (2018). Evaluating author name disambiguation for digital libraries: A case of DBLP. *Scientometrics*, *116*(3), 1867–1886. doi:10.100711192-018-2824-5

Kolassa, C., Riehle, D., & Salim, M. (2013). A Model of the Commit Size Distribution of Open Source. *Proc. the 39th Int'l Conf. Current Trends in Theory and Practice of Comput. Sci. (SOFSEM'13),* (pp. 52-66). IEEE. 10.1007/978-3-642-35843-2_6

Koren, I., & Klamma, R. (2018). The exploitation of OpenAPI documentation. Proc. Companion Web Conf. (WWW), (pp. 781-787). IEEE.

Kotusev, S., Kurnia, S., & Dilnutt, R. (2023). Enterprise architecture artifacts as boundary objects: An empirical analysis. *Information and Software Technology*, *155*, 107108. doi:10.1016/j.infsof.2022.107108

Kouters, E., Vasilescu, B., Serebrenik, A., & Van Den Brand, M. G. (2012). Who's who in Gnome: Using LSA to merge software repository identities. *2012 28th IEEE International Conference on Software Maintenance (ICSM),* (pp. 592-595). IEEE. 10.1109/ICSM.2012.6405329

Krishna, K. V. (2022). Speech Emotion Recognition Using Machine Learning.. *2022 6th International Conference on Computing Methodologies and Communication (ICCMC),* (pp. 1014–18*. IEEE Xplore 10.1109/ICCMC53470.2022.9753976

Kukulska-Hulme, A. (2016). *Personalization of Language Learning through Mobile Technologies.* Cambridge University Press.

Kukulska-Hulme, A., & Shield, L. (2008). An overview of mobile assisted language learning: From content delivery to supported collaboration and interaction. *ReCALL*, *20*(3), 271–289. doi:10.1017/S0958344008000335

Kumar, G., & Samanta, D. (2022). Effective View of Swimming Pool Using Autodesk 3ds Max: 3D Modelling and Rendering. In *Internet of Things and Its Applications: Select Proceedings of ICIA 2020,* (pp. 47–53). Springer Nature. 10.1007/978-981-16-7637-6_5

Kumar, V., Kumar, A., Sharma, A. K., & Singh, D. (2016). Implementation of MVC (Model-View-Controller) design architecture to develop web based Institutional repositories: A tool for Information and knowledge sharing. *Indian Research Journal of Extension Education*, *16*(3), 1–9.

Laaz, N., & Mbarki, S. (2016). Integrating IFML models and owl ontologies to derive UIs Web-Apps. *Proc. Int. Conf. Inf. Technol. Org. Develop. (IT4OD)*. IEEE.

Laera, F. (2020, November). Augmented Reality for Easy Sailing. In *2020 IEEE International Symposium on Mixed and Augmented Reality Adjunct (ISMAR-Adjunct)* (pp. 312-313). IEEE. 10.1109/ISMAR-Adjunct51615.2020.00088

Landis, J. R., & Koch, G. G. (1977). An application of hierarchical kappa-type statistics in the assessment of majority agreement among multiple observers. *Biometrics*, *33*(2), 363–374. doi:10.2307/2529786 PMID:884196

Langtangen, H. P. (2016). *A Primer on Scientific Programming with Python*. Springer.

Latif, Z., Sharif, K., Li, F., Karim, M., Biswas, S., & Wang, Y. (2020). Journal of Network and Computer Applications A comprehensive survey of interface protocols for software defined networks. *Journal of Network and Computer Applications*, *156*(February), 102563. doi:10.1016/j.jnca.2020.102563

Lee, D. R. (2022). *FDIC: Supervisory Insights - Authentication in Internet Banking: A Lesson in Risk Management*. Fdic.gov. www.fdic.gov/regulations/examinations/supervisory/insights/siwin07/article05_authentication.html

Lee, H., & Chuvyrov, E. (2012). Beginning Windows Phone App Development. *Apress*. https://appinventiv.com/blog/mobile-app-architecture-explained/

Lee, Y.-C. "Analysis of Service Experience of the Community Sports Center Mobile Apps Based on a Service Design Perspective", Master Thesis, 2019, [online] Available: https://reurl.cc/dV7r06

Leff, A., & Rayfield, J. T. (2001). Web-application development using the model/view/controller design pattern. *Proceedings Fifth IEEE International Enterprise Distributed Object Computing Conference*, 118-127.

Leffingwell, D., & Widrig, D. (2003). *Managing Software Requirements: A Use Case Approach*. Addison-Wesley.

Legner, C. (2016, September). Mobile Business Application for Service and Maintenance Processes: Using Ex Post Evaluation by End-Users as Input for Iterative Design. [ScienceDirect, https://doi.org/10.1016/j.im.2016.03.001.]. *Information & Management*, *53*(6), 817–831.

Levenshtein, V. I. (1966). Binary codes capable of correcting deletions insertions and reversals. *Soviet Physics, Doklady*, *10*(8), 707–710.

Li, K., Tao, M., & Chen, Z. (2019). A computation-communication tradeoff study for mobile edge computing networks. *2019 IEEE International Symposium on Information Theory, (ISIT)* (pp. 2639-2643). doi:10.1109/ISIT.2019.8849601

Li, Q., Cai, W., Wang, X., Zhou, Y., Feng, D. D., & Chen, M. (2014, December). Medical image classification with convolutional neural network. In *2014 13th international conference on control automation robotics & vision (ICARCV)* (pp. 844-848). IEEE. 10.1109/ICARCV.2014.7064414

Lin, P.-W. (2020). Research on the Marketing Strategy of Small Gyms: A Case Study of Small Gyms in Metropolitan District of Tainan City [Master Thesis, National Kaohsiung Normal University.] https://reurl.cc/2b8GaO

LinuxHint. (n.d.). [Programming]. Linux Hints. https://linuxhint.com

Lin, Z., Lin, M., Wang, J. B., de Cola, T., & Wang, J. (2019). Joint Beamforming and Power Allocation for Satellite-Terrestrial Integrated Networks with Non-Orthogonal Multiple Access. *IEEE Journal of Selected Topics in Signal Processing*, *13*(3), 657–670. doi:10.1109/JSTSP.2019.2899731

Lior Rokach, Maimon, Oded Z. (2015). *Data mining with decision trees: theory and applications*,,16~41, World Scientific, Cop.

Liu, Z., Gao, X., & Long, X. (2010). Adaptive random testing of mobile application. In *Computer engineering and technology (iccet), 2010 2nd international conference on* (*Vol. 2*, pp. V2-297).

Liu, C., & Kirubakaran, S.Daniel, A. J. (2022, April). Deep Learning Approach for Emotion Recognition Analysis in Text Streams. [IJTHI]. *International Journal of Technology and Human Interaction*, *18*(2), 1–21. www.igi-global.com. doi:10.4018/IJTHI.313927

Liu, D., Subramaniam, K., Eberlein, A., & Far, B. H. (2004). Natural language requirements analysis and class model generation using UCDA. *Proc. Int. Conf. Ind. Eng. Appl. Appl. Intell. Syst.*, (pp. 295-304).

Liu, S. (2020). [John Wiley & Sons.]. *Engineering Autonomous Vehicles and Robots*, *19*, 21.

Liu, X. K., & Cheng, G. G. (2013). Analysis and implementation of ASP.NET and PHP frameworks based on MVC Architecture. *Advanced Materials Research*, *798-799*, 749–752.

Li, W., Cao, P., Zhao, D., & Wang, J. (2016). Pulmonary nodule classification with deep convolutional neural networks on computed tomography images. *Computational and Mathematical Methods in Medicine*, *2016*, 2016. doi:10.1155/2016/6215085 PMID:28070212

Lozano, M. D., Gonzalez, P., & Ramos, I. “User interface specification and modelling in an object-oriented environment for automatic software development”, Proc. Int. Conf. Technol. Object-Oriented Lang. Syst. (TOOLS), pp. 373-381, 2000.

Lucia-Palacios, L. (2016, July). Complementary IT Resources for Enabling Technological Opportunism. [ScienceDirect, https://doi.org/10.1016/j.im.2016.02.002.]. *Information & Management*, *53*(5), 654–667.

Luczak-Rösch, M., Coskun, G., Paschke, A., Rothe, M., & Tolksdorf, R. (2010). Svont-version control of owl ontologies on the concept level. *GI Jahrestagung*, *176*(2), 79–84.

Lutz, M. (2006). *Programming Python*. O'Reilly Media, Inc.

Lutz, M. (2013). *Learning Python*. O'Reilly Media, Inc.

Ma, Y., Wu, Y., & Xu, Y. (2014). Dynamics of Open-Source Software Developer's Commit Behavior: An Empirical Investigation of Subversion. Proceedings of the 29th Annual ACM Symposium on Applied Computing (SAC' 14), (pp. 1171-1173). ACM. 10.1145/2554850.2555079

Ma, H., & Yarosh, S. (2021). A Review of Affective Computing Research Based on Function-Component-Representation Framework. [IEEE Xplore]. *IEEE Transactions on Affective Computing*, 1–1. doi:10.1109/TAFFC.2021.3104512

Majeed, A., & Rauf, I. (2018). MVC Architecture: A Detailed Insight to the Modern Web Applications Development. *Peer Review Journal of Solar & Photoenergy Systems*, *1*(1), 1–7.

Ma, L. (2017). Independent Travel System Based on Intelligent Solution. *Computer Systems & Applications*, *26*(3), 57–62.

Maqbool, B., Azam, F., Anwar, M. W., Butt, W. H., Zeb, J., Zafar, I. (2019). A Comprehensive investigation of BPMN models generation from textual requirements—Techniques tools and trends. *Proc. ICISA*, (vol. 514). IEEE.

Martin, R. C. (2017). *Clean Architecture: A Craftsman's Guide to Software Structure and Design* (1st ed.). Prentice Hall Press.

Martin, R. C., & Code, C. (2007). *A Handbook of Agile Software Craftsmanship*.

Masood, A., Sheng, B., Li, P., Hou, X., Wei, X., Qin, J., & Feng, D. (2018). Computer-assisted decision support system in pulmonary cancer detection and stage classification on CT images. *Journal of Biomedical Informatics*, *79*, 117–128. doi:10.1016/j.jbi.2018.01.005 PMID:29366586

Mathews, N. S., Chimalakonda, S., & Jain, S. (2021, October). AiR: An Augmented Reality Application for Visualizing Air Pollution. In *2021 IEEE Visualization Conference (VIS)* (pp. 146-150). IEEE. 10.1109/VIS49827.2021.9623287

Matviienko, A., Müller, F., Schön, D., Seesemann, P., Günther, S., & Mühlhäuser, M. (2022, April). BikeAR: Understanding Cyclists' Crossing Decision-Making at Uncontrolled Intersections using Augmented Reality. In *CHI Conference on Human Factors in Computing Systems* (pp. 1-15). 10.1145/3491102.3517560

Meedeniya, D., Perera, I., & Bowles, J. (2015). Transformation and composition of software design models for Model Driven Development. Proc. Int. Conf. Ind. Inf. Syst. (ICIIS). IEEE.

Meister, P., Miller, J., Wang, K., Dorneich, M. C., Winer, E., Brown, L. J., & Whitehurst, G. (2022). Designing Three-Dimensional Augmented Reality Weather Visualizations to Enhance General Aviation Weather Education. *IEEE Transactions on Professional Communication*, *65*(2), 321–336. doi:10.1109/TPC.2022.3155920

Merris, R. (2001). *Graph theory*. John Wiley.

Messaoudi, M. D., Menelas, B.-A. J., & Mcheick, H. (2022). Review of Navigation Assistive Tools and Technologies for the Visually Impaired. *Sensors (Basel)*, *22*(20), 7888. doi:10.339022207888

Miah, M. B. A., & Yousuf, M. A. (2015, May). Detection of lung cancer from CT image using image processing and neural network. In *2015 International conference on electrical engineering and information communication technology (ICEEICT)* (pp. 1-6). IEEE. 10.1109/ICEEICT.2015.7307530

Microsoft. (2013, June 16). *Visual Studio Resources*. Microsoft. https://msdn.microsoft.com/en-us/vstudio/cc136611

Middelweerd, A., Mollee, J. S., van der Wal, C. N., Brug, J., & te Velde, S. J. (2014). Apps to promote physical activity among adults: A review and content analysis. [Google Scholar] [CrossRef] [PubMed]. *The International Journal of Behavioral Nutrition and Physical Activity*, *11*(1), 97. doi:10.118612966-014-0097-9 PMID:25059981

Mirjalili, S. (2019). *PYTHON MACHINE LEARNING - THIRD EDITION: machine learning and deep learning with python, scikit... -Learn, and tensorflow 2*. Packt Publishing Limited.

Mishra, A., Singh, C., Dwivedi, A., Singh, D., & Biswal, A. K. (2021, October). Network Forensics: An approach towards detecting Cyber Crime. In *2021 International Conference in Advances in Power, Signal, and Information Technology (APSIT)* (pp. 1-6). IEEE. 10.1109/APSIT52773.2021.9641399

MIT. (n.d.). *Publications List Pre 2020*. MIT Media Lab, https://www.media.mit.edu/groups/affective-computing/publications-list/.

Mohajeri, N., Raste, R., & Daneshvar, S. (2011, July 6 - 8). An Obstacle Detection System for Blind People. *Proceedings of the World Congress on Engineering 2011*. https://www.iaeng.org/publication/WCE2011/WCE2011_pp1602-1605.pdf

Mohapatra, S., Rout, S., Tripathi, V., Saxena, T., & Karuna, Y. (2018, May 11-12). *Smart Walking Stick for Blind Integrated with SOS Navigation System*. 2018 2nd International Conference on Trends in Electronics and Informatics (ICOEI), Tirunelveli, India. https://ieeexplore.ieee.org/document/8553935

Mollahosseini, A., Hasani, B., & Mahoor, M. H. (2017). Affectnet: A database for facial expression, valence, and arousal computing in the wild. *IEEE Transactions on Affective Computing*, *10*(1), 18–31. doi:10.1109/TAFFC.2017.2740923

Monkam, P., Qi, S., Ma, H., Gao, W., Yao, Y., & Qian, W. (2019). Detection and classification of pulmonary nodules using convolutional neural networks: A survey. *IEEE Access: Practical Innovations, Open Solutions*, *7*, 78075–78091. doi:10.1109/ACCESS.2019.2920980

Montalvo, A. (2022, December). Model-Driven System-Level Validation and Verification on the Space Software Domain. [Springer Link, https://doi.org/10.1007/s10270-021-00940-8.]. *Software & Systems Modeling*, *21*(6), 2367–2394.

Montenegro, A. A., Carvalho, P. C. P., Velho, L., & Gattass, M. *Space carving with a hand-held camera.* Proceedings of the XVII Brazilian Symposium on Computer Graphics and Image Processing, (pp. 396~403). John Wiley & Sons. 10.1109/SIBGRA.2004.1352986

Moryossef, A., Tsochantaridis, I., Aharoni, R., Ebling, S., & Narayanan, S. (2020). Real-time sign language detection using human pose estimation. In Computer Vision–ECCV 2020 Workshops: UK.

Mukherjee, P. K. (2022). Artificial Intelligence Based Smart Government Enterprise Architecture (AI-SGEA) Framework. *Artificial Intelligence: First International Symposium, ISAI 2022*. Cham: Springer Nature Switzerland.

Nagavelli, U., Samanta, D., & Thomas, B. (2022). ML-Based Prediction Model for Cardiovascular Disease. In *Emerging Technologies in Data Mining and Information Security: Proceedings of IEMIS 2022*, (volume 2, pp. 91–98). Springer Nature.

Nandy, A., Prasad, J. S., Mondal, S., Chakraborty, P., & Nandi, G. C. (2010). Recognition of isolated indian sign language gesture in real time. In *Information Processing and Management: International Conference on Recent Trends in Business Administration and Information Processing, BAIP 2010*, (pp. 102~107).

Nanumyan, V., Gote, C., & Schweitzer, F. (2020). Multilayer network approach to modeling authorship influence on citation dynamics in physics journals. *Physical Review E, 102*, (pp. 032303). https://link.aps.org/doi/10.1103/PhysRevE.102.032303

Nazir, F., Butt, W. H., Anwar, M. W., & Khattak, M. A. K. (2017). The applications of natural language processing (NLP) for software requirement engineering—A systematic literature review. *Proc. 8th Int. Conf. Inf. Sci. Appl. (ICISA)*,(vol. 425, pp. 485-493). IEEE.

Newman, M. E. (2004). *Who is the best connected scientist? A study of scientific coauthorship networks in Complex Networks*. Springer.

Neyveli, V. R. N., Sivakumar, S. S., Arunagiri, D., Arumugam, C., & Veeramani, A. M. (2019). An approach to estimate the size of Web application using IFML user interface model. *Proc. Amity Int. Conf. Artif. Intell. (AICAI)*, (pp. 292-295). IEEE.

Nieto, V., Castro, V., Lopez, F., Ferro, R., & Gonzalez, C. (2016, January). Model driven architecture software and interaction flow modelling language for tourism data acquisition in Colombia. *Communications in Computer and Information Science*, *657*, 368–379.

Nikam, A. S., & Ambekar, A. G. (2016). *Sign language recognition using image based hand gesture recognition techniques. In 2016 online international conference on green engineering and technologies*. IC-GET.

Nouman, M., Azam, M., Saleh, A. M., Alsaeedi, A., & Abuaddous, H. Y. (2023). A systematic review of non-functional requirements mapping into architectural styles. *Bulletin of Electrical Engineering and Informatics*, *12*(2), 1226–1236. doi:10.11591/eei.v12i2.4081

Ntagiou, E. V., Klug, J., Luntama, J. P., & Sarkarati, M. (2022). Virtual Reality in Support of Space Weather Forecasting. In *Space Operations* (pp. 325–343). Springer. doi:10.1007/978-3-030-94628-9_15

Object Management Group IFML Specifications Version 1 . (2019). OMG.org. http://www.omg.org/spec/IFML/1.0/

Object Management Group Portals . (2019). OMG Wiki. https://www.omgwiki.org/

Oo, M. M., Kamolphiwong, S., Kamolphiwong, T., & Vasupongayya, S. (2019). *Advanced Support Vector Machine- (ASVM-) Based Detection for Distributed Denial of Service (DDoS) Attack on Software Defined Networking (SDN). 2019*. Semantic Scholar.

Oulasvirta, A., Wahlström, M., & Ericsson, K. A. (2011). What does it mean to be good at using a mobile device? an investigation of three levels of experience and skill. *International Journal of Human-Computer Studies, 69*(3), 155–169.

Oz, H. (2013). Prospective English teachers' ownership and usage of mobile device as m-learning tools. *Procedia: Social and Behavioral Sciences, 141*, 1031–1041. doi:10.1016/j.sbspro.2014.05.173

Pang, C.-Y. (2022). Applying Software Engineering Design Principles to Agile Architecture. In *Research Anthology on Agile Software, Software Development, and Testing, edited by Information Resources Management Association* (pp. 330–355). IGI Global. doi:10.4018/978-1-6684-3702-5.ch016

Pardalos, P. M., Piero Conca, Giuffrida, G., Nicosia, G., & Springerlink. (2016). Machine Learning, Optimization, and Big Data. *Second International Workshop, MOD 2016*. Springer International Publishing.

Parmar, A. (2021). *A Comprehensive Guide on Banking Software Development*. Prismetric. www.prismetric.com/guide-to-banking-software-development/

Patel, S. J., & Pancholi, P. D. (2018). Implementation and Comparison of MVC Model in ASP.net Framework and PHP Framework. *International Journal of Research and Analytical Reviews, 5*(4), 827–835.

Pathar, R., & Human Emotion Recognition Using Convolutional Neural Network in Real Time." 2019 1st International *Conference on Innovations in Information and Communication Technology (ICIICT), 2019*, (pp. 1–7). IEEE Xplore 10.1109/ICIICT1.2019.8741491

Pearce, Q. L. (2011). *Artificial Intelligence, 22*, 29. Lucent Books.

Pedregosa, F., Varoquaux, G., Gramfort, A., Michel, V., Thirion, B., & Grisel, O. (2011). Scikit-learn: Machine learning in Python. *Journal of Machine Learning Research, 12*, 2825–2830.

Petale, S., Member, G. S., & Thangaraj, J. (2020). Link Failure Recovery Mechanism in Software Defined Networks. *Link Failure Recovery Mechanism in Software Defined Networks., 8716*(c), 1–8. doi:10.1109/JSAC.2020.2986668

Phillips, A. C., Der, G., & Carroll, D. (2010). Self-reported health, self-reported ðtness, and all-cause mortality:Prospective cohort study. *British Journal of Health Psychology, 15*(2), 337–346. doi:10.1348/135910709X466180 PMID:19619405

Platt, E. L. (2019). *Network Science with Python and NetworkX Quick Start Guide*. Packt Publishing Ltd.

Pop, D.-P., & Altar, A. (2014). Designing an MVC model for rapid web application development. *Procedia Engineering, 69*, 1172–1179.

Potasso-Justino. Lucas. "What Is Iterative Development and How Does It Work? " Planbox, 16 Apr. 2019, https://www.planbox.com/what-is-iterative-development-and-how-does-it-work/

Priyadarsini, M., & Bera, P. (2021). *Software defined networking architecture, traffic management, security, and placement : A survey. 192*(June 2020).

Prophet, J., Kow, Y. M., & Hurry, M. (2018, July). Cultivating environmental awareness: Modeling air quality data via augmented reality miniature trees. In *International Conference on Augmented Cognition* (pp. 406-424). Springer, Cham. 10.1007/978-3-319-91470-1_33

Purushothaman, R., & Perry, D. E. (2005). Toward Understanding the Rhetoric of Small Source Code Changes. *IEEE Transactions on Software Engineering, 31*(6), 511–526. doi:10.1109/TSE.2005.74

Qiu, C., Yao, H., Richard Yu, F., Xu, F., & Zhao, C. (2019). Deep Q-Learning Aided Networking, Caching, and Computing Resources Allocation in Software-Defined Satellite-Terrestrial Networks. *IEEE Transactions on Vehicular Technology, 68*(6), 5871–5883. doi:10.1109/TVT.2019.2907682

Rafique, W., Qi, L., Yaqoob, I., Imran, M., Rasool, R. U., & Dou, W. (2020). Complementing IoT Services through Software Defined Networking and Edge Computing: A Comprehensive Survey. *IEEE Communications Surveys and Tutorials, 22*(3), 1761–1804. doi:10.1109/COMST.2020.2997475

Rahman, M. A., Sadi, M. S., Islam, M. M., & Saha, P. (2019, November 28-30). *Design and Development of Navigation Guide for Visually Impaired People*. 2019 IEEE International Conference on Biomedical Engineering, Computer and Information Technology for Health (BECITHCON), Dhaka, Bangladesh. https://ieeexplore.ieee.org/document/9063201

Rahy, S., & Bass, J. M. (2022). Managing non-functional requirements in agile software development. *IET Software, 16*(1), 60–72. doi:10.1049fw2.12037

Rajamani, S. K., & Iyer, R. S. (2022). Development of an Android Mobile Phone Application for Finding ClosedLoop, Analytical Solutions to Dense Linear, Algebraic Equations for The Purpose of Mathematical Modelling in Healthcare and Neuroscience Research. *NeuroQuantology: An Interdisciplinary Journal of Neuroscience and Quantum Physics, 20*, 4959–4973. doi:10.6084/m9.figshare.c.6156024.v1

Ranjan, R. (2022). The Mobile App Architecture Guide for 2022, Insights - Web and Mobile Development Services and Solutions. *Net Solutions*. https://www.netsolutions.com/insights/mobile-app-architecture-guide/

Rao, B. N. K. & Rao, B. K. (2019). Block chain Based Implementation of Electronic Medical Health Record. *International Journal of Innovative Technology and Exploring Engineering, 8*(8).

Rao, B. N. K. & Rao, B. K. (2019). Clustering based Test Suite Selection for Ranking of Program Execution Sequence using Improved Precision in Regression testing. *International Journal of Innovative Technology and Exploring Engineering*, *8*(7).

Rao, B. N. K. & Rao, B. K. (2019). Defect Detection in Printed Board Circuit using Image Processing. *International Journal of Innovative Technology and Exploring Engineering*, *9*(2).

Razavi, M. (2020). Neural *Regenerative Nanomedicine*, *4*, 9. Academic Press.

Real, S., & Araujo, A. (2019). Navigation Systems for the Blind and Visually Impaired: Past Work, Challenges, and Open Problems. *Sensors (Basel)*, *19*(15), 3404. doi:10.339019153404

Rehmani, M. H., Davy, A., Jennings, B., & Assi, C. (2019). Software Defined Networks-Based Smart Grid Communication: A Comprehensive Survey. *IEEE Communications Surveys and Tutorials*, *21*(3), 2637–2670. doi:10.1109/COMST.2019.2908266

Rhazali, Y., Hadi, Y., & Mouloudi, A. (2016). A model transformation in MDA from CIM to PIM represented by Web models through SoaML and IFML. *Proc. 4th IEEE Int. Colloq. Inf. Sci. Technol. (CiSt)*, (pp. 116-121). IEEE.

Riesener, M. (2021). Methodology for Iterative System Modeling in Agile Product Development. [ScienceDirect, https://doi.org/10.1016/j.procir.2021.05.101.]. *Procedia CIRP*, *100*(Jan), 439–444.

Ries, M., Nemethova, O., & Rupp, M. (2008). Video Quality Estimation for Mobile H.264/AVC Video Streaming. *Journal of Communication*, *3*, 41–50.

Roberto, R. (2019). Geometrical and Statistical Incremental Semantic Modeling on Mobile Devices. [ScienceDirect, https://doi.org/10.1016/j.cag.2019.09.003.]. *Computers & Graphics*, *84*(Nov), 199–211.

Robertson, D., Ross, J. W., & Weill, P. (2006). *Everything There is to Know About Automated Testing*. Global App Testing. https://www.globalapptesting.com/blog/everything-there-is-to-know-about-automated-testing

Robertson, J., & Robertson, S. (2000). Volere. [Google Scholar]. *Requir. Specif. Templates*, *16*, 1–80.

Robinson, P., & el Kaliouby, R. (2009, December). Computation of Emotions in Man and Machines. [PubMed Central]. *Philosophical Transactions of the Royal Society of London. Series B, Biological Sciences*, *364*(1535), 3441–3447. doi:10.1098/rstb.2009.0198 PMID:19884138

Robles, G., & Gonzalez-Barahona, J. M. (2005). Developer identification methods for integrated data from various sources. *Software Engineering Notes*, *30*(4), 1–5. doi:10.1145/1082983.1083162

Robson, C., & McCartan, K. (2016). *Real world research*. John Wiley & Sons.

Ronchetti, F., Quiroga, F., Estrebou, C. A., & Lanzarini, L. C. (2016). Handshape recognition for argentinian sign language using probsom. *Journal of Computer Science and Technology*, 16.

Roubi, S., Erramdani, M., & Mbarki, S. (2015). A model driven approach to generate graphical user interfaces for rich Internet applications using interaction flow modeling language. Proc. ISDA. IEEE.

Rudovic, O., Zhang, M., Schuller, B., & Picard, R. (2019). Multi-modal Active Learning From Human Data: A Deep Reinforcement Learning Approach. In *2019 International Conference on Multimodal Interaction (ICMI '19)*, ACM, 10.1145/3340555.3353742

Sakshi & Kukreja, V. (2022). A Dive in White and Grey Shades of ML and Non-ML Literature: A Multivocal Analysis of Mathematical Expressions. *Artificial Intelligence Review*. Springer. doi:10.1007/s10462-022-10330-1

Salini, A., Malavolta, I., & Rossi, F. (2016). Leveraging Web analytics for automatically generating mobile navigation models. *Proc. IEEE Int. Conf. Mobile Services (MS)*, (pp. 103-110). IEEE.

Samanta, D. (2022). *3D Modeling Using Autodesk 3ds Max With Rendering View*. IGI Global.

Samantaray, M., Biswal, A. K., Singh, D., Samanta, D., Karuppiah, M., & Joseph, N. P. (2021, December). Optical Character Recognition (OCR) based Vehicle's License Plate Recognition System Using Python and OpenCV. In *2021 5th International Conference on Electronics, Communication and Aerospace Technology (ICECA)* (pp. 849-853). IEEE.

Sangeethalakshmi, K., Shanthi, K. G., Raj, A. M., Muthuselvan, S., Taha, P. M., & Shoaib, S. M. (2021). Hand gesture vocalizer for deaf and dumb people. *Materials Today: Proceedings*.

Sannino, G., Graziani, A., de Pietro, G., & Pratola, R. (2016). *Wellness & LifeStyle Server: A platform for anthropometric and LifeStyle data analysis*. In Proceedings of the International Conference on P2P, Parallel, Grid, Cloud and Internet Computing, Asan, Korea.

Saquib, N., Kazi, R. H., Wei, L. Y., & Li, W. (2019, May). Interactive body-driven graphics for augmented video performance. In *Proceedings of the 2019 CHI Conference on Human Factors in Computing Systems* (pp. 1-12). 10.1145/3290605.3300852

Saquib, N., & Rahman, A. (2020). Application of machine learning techniques for real-time sign language detection using wearable sensors. In *Proceedings of the 11th ACM Multimedia Systems Conference*, (pp. 178~189).

Sarawagi, S., & Bhamidipaty, A. (2002). Interactive deduplication using active learning. *Proceedings of the Eighth ACM SIGKDD International Conference on Knowledge Discovery and Data Mining*, (pp. 269-278). ACM. 10.1145/775047.775087

Sarigol, E., Garcia, D., Scholtes, I., & Schweitzer, F. (2017). Quantifying the effect of editor-author relations on manuscript handling times. *Scientometrics, 113*(1), pp. 609-631. https://link.springer.com/article/10.1007/s11192-017-2309-y

Sarigol, E., Pfitzner, R., Scholtes, I., Garas, A., & Schweitzer, F. (2014). Predicting scientific success based on coauthorship networks. *EPJ Data Science, 3*(9). http://www.epjdatascience.com/content/3/1/9

Sarker, I. H. (2021, March). Machine Learning: Algorithms, Real-World Applications and Research Directions. [Springer]. *SN Computer Science, 2*(3), 160. doi:10.100742979-021-00592-x PMID:33778771

Sassi, M. S. H., & Fourati, L. C. (2021, October). Deep learning and augmented reality for IoT-based air quality monitoring and prediction system. In *2021 International Symposium on Networks, Computers and Communications (ISNCC)* (pp. 1-6). IEEE. 10.1109/ISNCC52172.2021.9615639

Satheesh, N., Rathnamma, A. M., Rajeshkumar, G., Sagar, P. V., Dadheech, P., Dogiwal, S. R., Velayutham, P., & Sengan, S. (2020). Flow-based Anomaly Intrusion Detection using Machine Learning Model with Software Defined Networking for OpenFlow Network. *Microprocessors and Microsystems, 103285*, 103285. doi:10.1016/j.micpro.2020.103285

Saxena, A, Rani, U., Agarwal, I., & Jangade. R. (2021). Introduction to Big Data, Chapter 1 Introduction to Big Data. In Ankur Saxena, Nicolas Brault, & Shazia Rashid (Eds.), Big Data and Artificial Intelligence for Healthcare Applications, (p. 3-21). Taylor and Francis..

Schwartz, M. (1995).Network management and control issues in multimedia wireless networks. *IEEE Personal Communications, 2*(3), 8-16.

Schwartz, M. (2016). *Internet of Things with ESP8266: build amazing Internet of Things projects using the ESP8266 Wi-Fi chip*. Packt Publishing.

Sejnowski, T. J. (2018). *The deep learning revolution*. The MIT Press. doi:10.7551/mitpress/11474.001.0001

Seo, S., Kang, D., & Park, S. (2018). Real-time adaptable and coherent rendering for outdoor augmented reality. *EURASIP Journal on Image and Video Processing, 2018*(1), 1–8. doi:10.118613640-018-0357-8

Serban, A., & Visser, J. (2022). Adapting software architectures to machine learning challenges. *2022 IEEE International Conference on Software Analysis, Evolution and Reengineering (SANER)*. IEEE. 10.1109/SANER53432.2022.00029

Sermet, Y., & Demir, I. (2020). Virtual and augmented reality applications for environmental science education and training. In *New Perspectives on Virtual and Augmented Reality* (pp. 261–275). Routledge. doi:10.4324/9781003001874-17

Sezer, S., Scott-Hayward, S., Chouhan, P., Fraser, B., Lake, D., Finnegan, J., Viljoen, N., Miller, M., & Rao, N. (2013). Are we ready for SDN? Implementation challenges for software-defined networks. *IEEE Communications Magazine, 51*(7), 36–43. doi:10.1109/MCOM.2013.6553676

Shah, F., Shah, M. S., Akram, W., Manzoor, A., Mahmoud, R. O., & Abdelminaam, D. S. (2021). Sign language recognition using multiple kernel learning: A case study of Pakistan sign language. *IEEE Access: Practical Innovations, Open Solutions, 9*, 67548–67558.

Shamil, F. R. (n.d.). Incremental Model, Advantages, Examples, Case Study. *T4Tutorials.Com.* https://t4tutorials.com/incremental-model-in-software-engineering-advantages-disadvantages-examples-case-study-of-incremental-model/

Shamil, F. R. (n.d.). *Iterative Model in Software Engineering Advantages Disadvantages.* T4Tutorials.Com. https://t4tutorials.com/what-is-iterative-model-software-engineering/

Sharifian, M., Abdolvand, N., & Harandi, S. R. (2021). Context-Based Expert Finding in Online Communities Using Ant Colony Algorithm. *Journal of Information Systems and Telecommunication, 8*(30), 130–139. doi:10.29252/jist.8.30.130

Sharma, S., & Kumar, K. (2021). ASL-3DCNN: American Sign Language recognition technique using 3-D convolutional neural networks. *Multimedia Tools and Applications, 80*(17), 26319~26331.

Shi, S. Y. (2018). The design and implementation of Bank Online Shopping mall system based on Cordova Technology, [Master's thesis, Lanzhou Jiaotong University].

Shih-Hsun, Ho. (2017). The Study of Fitness Applications based on Competence Set Analysis. [Master Thesis, National Chung Hsing University. https://reurl.cc/bzG8dM

Shrestha, A., & Mahmood, A. (2019). *Review of Deep Learning Algorithms and Architectures.* IEEE. doi:10.1109/ACCESS.2019.2912200

Sibarani, E. M., Hutagaol, A., Simarmata, D., & Manihuruk, J. (2014). Actor and use case extraction from text based requirement specification. Proc. Int. Conf. Image Process. Comput. Ind. Eng. (ICICIE). IEEE.

Sillitto H., Martin J., McKinney D., Griego R., Dori D., Krob D., Godfrey P., Arnold E., Jackson S. (2019). Systems Engineering and Systems Definitions.

Sims, E. (2006). *Deep learning*. SpecialistSchools,and Academies Trust.

Sinatra, R., Wang, D., Deville, P., Song, C., & Barabási, A.-L. (2016). Quantifying the evolution of individual scientific impact. *Science, 354*(6312), aaf5239. doi:10.1126cience.aaf5239 PMID:27811240

Singh, D. (2022). Search Engine Optimization for Digital Marketing to Raise the Rank, Traffic, and Usability of the Website. In *Human-Centric Smart Computing: Proceedings of ICHCSC 2022,* (pp. 217–226). Springer Nature Singapore Singapore.

Singh, D., Prusty, S. K., Sarangi, S. S., Sahoo, S., & Biswal, A. K. (2020). Anxiety, Psychological Effects and Prevention during COVID-19 in India.

Singh, D., Shivamurthaiah, M., Samanta, D., Bhattacharya, A., & Dutta, S. (2022). Applicability of Search Engine Optimization for WordPress (WP) Website. In *Human-Centric Smart Computing: Proceedings of ICHCSC 2022,* , (pp. 227–237). Springer Nature.

Singh, V., Aggarwal, A., Kumar, N., & Saini, A. K. (2020). A Novel Approach for Pre-Validation Auto Resiliency & Alert Notification for SVN To Git Migration Using Iot Devices. *PalArch's Journal of Arch. of Egypt/Egyptology, 17*(9), 7131-7145.

Singha, J., & Das, K. (2015). Automatic Indian Sign Language recognition for continuous video sequence. *ADBU Journal of Engineering Technology, 2*(1).

Singh, D., Bhanipati, J., Biswal, A. K., Samanta, D., Joshi, S., Shukla, P. K., & Nuagah, S. J. (2021). Approach for Collision Minimization and Enhancement of Power Allocation in WSNs. *Journal of Sensors*. doi:10.1155/2021/7059881

Singh, D., Biswal, A. K., Samanta, D., Singh, D., & Lee, H. N. (2022). Juice Jacking: Security Issues and Improvements in USB Technology. *Sustainability, 14*(2), 939. doi:10.3390u14020939

Singh, S. (2020). MVC Framework: A Modern Web Application Development Approach and Working. *International Research Journal of Engineering and Technology, 77*(1), 51–55.

Singh, S., & Iyer, J. (2016). Comparative Study of MVC (Model View Controller) Architecture with respect to Struts Framework and PHP. *International Journal on Computer Science and Engineering, 5*(3), 142–150.

Singh, V., Aggarwal, A., Kumar, A., & Sanwal, S. (2019). The Transition from Centralized (Subversion) VCS to Decentralized (Git) VCS: A Holistic Approach. *Journal of Electrical and Electronics Engineering (Oradea), 12*(1), 7–15.

Singh, V., Alshehri, M., Aggarwal, A., Alfarraj, O., & Sharma, P. (2021). A holistic proactive and novel approach for pre during and post migration validation from subversion to git. *Computers Materials & Continua, 66*(3), 2359–2371. doi:10.32604/cmc.2021.013272

Siri, W. E. (1956). The gross composition of the body. [Google Scholar]. *Advances in Biological and Medical Physics, 4*, 513. doi:10.1016/B978-1-4832-3110-5.50011-X PMID:13354513

Sivanantham, K., Kalaiarasi, I., & Leena, B. (2022). Brain Tumor Classification Using Hybrid Artificial Neural Network with Chicken Swarm Optimization Algorithm in Digital Image Processing Application. *Advance Concepts of Image Processing and Pattern Recognition: Effective Solution for Global Challenges, 91*.

Sivanantham, K. (2022). Deep Learning-Based Convolutional Neural Network with Cuckoo Search Optimization for MRI Brain Tumour Segmentation. In *Computational Intelligence Techniques for Green Smart Cities* (pp. 149–168). Springer. doi:10.1007/978-3-030-96429-0_7

Smart Health 2.0. (2014). *Home.* Smart Health 2.0. http://www.smarthealth2.com/

Snehi, M., & Bhandari, A. (2021). Vulnerability retrospection of security solutions for software-defined Cyber-Physical System against DDoS and IoT-DDoS attacks. *Computer Science Review*, *40*, 100371. doi:10.1016/j.cosrev.2021.100371

Solanki, A., Nayyar, A., & Naved, M. (2021). *Generative Adversarial Networks for Image-to-Image Translation*, *10*, 23. Academic Press.

Soni, A., Khare, A., Balaji, P. D., Verma, S., Rani, K. A., & Gowrishankar, S. (2022, October). Pest Identification and Control using Deep Learning and Augmented Reality. In *2022 International Conference on Edge Computing and Applications (ICECAA)* (pp. 1-7). IEEE. 10.1109/ICECAA55415.2022.9936053

Stamile, C., Marzullo, A., & Deusebio, E. (2021). Graph *Machine Learning*, *11*, 26. Packt Publishing Ltd.

Subari, A., Manan, S., & Ariyanto, E. (2021). Implementation of MVC (model-view-controller) architecture in online submission and reporting process at Official Travel Warrant Information System based on web application. *Journal of Physics: Conference Series*, *1918*(4), 042145.

Sullivan, W. (2017). *Machine learning for beginner's guide algorithms: decision tree & random forest introduction.* Healthy Pragmatic Solutions.

Sumathy, B., Pavithran, K. M., Nizam, N., & Surya, V. A. (2020 October 2-3). *Smart Guidance System for Blind with Wireless Voice Playback.* IOP Conference Series: Materials Science and Engineering, Volume 1012, International Conference on Robotics, Intelligent Automation and Control Technologies (RIACT 2020), Chennai, India. https://iopscience.iop.org/article/10.1088/1757-899X/1012/1/012045/pdf

Svajlenko, J., & Roy, C. K. (2016). A machine learning based approach for evaluating clone detection tools for a generalized and accurate precision. *International Journal of Software Engineering and Knowledge Engineering, 26*, pp. 1399-1429.

T. Nguyen, T. (2015). Structural modeling from depth images. *IEEE Trans Visual Comput Gr.* IEEE.

Taherian-fard, E., Niknam, T., Sahebi, R., Javidsharifi, M., Kavousi-fard, A., & Aghaei, J. (2022). A Software Defined Networking Architecture for DDoS-Attack in the Storage of Multimicrogrids. *IEEE Access: Practical Innovations, Open Solutions*, *10*(July), 83802–83812. doi:10.1109/ACCESS.2022.3197283

Tang, S. W., Dai, A. L., & Chen, Y. Z. (2015). Research on Cross platfrom APP for Minxi's Pig Disease Base on CORDOVA and Its Implementation. *Journal of Wuyi University*, *34*(12), 60–64.

Tao, J., & Tan, T. (2005). Affective Computing: A Review. In J. Tao & ... (Eds.), *Affective Computing and Intelligent Interaction* (pp. 981–995). Springer. doi:10.1007/11573548_125

Tasaka, S., & Watanabe, Y. (2007, November). Real-Time Estimation of User-Level QoS in Audio-Video IP Transmission by Using Temporal and Spatial Quality. *IEEE Globecom Workshops*, 2661–2666. doi:10.1109/GLOCOM.2007.505

Tawsif, K., & ... . (2022, August). A Systematic Review on Emotion Recognition System Using Physiological Signals: Data Acquisition and Methodology. [ijournalse.org]. *Emerging Science Journal*, *6*(5), 1167–1198. doi:10.28991/ESJ-2022-06-05-017

Thakur, R. N., & Pandey, U. S. (2019). A Study Focused on Web Application Development using MVC Design Pattern. *International Research Journal of Engineering and Technology*, *6*(8), 237–244.

Thakur, S. K., Singh, D. P., & Choudhary, J. (2020). Lung cancer identification: A review on detection and classification. *Cancer and Metastasis Reviews*, *39*(3), 989–998. doi:10.100710555-020-09901-x PMID:32519151

The Mobile App Architecture Guide. (n.d.). Net Solutions. https://www.netsolutions.com/insights/mobile-app-architecture-guide/

Tiwari, S., Ameta, D., & Banerjee, A. (2019). An approach to identify use case scenarios from textual requirements specification. *Proc. ISEC*. IEEE.

Tripathi, K., & Nandi, N. B. G. (2015). Continuous Indian sign language gesture recognition and sentence formation. *Procedia Computer Science*, *54*, 523–531.

Tyagi, V., & Kumar, A. (2017). Internet of things and social networks: a survey. *International Conference on Computing, Communication and Automation, ICCCA*, (pp. 1268-1270). IEEE. doi:10.1109/CCAA.2017.8230013

Umuhoza, E., Brambilla, M., Cabot, J., & Bongio, A. (2015). Automatic code generation for cross-platform multi-device mobile apps: Some reflections from an industrial experience. Proc. 3rd Int. Workshop Mobile Develop. Lifecycle, (pp. 37-44). IEEE.

Usman, M., Ali, I., Khan, S., & Khurram, M. (2019). Journal of Network and Computer Applications A survey on software defined networking enabled smart buildings : Architecture, challenges and use cases. *Journal of Network and Computer Applications*, *137*(March), 62–77. doi:10.1016/j.jnca.2019.04.002

van Kessel, K. (2021). Design Considerations for a Multiple Sclerosis Fatigue Mobile App MS Energize: A Pragmatic Iterative Approach Using Usability Testing and Resonance Checks. *Internet Interventions: the Application of Information Technology in Mental and Behavioural Health*, *24*(Apr), 100371.

Vanderplas, J. T. (2017). *Python data science handbook: Essential tools for working with data*. O'reilly, Cop.

Varoquaux, G., Buitinck, L., Louppe, G., Grisel, O., Pedregosa, F., & Mueller, A. (2015). Scikit-learn. *GetMobile: Mobile Computing and Communications*, *19*(1), 29–33. doi:10.1145/2786984.2786995

Velasco-Elizondo, P. (2022). Software Architecture: Developing Knowledge, Skills, and Experiences. In *Research Anthology on Agile Software, Software Development, and Testing, edited by Information Resources Management Association* (pp. 17–39). IGI Global. doi:10.4018/978-1-6684-3702-5.ch002

Vemuri, S., Chala, S., & Fathi, M. (2017). Automated use case diagram generation from textual user requirement documents. *Proc. IEEE 30th Can. Conf. Electr. Comput. Eng. (CCECE),* (pp. 1-4). IEEE.

Ventura, S. L., Nugent, R., & Fuchs, E. R. (2015). Seeing the non-stars: (Some) sources of bias in past disambiguation approaches and a new public tool leveraging labeled records. *Research Policy*, *44*(9), 1672–1701. doi:10.1016/j.respol.2014.12.010

Vissicchio, S., & Vanbever, L. (2014).. . *Opportunities and Research Challenges of Hybrid Software Defined Networks.*, *44*(2), 70–75.

Wade, C. (2020). *Hands-on gradient boosting with XGBoost and scikit-learn: Perform accessible machine learning and extreme gradient boosting with python*. Packt Publishing.

Wang, M. (2022). Users' Experience Matter: Delay Sensitivity-Aware Computation Offloading in Mobile Edge Computing. *Digital Communications and Networks*. ScienceDirect. doi:10.1016/j.dcan.2022.08.005

Wang, H. U. A., Rafique, W., & Anwar, Z. (2019). Cyberpulse : A Machine Learning Based Link Flooding Attack Mitigation System for Software Defined Networks. *IEEE Access: Practical Innovations, Open Solutions*, *7*, 34885–34899. doi:10.1109/ACCESS.2019.2904236

Wang, J. (2022, December). A Review on Extreme Learning Machine. *Multimedia Tools and Applications*, *81*(29), 41611–41660.

Wang, X. (2021). Design and Implementation of Cross Platform IntelligentTravel APP Based on Cordova. *Railway Signalling & Communication Engineering*, *18*(7), 40–44.

Wang, X., Xia, M., Cai, H., Gao, Y., & Cattani, C. (2012). Hidden-markov-models-based dynamic hand gesture recognition. *Mathematical Problems in Engineering*.

Wankhade, M., Rao, A. C. S., & Kulkarni, C. (2022, October). A Survey on Sentiment Analysis Methods, Applications, and Challenges. [Springer]. *Artificial Intelligence Review*, *55*(7), 5731–5780. doi:10.100710462-022-10144-1

Watters, A. (2021). *Teaching Machines*, *19*, 24 MIT Press..

Wei, Y., Kou, X., & Lee, M. (2013, October 20-23). *Development of a guide-dog robot system for the visually impaired by using fuzzy logic based human-robot interaction approach.* 2013 13th International Conference on Control, Automation and Systems (ICCAS 2013), Gwangju, South Korea. https://ieeexplore.ieee.org/document/6703877

Weisberg, S. (2013). *Applied Linear Regression*. John Wiley & Sons.

What Is Multimodal Emotion Recognition (2021). IGI Global. https://www.igi-global.com/dictionary/multimodal-emotion-recognition/19631.

What is Network Virtualization? Everything You Need to Know. (n.d.). TechTarget. https://www.techtarget.com/searchnetworking/What-is-network-virtualization-Everything-you-need-to-know

Wiese, I. S., Da Silva, J. T., Steinmacher, I., Treude, C., & Gerosa, M. A. (2016). *"Who is who in the mailing list? Comparing six disambiguation heuristics to identify multiple addresses of a participant", 2016 IEEE international conference on software maintenance and evolution.* IC-SME.

Wilson, G., Pereyda, C., Raghunath, N., de la Cruz, G., Goel, S., Nesaei, S., Minor, B., Schmitter-Edgecombe, M., Taylor, M. E., & Cook, D. J. (2019). Robot-enabled support of daily activities in smart home environments. *Cognitive Systems Research, 54*, 258–272. doi:10.1016/j.cogsys.2018.10.032

Winkler, S., & Dufaux, F. (2003). Video Quality Evaluation for Mobile Applications. *Proc. of SPIE Conference on Visual Communications and Image Processing*. IEEE.

Winkler, W. E. (1990). String comparator metrics and enhanced decision rules in the Fellegi-Sunter model of record linkage. *Proceedings of the Section on Survey Research Methods*, (pp. 433-444). ACM.

Wong, W. W. S., & Sousa, E. S. (1999). A CDMA mixed cellular architecture with space division duplex (SDD) for symmetric/asymmetric wireless communications. WCNC *1999 IEEE Wireless Communications and Networking Conference (Cat. No.99TH8466),* (pp. 1338-1342). IEEE. 10.1109/WCNC.1999.796955

World Health Organization. (2008). *2008–2013 Action Plan for the Global Strategy for the Prevention and Control and Diabetes.* World Health Organization.

Wu, D. (2001). Streaming Video over the Internet: Approaches and Directions. *IEEE Transactions on Circuits and Systems for Video Technology*, (Mar), 282–300.

Xhu, X., & Godberg, A. B. (2009). *Introduction to semi-supervised learning*. Morgan & Claypool.

Xuebin, J. (2017). Research on Computer Network Virtual Laboratory based on asp.net. *2017 International Conference on Computer Systems, Electronics and Control (ICCSEC)*, 464–467.

Yamagishi, K., & Hayashi, T. (2006, November). Opinion Model for Estimating Video Quality of Videophone Services. *IEEE Globecom Workshops*, 1–5. doi:10.1109/GLOCOM.2006.458

Yang, L., & Shami, A. (2020). On hyperparameter optimization of machine learning algorithms: Theory and practice. *Neurocomputing*, *415*, 295–316. doi:10.1016/j.neucom.2020.07.061

Yan, W. (2021). *"Research on Emotions Based on Brain-Computer Technology: A Bibliometric Analysis and Research Agenda." Frontiers in Psychology* (Vol. 12). Frontiers.

Yao, Y. & Liu, W. (2008). *Research on Android architecture and application development.* Computer system application.

Yasui, T. (2005). Influences of network latency and packet loss on consistency in networked racing games. ACM SIGCOMM workshop on Network and System Support for Games. NetGames.

Ye, J.-H. (2017). *Development of App for Strength Fitness Products* [Master Thesis, National Kaohsiung University of Applied Sciences https://reurl.cc/KxXY2g

Ye, Y., Tian, Y., Huenerfauth, M., & Liu, J. (2018). Recognizing American Sign Language gestures from within continuous videos. In *Proceedings of the IEEE Conference on Computer Vision and Pattern Recognition Workshops*, (pp. 2064-2073). IEEE. 10.1109/CVPRW.2018.00280

Yigitbas, E. & S. Sauer, S. (2016). Engineering context-adaptive UIs for task-continuous cross-channel applications. In *Human-Centered and Error-Resilient Systems Development*. Springer.

Yoon, W.-J., & Park, K.-S. (2007). A Study of Emotion Recognition and Its Applications.. Modeling Decisions for Artificial Intelligence,, pp. 455–62. Springer doi:10.1007/978-3-540-73729-2_43

Yousaf, N., Azam, F., Butt, W. H., Anwar, M. W., & Rashid, M. (2019). Automated model-based test case generation for Web user interfaces (WUI) from interaction flow modeling language (IFML) models. *IEEE Access: Practical Innovations, Open Solutions*, *7*, 67331–67354.

Yuan, X. Y. (2018). Problems and countermeasures of catering management in tourist attractions. *Yangtze River Series, 40*(26).

Yu, D. S., Lu, Q., & Leng, Y. J. (2017). Design and implementation of self service tourism management system based on B/S mode. *Inner Mongolia Science Technology & Economy, 21*(19), 61–62.

Yu, X., Jiang, M., & Liu, A. (2022). Design and Application of Self-service Cultural Tourism Information Platform in the Mobile Information Age. *2022 International Conference on Information System, Computing and Educational Technology (ICISCET)*. 10.1109/ICISCET56785.2022.00059

Zaccone, G. (2017). *Deep learning with TensorFlow*. Uk Packt Publishing.

Zein, S., Salleh, N., & Grundy, J. (2016). A systematic mapping study of mobile application testing techniques. *Journal of Systems and Software*, *117*, 334–356.

Zephyr Techonology. (2016). *New Zephyr BioPatch Monitoring Device for Human Performance*. Zephyr Techonology. [Google Scholar]

Zhai, P., Song, Y., Zhu, X., Cao, L., Zhang, J., & Yang, C. (2020). Distributed Denial of Service Defense in Software Defined Network Using OpenFlow. *ICCC*, 1274–1279.

Zhang, C., Yang, X., & Tian, Y. (2013). Histogram of 3d facets: A characteristic descriptor for hand gesture recognition. In *2013 10th IEEE international conference and workshops on automatic face and gesture recognition (FG)*, (pp. 1~8). IEEE.

Zhang, L. (2017). Research on application development based on Android platform architecture. *Computer programming skills and maintenance*, (06), 52-53 + 85.

Zhang, Z. (2016). Research on Android architecture and application development. *Inner Mongolia Science and technology and economy,* (17), 72-73 + 78.

Zhang, H. C., & Li, D. J. (2021). The theoretical interpretation and test of the integration of leisure sports with tourism phenomenon *Journal of South-Central Minzu University*, *40*(5), 543–550.

Zhang, Q., Zhang, L., & Qiang, Y. S. (2013). Study on intelligent travelling self service system based on Android. *Internet of Things Technologies*, *3*(8), 53–55.

Zhang, Y., & Rajapakse, J. C. (2009). *Machine Learning in Bioinformatics*, *74*, 84.

Zhao, J. (2017). Research on Android system architecture and application development. *Automation and instrumentation*, (01), 86-87 + 90.

Zhong, Z., Guo, J., Yang, W., Xie, T., Lou, J.-G., & Liu, T. (2018). Generating regular expressions from natural language specifications: Are we there yet? *Proc. Workshop NLP Softw. Eng. (NL4SE)*. IEEE.

Zhou, G., Liu, Z., & Fu, C. (n.d.). *An Efficient Design of Intelligent Network Data Plane*. Research Gate.

Zhou, B., Zang, R., Zhang, M., Song, P., Liu, L., Bie, F., & Gao, S. (2022). Worldwide burden and epidemiological trends of tracheal, bronchus, and lung cancer: A population-based study. *EBioMedicine*, *78*, 103951. doi:10.1016/j.ebiom.2022.103951 PMID:35313216

Zhou, Q. D. (2019). Construction of smart tourism management platform based on mobile Internet technology. *Electronic Technology & Software Engineering*, *8*(3), 2–2.

Zingg, C. Nanumyan, V., & Schweitzer, F., (2020).Citations driven by social connections? A multilayer representation of coauthorship networks. *Quantitative Science Studies, 1*(4), pp. 1493-1509. doi: 1. doi:0.1162/qssa00092

About the Contributors

Debabrata Samanta is presently working as Assistant Professor, Rochester Institute of Technology, Kosovo, Europe. He obtained his B.Sc. (Physics Honors), from Calcutta University; Kolkata, India. He obtained his MCA, from the Academy of Technology, under WBUT, West Bengal. He obtained his PhD in Computer Science and Engg. from National Institute of Technology, Durgapur, India, in the area of SAR Image Processing. His areas of interest are SAR Image Analysis, Video surveillance, Heuristic algorithm for image classification, Deep Learning Framework for Detection and Classification, Blockchain, Statistical Modelling, Wireless Adhoc Network, Natural Language Processing, V2I Communication. He is the owner of 22 Indian Patents. He has authored and coauthored over 207 research papers in international journal (SCI/SCIE/ESCI/Scopus) and conferences including IEEE, Springer and Elsevier Conference proceedings. He has received "Scholastic Award" at 2nd International conference on Computer Science and IT application, CSIT-2011, Delhi, India. He has published 9 books, available for sale on Amazon and Flipkart.

***

Jayalakshmi A. is working as a Assistant Professor, Department of Computer Applications, Hindusthan College of Arts & Science, Coimbatore, Tamilnadu, India.

Deti Baholli is a student of the Department of Computer Information of Technology, Rochester Institute of Technology Kosovo.

Herolind Balaj is a student of the Department of Computer Information of Technology, Rochester Institute of Technology Kosovo.

Tapan Behera is a highly skilled Architect and Analytics expert who has over the years channelled his expertise within the healthcare, market research space. Tapan advises Business and technology leaders in the transformation of both their organization and technology platform. He has worked with several Fortune 500 healthcare, market research and product-based companies such as Forrester Research, Walgreens, Cognizant, Hewlett-Packard, DXC and HCL.

Alma Beluli is a student of the department of Computing and Information Technologies, Rochester Institute of Technology Kosovo.

Shkurte Berisha is a final-year student at Rochester Institute of Technology, Kosovo

Pratibha Bharati is currently pursing B.Tech. in Computer Science and Engineering from NIST Institute of Science and Technology (Autonomous), Berhampur, Odisha.

Anil Kumar Biswal is currently pursing PhD in Computer Science and Engineering from ITER, Siksha 'O' Anusandhan (Deemed to be University). He has also an extensive record of teaching for the past 8 years and his research interests include Wireless Sensor Networks, Cloud Computing and IoT. He has worked on various research projects that solve some social related issues like traffic with street light monitoring process, drowsiness detection system, designing women safety App, etc.

Banupriya C. V. is currently working as Assistant Professor, Department of Computer Science, PSG College of Arts and Science, Coimbatore, Tamilnadu, India.

Leon Cana is a student of the Department of Computer Information of Technology, Rochester Institute of Technology Kosovo.

Erita Çunaku is a final-year student at Rochester Institute of Technology, Kosovo.

Muskan Garg is working as postdoctoral research fellow at Mayo Clinic, Rochester, Minnesota. She was previously working as Postdoctoral research associate at University of Florida and as an assistant professor in Thapar institute of engineering and technology, Patiala. Her research focuses on the problems in natural language processing, information retrieval, and social media analysis. She received her Masters and PhD from Panjab University, India. Prior to TIET, she has worked as an assistant professor in Amity school of engineering and technology at Amity University. Her focus is on research and development of cutting edge NLP approaches to solving problems of national and international importance and on initiation and broadening a new program in natural language processing (including a new NLP course series). Her current research interests are causal inference, mental health on social media, event detection and sentiment analysis. She contributes as a reviewer in The ACM Transactions on Asian and Low-Resource Language Information Processing (TALLIP) and Expert Systems with Applications.

Klinton Gashi is a student of the Department of Computer Information of Technology, Rochester Institute of Technology Kosovo.

Drin Hoti is a student of the Department of Computer Information of Technology, Rochester Institute of Technology Kosovo.

Radha Srinivasan Iyer is a Physiotherapist with a master's degree in Neurosciences from the prestigious D.Y Patil University, Pune. She frequently collaborates with her spouse Santhosh Kumar Rajamani in his research work and publications.

Akshat Jain is the final year student of BCA, Christ University, India. He has published 2 research papers in IEEE conferences.

Art Jashari is a student of the Department of Computer Information of Technology, Rochester Institute of Technology Kosovo.

Marimuthu Karuppiah received his B.E. degree in Computer Science and Engineering from Madurai Kamaraj University, Madurai, India in 2003, M.E. degree in Computer Science and Engineering from Anna University, Chennai, India in 2005, Ph.D. degree in Computer Science and Engineering from VIT University, Vellore, India in 2015. Presently, he is a Professor in Department of Computing Science and Engineering, Presidency University, Bengaluru, India. He has published more than 30 research papers in SCI indexed journals. Also, he has published more than 40 research papers in SCOPUS indexed journals and international conferences. He is a life member of Cryptology Research Society of India (CRSI) and Computer Society of India (CSI), senior member of IEEE and member of ACM. His main research interests include cryptography and wireless network security, in particular, authentication and encryption schemes.

Jyoti Khandelwal is an Assistant Professor in University of Engineering & Management, Jaipur, and Ph.D. research scholar at Manipal University in Jaipur. She received her master's degree in audio watermarking in 2014. Her current field of research is image steganography. She is working on an integrated field of AI and ML to enhance the information security process.

Veton Klinaku is a student of the Department of Computer Information of Technology, Rochester Institute of Technology Kosovo.

Ankit Kumar is currently pursing B.Tech. in Computer Science and Engineering from NIST Institute of Science and Technology (Autonomous), Berhampur, Odisha.

Narendra Kumar Rao is currently working as Professor and Program Head, AIML in the School of Computing, Mohan Babu University, Tirupati. He is a Member of the Research Advisory Committee and also a Doctoral Supervisor Cluster Head at the School of Computing, Mohan Babu University. His research interests include Software Testing, Embedded Systems and Machine Learning. He has been part of several International Conferences as Program Coordinator and Conference Chair. He was the convener for International Conference on Data Analytics, Intelligent Computing and Cyber Security, and was also an editor for two Proceedings by Springer published. He is associated with IEEE, ACM, CSTA and IAENG. He won Best Faculty Recognition, the Nava Bharat Nirman Award by the Information Technology Association of AP & India Servers, and the Best Researcher Award by Integrated Research Group(IRG), Chennai for his research work. Apart from this, he was awarded Trial Blazer-Highest Domain Level Award from Embedded Systems President, STAR-Highest Business Unit Level Award from Vice-President, FIMC-Highest Project Level Award from Sr.Project Manager at Wipro Technologies. He received his PhD in CSE from Jawaharlal Nehru Technological University, Hyderabad, India. He has published 22 publications in reputed Journals and presented 30 papers in indexed conferences to date.

Islabudeen M. is currently working as an Associate Professor in the School of Computer Science Engineering and Information Science, Presidency University, Bengaluru, India. He received his B.E. Degree in Computer Science and Engineering from Madurai Kamaraj University, Madurai, India in 2001 and M.E. Degree in Computer Science and Engineering and Ph.D. Degree in Information and Communication Engineering from Anna University, Chennai, India in 2008 and 2021, respectively. He has more than 21 years of academic experience. His research interests include Cryptography and Network Security, Blockchain, Wireless Networks and Machine Learning. He is a life member of the Computer Society of India (CSI) and Indian Society for Technical Education (ISTE). He has served as an active reviewer and chair for many reputed international conferences and journals including IEEE, Elsevier and Springer. He has more than 20 publications in reputed international conferences and refereed journals.

Karthikeyan M. P. is presently working as Assistant Professor, School of Computer Science and IT, Jain (Deemed to be University), Bangalore, India. His area of Specialization Machine Learning, Deep Learning and Digital Marketing. He is Corporate Trainer at Symbiosis International Skill University Pune. He obtained his Bachelors in Computer Science, from Sree Saraswathi Thyayagaraja College, Pollachi. He obtained his MCA, from Sree Saraswathi Thyayagaraja College, Pollachi. He Pursuing his PhD in Computer Science from Sri SRNM College, Sattur in the area of Query Optimization in Deep Reinforcement Learning. He Completed Industry Integrated Course Data Science and Machine Learning from Nunnari Labs, Coimbatore. He has acted as resource person for various DBT Sponsored workshop on “Big data Analytics using python programming & R Programming”, and Blockchain Technology. He has authored and coauthored over 25 research papers in international journal (SCI/SCIE/ESCI/Scopus) and conferences including IEEE, Springer and Elsevier Conference proceeding. He is a co-author of 5 books and 10 books Chapters. He has presented various papers at international conferences. He has Completed more than 50 Online Course in Coursera, LinkedIn, etc. He is an IEEE Member, Robotics Club Member and Member of Computer Society of India (CSI). He is a Board Studies Member for Various Autonomous Institutions and Question Paper Setter for Various Colleges. He acted as a Resource Person for Several Institutions.

Monika Maloku is a student of the Department of Computer Information of Technology, Rochester Institute of Technology Kosovo.

Rafaela Marku is a student of the Department of Computer Information of Technology, Rochester Institute of Technology Kosovo.

Jona Ndrecaj is a final-year student at Rochester Institute of Technology, Kosovo.

Devasis Pradhan is currently working as an Assistant Professor lResearch Coordinator in the Department of Electronics & Communication Engineering at Acharya Institute of Technology, Bengaluru, Karnataka from 2017 onwards. His current research includes the effectiveness of 5G Green Communications, mmWave antenna design, UWB antennas, and its implementation. Presently he is continuing his Ph.D. at the National Institute of Technology, Rourkela, Odisha in the department of Electrical Engineering. . He has published 76 + research papers and 8 papers submitted to IEEE Access and Reputed Book and 4 books with a reputed publishing house. He is a co-editor in the Editorial Board & Peer-review of 8 International Journals and a committee member in the much-reputed organization. He has authored and co-authored 15 Book Chapters. He has been a Technical Committee Member and Reviewer for a reputed internal conference such as IEEE, Grenz Society, IFERP, etc; He has received 7 National Awards in the field of Academic and Research work from various governing bodies associated with the government of India. He is also an active member of ISTE, IEEE, ATMS, and other professional associations toward professional growth.

Alisa Qatipi is a student of the Department of Computer Information of Technology, Rochester Institute of Technology Kosovo.

Kowsalya R. is working as an Assistant Professor, PG Department of Computer Science, PSGR Krishnammal College for Women, Coimbatore, Tamilnadu, India.

Santhosh Kumar Rajamani is a medical teacher, who is enthusiastic about developments in the field of Artificial intelligence and applications of AI and Machine Learning in Healthcare. He has collaborated in many AI healthcare projects and published articles and patents in the field of Medical applications of Artificial Intelligence and Machine Learning. He is a Surgeon by profession and a naturalist by passion.

Rajesh Kanna Rajendran is an Assistant Professor at CHRIST(Deemed to be University), Bangalore, India. He Holds a PhD degree in Computer Science in Mobile adhoc Network at Bharathiar University, Coimbatore, India. His research areas are Mobile Adhoc Networks, VANET, Machine Learing and medical image analysis. He is acted as Coordinator for centre for e-learning and NAAC Criterion II.He published 27 Journal articles in various international peer reviewed journals, 3 Book publications and 35 presentations in various international level conference held in india and abroad.He received morethan 15 lakhs for researchand consultancy for research promotion. He can be contacted at email: drrajeshkannacbe@gmail.com

Swati Ranjan is currently pursing B.Tech. in Computer Science and Engineering from NIST Institute of Science and Technology (Autonomous), Berhampur, Odisha.

Blendi Rrustemi is a student of the Department of Computer Information of Technology, Rochester Institute of Technology Kosovo.

Milan Samantaray is currently pursing PhD in Computer Science and Engineering from C. V. Raman Global Univertsity. He has also an extensive record of teaching for the past 8 years and his research interests include Cyber Security, Adhoc Networking and IoT.

Ramani Selvanambi is an Associate Professor in the School of Computing Science and Engineering at Vellore Institute of Technology (VIT), Vellore, India. He received his B.E in Computer Science and Engineering from Madras University and M. Tech Computer Science and Engineering from Bharathidasan University, and Ph.D. in Computer Science and Engineering from Vellore Institute of Technology (VIT), Vellore. He has 13+ years of experience in teaching and 2 years of experience in the Consultancy and Software Industry. He has published about 40 research papers in International Journals on Machine Learning, Nature-Inspired Algorithms, Cyber Security and Health Care. His research interest includes Data Mining, Machine learning, Database Systems, Optimization Techniques, and Cyber Security. He is a Senior member of IEEE and member in other technical societies.

Mohana Priya T. is Assistant Professor at CHRIST(Deemed to be University), Bangalore, India. She Holds a PhD degree in Computer Science in Machine learning at Bharathiar University, Coimbatore, India. Her research areas are Machine Learing and medical image analysis. She is acted as Coordinator for curriculum development cell and Software development cell. She published 10 Journal articles in various international peer reviewed journals, 2 Book publications and 15 presentations in various international level conference held in India and abroad.

Yllka Totaj is a student of the Department of Computer Information of Technology, Rochester Institute of Technology Kosovo.

Hla Tun currently works as the Pro-Rector (Research) at the Research Department of Yangon Technological University. Hla does research in Telecommunication Engineering, Control Systems Engineering and Electronic Engineering. Their current project is 'Biomedical Signal Processing' and Semiconductor Electronics Measurement System.

Deni Turku is a student of the Department of Computer Information of Technology, Rochester Institute of Technology Kosovo.

Lumi Zhubi is a student of the Department of Computer Information of Technology, Rochester Institute of Technology Kosovo.

Index

T

U

V

W

www.ingramcontent.com/pod-product-compliance
Lightning Source LLC
LaVergne TN
LVHW080610300926
849162LV00005B/1110